Fluoroplastics

Volume 1: Non-Melt Processible Fluoroplastics
The Definitive User's Guide and Databook

Sina Ebnesajjad

Plastics Design Library

Copyright © 2000, Plastics Design Library. All rights reserved.
ISBN 1-884207-84-7
Library of Congress Card Number 99-086655

Published in the United States of America, Norwich, NY, by Plastics Design Library, a division of William Andrew Inc.

Information in this document is subject to change without notice and does not represent a commitment on the part of Plastics Design Library. No part of this document may be reproduced or transmitted in any form or by any means, electronic or mechanical, including photocopying, recording, or any information retrieval and storage system, for any propose without the written permission of Plastics Design Library.

Comments, criticisms, and suggestions are invited, and should be forwarded to Plastics Design Library.

Plastics Design Library and its logo are trademarks of William Andrew Inc.

Please Note: Although the information in this volume has been obtained from sources believed to be reliable, no warranty, expressed or implied, can be made as to its completeness or accuracy. Design, processing methods and equipment, environment and other variables affect actual part and mechanical performance. Inasmuch as the manufacturers, suppliers, William Andrew Inc. or Plastics Design Library have no control over those variables or the use to which others may put the material and, therefore, cannot assume responsibility for loss or damages suffered through reliance on any information contained in this volume. No warranty is given or implied as to application and to whether there is an infringement of patents is the sole responsibility of the user. The information provided should assist in material selection and not serve as a substitute for careful testing of prototype parts in typical operating environments before beginning commercial production.

Manufactured in the United States of America.

Library of Congress Cataloging-in-Publication Data

Ebnesajjad, Sina.
  Non-melt processible fluoroplastics : the definitive user's guide and databook / by Sina Ebnesajjad.
     p. cm.
  Includes bibliographical references and index.
  ISBN 1-884207-84-7
   1. Fluoropolymers--Handbooks, manuals, etc. I. Title.

QD412.F1 E36 2000
668.4'1--dc21

99-086655

Plastics Design Library, 13 Eaton Avenue, Norwich, NY 13815 Tel: 607/337/5080 Fax: 607/337/5090

*Dedicated to my mother, father and
beloved son Cyrus*

Acknowledgments

I owe the majority of my learning and experience in the field of fluoropolymers to my employment at the DuPont Company for the last seventeen years. DuPont Fluoroproducts has generously contributed to these volumes by providing a great deal of the data, research material, and supporting literature searches, securing reference material and the preparation of the manuscript.

A number of companies have provided me with information; they have been cited usually in the bibliography section at the end of each chapter. The following corporations have provided me with the data published in their commercial information bulletins: DuPont, Ausimont, Daikin, Dyneon, Imperial Chemical Industries, Elf Atochem and Asahi Glass. I sincerely appreciate the contribution that the data supplied by each company has made to this book.

The author is especially appreciative of Dr. Subash Gangal for his mentoring and teachings and his numerous encyclopedia review articles. I wish to thank the following individuals: Dr. Marianne Marsi, Dr. Louis Glascow and Dr. George Senkler for general support of this project. Mr. Yasef Adato, Dr. Ralph Aten, Dr. Robert J. Cavanaugh, Mr. David DeVoe, Mr. Pradip Khaladkar, Mr. Mathew Koenings, Ms. Sharon A. Libert, and Dr. Ron Uschold have kindly reviewed and made many helpful suggestions to the author. Thank you. Mr. Bill Digel has been most helpful with the writing of Ch. 17, "Applications."

I have made a great deal of use of data developed and organized by my colleagues and friends in the DuPont Company, S. A., in Geneva, Switzerland. I would like to specially acknowledge my friend Dr. Theodore Schroots for data, figures and tables that he has developed over the last three decades.

My friend Dr. Ted S. Wilks has masterfully conducted the literature search for this book. Mr. Tod Williams, Mr. Tom Johns, and Ms. Charlene Martin and her group have efficiently procured the reference material, including articles, books, and patents. Thank you.

My sincere thanks go to Mrs. Jeanne Roussel for converting the raw manuscript into a real book, with her endless care and patience. Jeanne, you have amazed me with your transformation of the raw material that I provided you.

Editorial suggestions and support by William Andrew Publishing have greatly upgraded the organization and the text. It would have been impossible for me to complete this project without the many helpful suggestions and moral support of William Woishnis, Editor-in-Chief of *Plastics Design Library*, at William Andrew Publishing.

Thanks to my son Cyrus Ebnesajjad, preparation of the manuscript at times approached what he calls *fun*.

Finally, there are not enough words for the author to thank many friends and colleagues, particularly Ms. Ghazale Dastgheib, who have provided general suggestions and unlimited encouragement to the author throughout this project.

Preface

The era of fluoropolymers began with the discovery of polytetrafluoroethylene in 1938. The history of fluoropolymers, as will be seen in this book, is steeped in both scientific curiosity and serendipity—the seemingly indispensable elements of most major discoveries and inventions. There has been an explosion of fluoropolymer applications in all facets of human affairs. Old and new technological advances have been made possible by their unique properties. From the Manhattan project in the early 1940's, to the Apollo missions in the 1970's, to integrated manufacturing in the late 1980's, industries have relied on fluoropolymers for their inertness and durability. New applications are being developed everyday after more than six decades since the discovery of this plastic family. Few materials have impacted the lives of peoples as extensively as fluoropolymers.

The aim of this volume and a second companion volume is to compile in one place a working knowledge of the polymer chemistry and physics of fluoropolymers with detailed descriptions of commercial processing methods. The books focus on providing a reference as well as a source for learning the basics for those involved in polymer manufacturing and part fabrication, as well as end-users of fluoropolymers and students. The two volumes have been arranged according to the processing methods for fluoropolymers, which is more or less the chronological order in which the important polymers were developed. Volume One concentrates mostly on polytetrafluoroethylene and polychlorotrifluoroethylene and their processing techniques, which are essentially non-melt-processes. Volume Two concentrates on melt-processable fluoropolymers. These include fluorocopolymers and specific other fluorine-containing polymers, which are by-and-large melt-processable.

These volumes emphasize the practical over theoretical. There are numerous sources for in-depth information about the polymerization and polymer science of fluoropolymers. The references listed at the end of each chapter serve as both bibliography and additional reading sources. Review papers are helpful as a starting point for finding additional sources for concentrated reading in a selected area.

None of the views or information presented in this book reflects the opinion of any of the companies or individuals that have contributed to the book. If there are errors, they are oversight on the part of the author. A note indicating the specific error to the publisher, for the purpose of correction of future editions, would be much appreciated.

Sina Ebnesajjad   2000

Table of Contents

Introduction 1

PART I

1 Fundamentals 3

    1.1   Introduction 3
    1.2   What are Fluoropolymers? 4
    1.3   Fundamental Properties of Fluoropolymers 4
    1.4   Developmental History of Fluoropolymers 5
    1.5   Examples of Uses of Fluoropolymers 6
    References 8

2 Fluoropolymers: Properties and Structure 9

    2.1   Introduction 9
    2.2   Impact of F and C-F Bond on the Properties of PTFE 9
    2.3   Disturbing the PTFE Structure: Perfluorinated Ethylene-Propylene Copolymer (FEP) and Polychlorotrifluoroethylene 11
        2.3.1   Disturbing the PTFE Structure: Perfluorinated Ethylene-Propylene Copolymer (FEP) .. 11
        2.3.2   Disturbing the PTFE Structure: Polychlorotrifluoroethylene 12
    2.4   Reaction Mechanism 12
    2.5   Effect of Solvents on Fluoropolymers 13
    2.6   Molecular Interaction of Fluoropolymers: Low Friction and Low Surface Energy 14
    2.7   Conformations and Transitions of Polytetrafluoroethylene 15
    2.8   Conformations and Transitions of Polychlorotrifluoroethylene (PCTFE) 16
    References 17

3 Operational Classification of Fluoropolymers 19

    3.1   Introduction 19
    3.2   TFE Homopolymers 22
    3.3   TFE Copolymers 23
    3.4   CTFE Polymers 23
    3.5   Vinylidene Fluoride Polymers 23
    3.6   Vinyl Fluoride Polymers 23
    3.7   Process Classification 23
    References 24

4 Fluoropolymer Monomers 25

    4.1   Introduction 25
    4.2   Synthesis of Tetrafluoroethylene 25

4.3	Properties of Tetrafluoroethylene	26
4.4	Synthesis of Hexafluoropropylene	27
4.5	Properties of Hexafluoropropylene	27
4.6	Synthesis of Perfluoroalkylvinylethers	28
4.7	Properties of Perfluoroalkylvinylethers	29
4.8	Synthesis of Chlorotrifluoroethylene	30
4.9	Properties of Chlorotrifluoroethylene	30
	References	31

5 Homofluoropolymer Polymerization and Finishing 33

5.1	Introduction	33
5.2	Polymerization Mechanism	33
5.3	Tetrafluoroethylene Polymers	33
5.4	Preparation of Polytetrafluoroethylene by Suspension Polymerization	35
5.5	Preparation of Polytetrafluoroethylene by Dispersion Polymerization	40
5.6	Chlorotrifluoroethylene Polymers and Polymerization	46
5.7	Characterization of Polytetrafluoroethylene	48
5.8	Characterization of Polychlorotrifluoroethylene	54
	References	56

6 Commercial Grades of Homofluoropolymers 59

6.1	Introduction	59
6.2	Granular PTFE	59
6.3	PTFE Dispersions	59
6.4	Fine Powder PTFE	59
6.5	PCTFE Dispersions	60
6.6	Polychlorotrifluoroethylene Polymers	60
6.7	Fluoropolymer Manufacturers	60
	References	73

PART II

7 Fabrication and Processing of Granular Polytetrafluoroethylene 76

7.1	Introduction		76
7.2	Resin Selection		76
7.3	Compression Molding		78
	7.3.1	Equipment	79
		7.3.1.1 Mold Design	79
		7.3.1.2 Presses	80
		7.3.1.3 Ovens	81
	7.3.2	Densification and Sintering Mechanism	81

	7.3.3	Billet Molding	82
		7.3.3.1 Preforming	82
		7.3.3.2 Sintering	92
		7.3.3.3 Cooling	95
	7.3.4	Sintering Cycles	97
	7.3.5	Hot Compression-Molding	97
7.4	Automatic Molding		97
7.5	Isostatic Molding		97
	7.5.1	Introduction to Isostatic Molding	103
	7.5.2	Comparison of Isostatic with Other Fabrication Techniques	103
	7.5.3	Basic Isostatic Process	105
	7.5.4	Complex Isostatic Molding	105
	7.5.5	Wet and Dry Bag Isostatic Molding	107
	7.5.6	Isostatic Mold Design	107
		7.5.6.1 Preform Shape	107
		7.5.6.2 Surface Finish of Preform	110
		7.5.6.3 Pressurization Direction	110
		7.5.6.4 Other Design Factors	111
		7.5.6.5 Mold Configuration and Dimensions	111
		7.5.6.6 Mold and Bag Sealing	113
		7.5.6.7 Tooling and Material: Rigid Part	115
		7.5.6.8 Tooling and Material: Flexible Bag	115
		7.5.6.9 Design Procedure	116
	7.5.7	PTFE Resin Selection	116
	7.5.8	Isostatic Processing of PTFE Resins	116
7.6	Ram Extrusion		118
	7.6.1	Introduction to Ram Extrusion	118
	7.6.2	Ram Extrusion: Basic Technology	119
	7.6.3	Ram Extrusion: Resin Feed	119
	7.6.4	Ram Extrusion: Compaction	121
	7.6.5	Ram Extrusion: Sintering and Cooling	121
		7.6.5.1 Ram Extrusion: Heat Requirements of Sintering	121
		7.6.5.2 Ram Extrusion: Pressure Requirements	123
		7.6.5.3 Ram Extrusion: Cooling	127
		7.6.5.4 Ram Extrusion: Die Skin Formation	127
	7.6.6	Ram Extrusion Equipment Design	128
		7.6.6.1 Drive System	130
		7.6.6.2 Extrusion Ram	130
		7.6.6.3 Extrusion Die	131
		7.6.6.4 Resin Feed System	131
	7.6.7	Other Ram Extrusion	131
	7.6.8	Typical Resin, Process and Property Data	132
References			134

8 Fabrication and Processing of Fine Powder Polytetrafluoroethylene 135

- 8.1 Introduction 135
- 8.2 Resin Handling and Storage 135
- 8.3 Paste Extrusion Fundamentals 136
- 8.4 Extrusion Aid or Lubricant 138
- 8.5 Wire Coating 138
- 8.6 Extrusion of Tubing 150
- 8.7 Unsintered Tape 157
- 8.8 Expanded PTFE Manufacturing 162
- 8.9 Fine Powder Resin Selection 165
- References 166

9 Fabrication and Processing of PTFE Dispersions 168

- 9.1 Introduction 168
- 9.2 Applications 168
- 9.3 Storage and Handling 169
- 9.4 Principles of Coating Technology 169
- 9.5 Dispersion Formulation and Characteristics 173
- 9.6 Glass Cloth Coating 175
- 9.7 Impregnation of Flax and Polyaramide 178
- 9.8 Coating Metal and Hard Surfaces 179
- 9.9 PTFE Yarn Manufacturing 180
- 9.10 Film Casting 180
- 9.11 Anti-Drip Applications 182
- 9.12 Filled Bearings 182
- 9.13 De-Dusting Powders 182
- 9.14 Other Applications 183
- References 183

10 Processing of Polychlorotrifluoroethylene 185

- 10.1 Introduction 185
- 10.2 Processing Considerations 185
- 10.3 Compression Molding 187
- 10.4 Injection Molding 187
- 10.5 Extrusion 189
- 10.6 Machining and Bonding 189
- 10.7 Processing PCTFE Dispersions 189
- References 190

11 Fluoroadditives 191

- 11.1 Introduction 191
- 11.2 Feedstock 191

11.3	Degradation of Polytetrafluoroethylene	191
11.4	Production Methods	193
	11.4.1 Production of Fluoroadditives by Thermal Cracking (Pyrolysis)	193
	11.4.2 Production of Fluoroadditives by Electron Beam Irradiation	193
	11.4.3 Grinding Irradiated PTFE	196
11.5	Commercial Products	197
11.6	Applications	199
	11.6.1 Thermoplastic Modification	200
	11.6.2 Elastomers	200
	11.6.3 Printing Inks	200
	11.6.4 Paints and Coatings	201
	11.6.5 Oils and Greases	201
	11.6.6 Dry Lubricant	203
11.7	Regulatory Compliance	203
	References	204

12 Filled Fluoropolymer Compounds ... 206

12.1	Introduction	206
12.2	Granular-based Compounds	206
	12.2.1 Fillers	206
	12.2.2 Polytetrafluoroethylene Selection	208
	12.2.3 Filled PTFE—Production Techniques	208
12.3	Fine Powder-Based Compounds	209
	12.3.1 Fillers and Compounding Methods	210
	12.3.2 Fabrication of Reinforced Gasketing Material	210
	12.3.2.1 Process Background	211
	12.3.2.2 Process Description	211
12.4	Co-Coagulated Compounds	214
12.5	Processing Compounds	214
12.6	Typical Properties of Filled Fluoropolymers	216
	12.6.1 Mechanical Properties	217
	12.6.2 Thermal Properties	222
	12.6.3 Electrical Properties	222
	12.6.4 Chemical Properties	222
	12.6.5 Tribological Properties of Filled PTFE	223
12.7	Commercial Products	227
	References	232

PART III

13 Chemical Properties of Fluoropolymers ... 233

13.1	Introduction	233

	13.2	Chemical Compatibility of Polytetrafluoroethylene	233
		13.2.1 Effect of Ozone with Polytetrafluoroethylene	236
		13.2.2 Oxygen Compatibility of PTFE	236
	13.3	Chemical Compatibility of Polychlorotrifluoroethylene	237
	13.4	Permeation Fundamentals	237
	13.5	Permeation Measurement and Data	239
	13.6	Environmental Stress Cracking	241
	References		242

14 Properties of Tetrafluoroethylene Homopolymers ... 243

- 14.1 Introduction ... 243
- 14.2 Influence of Processing ... 243
  - 14.2.1 Measurement of Flaws ... 244
- 14.3 Mechanical Properties ... 246
  - 14.3.1 Deformation Under Load (Creep) and Cold Flow ... 249
  - 14.3.2 Fatigue Properties ... 254
  - 14.3.3 Impact Strength ... 254
  - 14.3.4 Hardness ... 255
  - 14.3.5 Friction ... 255
  - 14.3.6 PV Limit ... 256
  - 14.3.7 Abrasion and Wear ... 256
- 14.4 Electrical Properties ... 258
- 14.5 Thermal Behavior ... 259
  - 14.5.1 Thermal Stability ... 259
  - 14.5.2 Thermal Expansion ... 259
  - 14.5.3 Thermal Conductivity and Heat Capacity ... 263
  - 14.5.4 Heat Deflection Temperature ... 263
- 14.6 Irradiation Behavior ... 264
- 14.7 Standard Measurement Methods ... 265
- Reference ... 266

15 Properties of Chlorotrifluoroethylene Homopolymers ... 267

- 15.1 Introduction ... 267
- 15.2 Crystallinity ... 267
- 15.3 Mechanical Properties ... 267
- 15.4 Electrical Properties ... 267
- 15.5 Thermal Properties ... 267
- 15.6 Irradiation Behavior ... 268
- 15.7 Properties of PCTFE Films ... 268
- References ... 275

16 Fabrication Techniques for Fluoropolymers ... 276

- 16.1 Introduction ... 276

16.2	Machining 276
	16.2.1	Sawing and Shearing 277
	16.2.2	Drilling, Tapping and Threading 277
	16.2.3	Skiving 277
16.3	Adhesive Bonding Methods 277
	16.3.1	Contact Adhesives 277
	16.3.2	Bonding Adhesives 278
	16.3.3	Sodium Etching 278
	16.3.4	Plasma Treatment 279
16.4	Welding and Joining 281
	16.4.1	Welding Technique 281
16.5	Thermoforming 283
16.6	Other Processes 283
Reference 284

17 Typical Applications of Fluoropolymers 285

17.1	Chemical Processing 285
17.2	Piping 285
17.3	Vessels 286
17.4	CPI Components 288
17.5	Seals and Gaskets 288
17.6	Self-Supporting Components 289
17.7	Trends in Using Fluoropolymers in Chemical Service 289
17.8	Semiconductor Processing 289
	17.8.1	Trends for the Use of Fluoropolymers in the Semiconductor Industry 290
17.9	Electrical and Mechanical 290
	17.9.1	Electrical Applications 290
	17.9.2	Mechanical Applications 293
17.10	Automotive 294

18 Safety, Disposal, and Recycling of Fluoropolymers 296

18.1	Introduction 296
18.2	Toxicology of Fluoropolymers 296
18.3	Thermal Properties of Fluoropolymers 296
18.4	Emission During Processing 297
18.5	Safety Measures 297
	18.5.1	Ventilation 297
	18.5.2	Processing and Fabrication 299
		18.5.2.1 Sintering 300
		18.5.2.2 Paste Extrusion 300
		18.5.2.3 Dispersion Coating 300
		18.5.2.4 Melt Processing 300
		18.5.2.5 Machining 300
		18.5.2.6 Soldering and Melt Stripping 301

| | | 18.5.2.7 | Welding Fluoropolymer .... 301 |
| | | 18.5.2.8 | Welding and Flame-Cutting Fluoropolymer-lined Metals .... 301 |

      18.5.3 Spillage Cleanup 301
      18.5.4 Equipment Cleaning and Maintenance 301
      18.5.5 Protective Clothing 301
      18.5.6 Personal Hygiene 301
      18.5.7 Fire Hazard 301
      18.5.8 Material Incompatibility 301

18.6 Food Contact and Medical Applications 302
18.7 Fluoropolymer Scrap and Recycling 302
18.8 Environmental Protection and Disposal Methods 303
Reference 304

Appendix I: Polytetrafluoroethylene 305

Appendix II: Polychlorotrifluoroethylene 311

Appendix III: Trademarks 319

Glossary 320

Index 347

Introduction

It is hard for most people to imagine what the world would be like without plastics. It is equally difficult for many design engineers to envision designing projects without the availability of fluoropolymers. Only expensive exotic metals can replace the chemical inertness of these plastics to highly corrosive and aggressive substances. Even though some engineering plastics can stand up to temperatures as well as fluoropolymers, they suffer major property losses beyond their glass transition temperatures. Engineering plastics lack the chemical inertness and low friction characteristics of fluoropolymers. No other material offers the low coefficient of friction of fluoropolymers.

Fluoropolymers have evolved over six decades. New monomers have allowed the synthesis of new polymers with new methods of processing. Most newer perfluoropolymers have the same basic properties as polytetrafluoroethylene (PTFE) but they offer new methods of processing. These new techniques have greatly increased the range of parts that can be fabricated from fluoropolymers at reduced cost. Today, fluoropolymers are processed by methods almost identical to those used a half century ago as well as by state of the art molding technologies.

This book is the first of two volumes about fluoropolymers. The division of the volumes is based on the processing techniques of commercial fluoropolymers. Homopolymers of tetrafluoroethylene (TFE), the main building block or "monomer" for polytetrafluoroethylene plastics are processed by non-traditional techniques. The extremely high melt viscosity of TFE polymers precludes their processing by conventional melt processing methods such as injection molding and melt extrusion. Copolymers of trace amounts of other perfluorinated monomers with TFE known as *Modified PTFE* are covered in this book because they are processed by the same techniques. Copolymers of TFE with extensive comonomer contents are handled by melt-processing methods. Volume One is mainly devoted to exploring the various homopolymers of tetrafluoroethylene and chlorotrifluoroethylene. Polychlorotrifluoroethylene (PCTFE) plastics, which can be processed by non-melt processing and some melt-processing techniques, have been covered in Volume One.

All aspects of the fluoropolymers including monomer synthesis, polymerization, properties, applications, part fabrication techniques, safety in handling, and recycling are discussed. Homopolymers and copolymers of vinylidene fluoride, all melt processable, have been covered in Volume Two because of the close resemblance of polyvinylidene fluoride in composition and in processing techniques to melt processable copolymers of tetrafluoroethylene. Polyvinyl fluoride (PVF) which is processed by melt processing of its dispersion in a polar solvent has also been covered in Volume Two.

Fluoropolymers have outstanding chemical resistance, low coefficient of friction, low dielectric constant, high purity, and broad use temperatures. Most of these properties are enhanced with an increase in the fluorine content of the polymers. For example, polytetrafluoroethylene, which contains four fluorine atoms per repeat unit, has superior properties compared to polyvinylidene fluoride, which has two fluorine atoms for each repeat unit. Generally, these plastics are mechanically weaker than engineering polymers. Their relatively low values of tensile strength, deformation under load or creep, and wear rate require the use of fillers and special design strategies. This book addresses ways of overcoming these shortcomings.

The primary objective in every section is to facilitate comprehension by a reader with a modest background in science or engineering. There are a number of measurement methods and techniques which are unique to fluoropolymers. In each case, these methods are defined for the reader. The book is divided into three parts; each part consists of several chapters. Each chapter has been written so it can be consulted independently of the others.

The first part of the book deals with definitions and fundamental subjects surrounding the polymerization of fluoropolymers. Basic subjects such as the identification of fluoropolymers, their key properties, and some of their everyday uses are addressed. The main monomer, tetrafluoroethylene, is extremely flammable and explosive. Consequently, safe polymerization of this monomer requires special equipment and technology. Molecular forces within these polymers are reviewed and connected to macro properties. Monomor and polymer synthesis techniques and properties are described. Part One ends with a detailed list of advertised commercial grades of various forms of PTFE and PCTFE.

Part Two describes the nature of and fabrication techniques for the various forms of fluoropolymers. The products of commercially common regimes of polymerization, slurry and dispersion are processed in different ways. The three product forms of PTFE, granular, fine powder and dispersion, of which the last two forms are products of dispersion polymerization, are described. Fabrication methods for the various forms of homopolymers are explained in detail. Fluoroadditives and filled-compounds are covered in Part Two. All important aspects of fluoroadditives and filled-compounds are collected in separate chapters.

Part Three describes the properties, characteristics, and applications of fluoropolymers. The reader involved in the processing of fluoropolymers is advised to carefully study the safety issues and disposal methods of fluoropolymers. In this book, Ch. 18, "Safety, Disposal, and Recycling of Fluoropolymers," covers both topics and provides additional references.

References are provided in each section to allow the interested reader further exploration of each topic. In this volume the author has drawn on his own and many colleagues' experiences over nearly two decades. We trust that readers will find this book helpful both as a text and as a design reference.

PART I
1 Fundamentals

1.1 Introduction

The era of fluoropolymers began with a small mishap, which did not go unnoticed by the ingenious and observant Dr. Roy Plunkett of DuPont Company.[1] In 1938, he had been at DuPont for two years, concentrating mostly on the development of fluorinated refrigerants. He was experimenting with tetrafluoroethylene (TFE) for synthesis of a useful refrigerant ($CClF_2$-CHF_2).[2] The effort was spurred by the desire to create safe, nonflammable, nontoxic, colorless, and odorless refrigerants. On the morning of April 6, 1938, when Plunkett checked the pressure on a full cylinder of TFE, he found none. However, the cylinder had not lost weight. Careful removal of the valve and shaking the cylinder upside down yielded a few grams of a waxy looking white powder—the first polymer of *tetrafluoroethylene*.[2]

Plunkett analyzed the white powder, which was conclusively proven to be *polytetrafluoroethylene* (PTFE). The slippery PTFE could not be dissolved in any solvent, acid, or base and upon melting formed a stiff clear gel without flow.[3] Later, research led to the discovery of processing techniques similar to those used with metal powders. At the time, the Manhattan Project was seeking new corrosion-resistant material for gaskets, packings, and liners for UF_6 handling. PTFE provided the answer and was used in production. The US government maintained a veil of secrecy over the PTFE project until well after the end of World War II.

Large-scale monomer synthesis and controlled polymerization were technical impediments to be resolved. Intensive studies solved these problems and small-scale production of Teflon® (trademark, 1944) began in 1947. In 1950, DuPont scaled up the commercial production of Teflon in the USA with the construction of a new plant in Parkersburg, West Virginia. In 1947, Imperial Chemical Industries built the first PTFE plant outside the US, in Western Europe. Since then, many more plants have been built around the globe. Over the last six decades, many forms of PTFE and copolymers of other monomers and TFE have been developed and commercialized.

The words of Plunkett himself best summarize the discovery of PTFE. He recounted the story of Teflon in a speech to the American Chemical Society at its April 1986 meeting in New York. "The discovery of polytetrafluoroethylene (PTFE) has been variously described as *(i)* an example of serendipity, *(ii)* a lucky accident and *(iii)* a flash of genius. Perhaps all three were involved. There is complete agreement, however, on the results of that discovery. It revolutionized the plastics industry and led to vigorous applications not otherwise possible."[2]

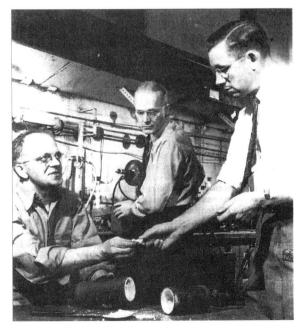

Figure 1.1 Reenactment of the discovery of polytetrafluoroethylene (Teflon®) in 1938 by DuPont scientist Dr. Roy Plunkett *(right)*. *(Courtesy DuPont.)*

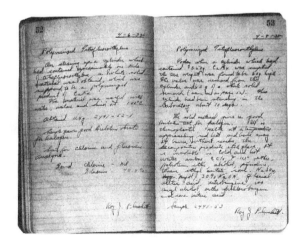

Figure 1.2 Photograph of two pages from the research notebook of Dr. Roy Plunkett recording the discovery of polytetrafluoroethylene on April 6, 1938. *(Courtesy DuPont.)*

1.2 What are Fluoropolymers?

Traditionally, a fluoropolymer or fluoroplastic is defined as a polymer consisting of carbon (C) and fluorine (F). Sometimes these are referred to as *perfluoropolymers* to distinguish them from partially fluorinated polymers, fluoroelastomers and other polymers that contain fluorine in their chemical structure. For example fluorosilicone and fluoroacrylate polymers are not referred to as fluoropolymers. An example of a linear fluoropolymer is tetrafluoroethylene polymer (PTFE):

```
  F   F   F   F
  |   |   |   |
— C — C — C — C —
  |   |   |   |
  F   F   F   F
```

A simplistic analogy would be to the chemical composition of polyethylene [(-CH$_2$-CH$_2$-)$_n$] where all the hydrogen atoms have been replaced by fluorine atoms. Of course, in practice PTFE and polyethylene are prepared in totally different ways. There are branched fluoropolymers such as fluorinated ethylene propylene polymer (FEP):

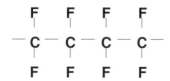

Oxygen (O) and chlorine (Cl) are present in the chemical structure of some commercial fluoropolymers. Examples include perfluoroalkoxy and polychlorotrifluoroethylene:

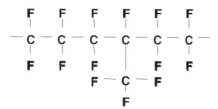

R_f is usually a perfluorinated group consisting of carbon and fluorine. Introduction of nonlinearity, oxygen and side chains, or chlorine invoke a variety of polymer properties which will be dealt with later in this book.

There is a second class of fluoropolymers called "partially fluorinated" in contrast to "perfluorinated polymers." These molecules include hydrogen (H) in addition to fluorine and carbon. Examples include polyvinylfluoride, polyvinylidenefluoride, and ethylene tetrafluoroethylene copolymer.

```
  H   H   H   H
  |   |   |   |
— C — C — C — C —      Polyvinyl fluoride
  |   |   |   |
  H   F   H   F
```

```
  H   F   H   F
  |   |   |   |
— C — C — C — C —      Polyvinylidenefluoride
  |   |   |   |
  H   F   H   F
```

```
  F   F   H   H   F   F
  |   |   |   |   |   |
— C — C — C — C — C — C —
  |   |   |   |   |   |
  F   F   H   H   F   F
```
Ethylene Tetrafluoroethylene Copolymer

Partially fluorinated fluoropolymers are significantly different from the perfluoropolymers with respect to properties and processing characteristics. For example, perfluoropolymers are more thermally stable but physically less hard than partially fluorinated polymers. The former has much higher "hardness" than the latter. Applications for both classes of polymers are discussed later in this book and in Vol. 2.

1.3 Fundamental Properties of Fluoropolymers

The basic properties of fluoropolymers arise from the atomic structure of fluorine and carbon and their covalent bonding in specific chemical structures. These properties are weakened as the chemical structure becomes less "perfluorinated," as in polyvinylidene fluoride. Because PTFE has a linear structure, it is a good subject for discussion of extreme properties. The backbone is formed of carbon-carbon bonds and carbon-fluorine bonds. Both are extremely strong bonds (C-C = 607 kJ/mole and C-F = 552 kJ/mole.)[4][5] The basic properties of PTFE stem from these two very strong chemical bonds. The PTFE molecule resembles a carbon rod completely blanketed with a sheath of fluorine atoms.[6]

The size of the fluorine atom allows the formation of a uniform and continuous sheath around the carbon-carbon bonds and protects them from attack, thus imparting chemical resistance and stability to the molecule. The fluorine sheath is also responsible for the low surface energy (18 dynes/cm)[7] and low coefficient of friction (0.05-0.8, static)[6] of PTFE. Another attribute of the uniform fluorine sheath is the electrical inertness (or non-polarity) of the PTFE molecule. Electrical fields impart only slight polarization in this molecule, so volume and surface resistivity are high. Table 1.1 summarizes the fundamental properties of PTFE.

The basic properties of PTFE provide beneficial attributes with high commercial value (Table 1.2).

Table 1.1. Fundamental Properties of PTFE

- High melting point, 342°C (648°F)
- High thermal stability
- Useful mechanical properties at extremely low and high temperatures
- Insolubility
- Chemical inertness
- Low coefficient of friction
- Low dielectric constant/dissipation factor
- Low water ab/adsorptivity
- Excellent weatherability
- Flame resistance
- Purity

Table 1.2. Useful Attributes of PTFE

- Stability
  - high continuous use temperature
  - excellent weatherability
  - excellent chemical resistance
  - excellent fire properties
- Low Surface Energy
  - good release properties
  - biological inertness
  - low friction
- Cryogenic Properties
  - retains flexibility
- Electrical Properties
  - low dielectric constant
  - low dissipation factor

1.4 Developmental History of Fluoropolymers

The development of fluoropolymers began with the invention of PTFE in 1938 and continued to 1992 when a soluble perfluoropolymer (Teflon® AF) was introduced. Table 1.3 summarizes the timeline for the development of fluoropolymers that have brought about major changes in properties and/or fabrication processes.

The discovery of PTFE was a major leap forward in material science. Yet, the new polymer could not be fabricated by melt-processing. The next two forms of PTFE, fine powder and dispersion, were also not melt-processable. The pursuit of a more easily processible polymer led to FEP, which could be melted in an extruder. Compared with PTFE the major disadvantage of FEP is its reduced thermal stability and lower maximum continuous use temperature (200°C) (Table 1.3). PFA, which was introduced in 1973, offers both melt-processing and the same upper continuous use temperature as PTFE (260°C).

ETFE addresses the need for a mechanically stronger polymer, albeit at a loss of fluoropolymer properties because of the presence of hydrogen in its molecule:

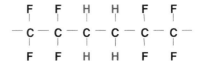

Compared to PTFE, ETFE has a lower continuous use temperature (150°C) less chemical resistance, and a higher coefficient of friction. Mechanical properties, including tensile strength, elongation at break, and tensile modulus is increased, leading to cut-through resistance in wire insulation.

Teflon® AF is an amorphous polymer which is soluble in certain halogenated solvents. It can be applied as a solution, followed by the removal of the solvent. The remaining coating will be as resistant to almost as many chemicals as PTFE. The thickness of the coating can range upward from less than a micrometer.

There are a number of other polymers in this family including polychlorotrifluoroethylene (PCTFE), polyvinylfluoride (PVF), polyvinylidenefluoride, ethylene chlorotrifluoroethylene (ECTFE), tetrafluoroethylene/hexafluoropropylene/vinylidene fluoride copolymers, perfluoroacrylates, fluorinated polyurethanes, and chlorotrifluoroethylene/vinylether copolymers.

Table 1.3. Commercialization Timeline of Major Fluoropolymers vs Key Processing/Application Trade-Offs

Fluoropolymer	Year Commercialized	Monomers	Trade-off +	Trade-off -
PTFE	1947	TFE	Continuous Use Temperature 260°C	Non-melt Processable
PCTFE	1953	CTFE	Melt Processable/Non-Melt Processable	Maximum Continuous Use Temperature 180°C
FEP	1960	TFE, HFP[3]	Melt-Processable	Maximum Continuous Use Temperature 200°C
PVF	1961	VF[1]	Thin film/Weatherable	Maximum Continuous Use Temperature 107°C
PVDF	1961	VDF[2]	Melt Processable	Maximum Continuous Use Temperature 150°C
ECTFE	1970	CTFE, E[4]	Hardness/Toughness	Maximum Continuous Use Temperature 150°C
PFA	1972	TFE, PAVE[5]	Melt Processable. Continuous Use Temperature 260°C	Low Molecular Weight
ETFE	1973	TFE, E	Hardness/Toughness	Maximum Continuous Use Temperature 150°C
Teflon®AF	1985	TFE, PDD[6]	Soluble in Special Halogenated Solvents	High Cost

[1] Vinyl fluoride ($CH_2=CHF$)
[2] Vinylidene fluoride ($CH_2=CF_2$)
[3] Hexafluoropropylene ($CF_2=CF-CF_3$)
[4] Ethylene ($CH_2=CH_2$)
[5] Perfluoroalkylvinylether ($CF_2=CF-O-R_f$)
[6] 2,2-Bistrifluoromethyl-4,5difluoro-1,3-dioxole:

1.5 Examples of Uses of Fluoropolymers

The consumption of PTFE has increased over the years as technological advancement has required the properties of fluoropolymers. The applications of PTFE, and fluoropolymers in general, span all facets of human life, from household uses to the aerospace and electronic industries. Basic properties (Table 1.1) of PTFE lead directly to its applications: chemical resistance, thermal stability, cryogenic properties, low coefficient of friction, low surface energy, low dielectric constant, high volume and surface resistivity, and flame resistance. Applications for fluoropolymers always exploit one or more of the properties (Table 1.4) that set them apart from other materials, particularly other plastics.

In the chemical process industry, for example, fluoropolymers are selected for their unmatched resistance to chemical attack. They serve as linings for carbon steel vessels, and for piping and other fluid handling components. They provide durable, low maintenance and economical alternatives to exotic metal alloys. In these applications, fluoropolymers also offer thermal stability for use at high temperatures. And because they do not react with process streams, they help prevent contamination of products.

Table 1.4. Major Applications and Some Uses of PTFE

INDUSTRY/ APPLICATION AREA	KEY PROPERTIES	TYPICAL USES
Chemical Processing	Chemical Resistance Good Mechanical Properties Thermal Stability Cryogenic Properties	Gaskets, Vessel Liners, Valve and Pipe Liners, Tubing, Coatings
Electrical & Communications	Low Dielectric Constant High Volume/Surface Resistivity High Dielectric Break-down Voltage Flame Resistance, Thermal stability	Wire and Cable Insulation, Connectors
Automotive & Office Equipment	Low Coefficient of Friction Good Mechanical Properties Cryogenic Properties, Chemical Resistance	Seals and Rings in Automotive Power Steering, Transmission, and Air-conditioning. Copier Roller and Food Processing Equipment Covering.
Houseware	Thermal Stability Low Surface Energy Chemical	Cookware Coatings.
Medical	Low Surface Energy Stability Excellent Mechanical Properties Chemical Resistance	Cardiovascular Grafts, Heart Patches, Ligament Replacement
Architectural Fabric	Excellent Weatherability Flame Resistance Low Surface Energy	Coated Fiberglass Fabric for Stadiums and Airport Roofs

Electrical properties of fluoropolymers are highly valuable in electronic and electrical applications. In data communications, for example, FEP is used to insulate cables installed in air-handling spaces (plenums) in office buildings. FEP provides the excellent dielectric properties these cables require to perform well at high data transmission rates as well as long-term stability so performance will not change over the life of the cabling system. Most importantly, FEP helps these cables meet strict building code requirements for low flame spread and low smoke generation.

Fluoropolymers are used to insulate wire for critical aerospace and industrial applications where chemical and thermal resistance is essential. They are also materials of construction for connectors for high-frequency cables and for thermocouple wiring that must resist high temperatures.

In the automotive, office equipment and other industries, mechanical properties of fluoropolymers are beneficial in low-friction bearings and seals that resist attack by hydrocarbons and other fluids. In food processing, the Food and Drug Administration (FDA) approved fluoropolymer grades are fabrication material for equipment due to their resistance to oil and cleaning materials, and their anti-stick and low friction properties.

In houseware, fluoropolymers are applied as non-stick coatings for cookware and appliance surfaces. These applications depend on thermal and chemical resistance as well as anti-stick performance. PTFE and ETFE are chosen to insulate appliance wiring that must withstand high temperatures.

Medical articles such as surgical patches and cardiovascular grafts rely on the long-term stability of fluoropolymers as well as their low surface energy and chemical resistance.

For airports, stadiums and other structures, glass fiber fabric coated with PTFE is fabricated into roofing and enclosures. This architectural fabric is supported by cables or air pressure to form a range of innovative structures. PTFE provides excellent resistance to weathering, including exposure to ultraviolet rays in sunlight, flame resistance for safety, and low surface energy for soil resistance and easy cleaning.

References

1. Plunkett, R. J., US Patent 2,230,654, assigned to DuPont Co., Feb. 4, 1941.
2. Plunkett, R. J., "The History of Polytetrafluoroethylene: Discovery and Development," in *High Performance Polymers: Their Origin and Development,* Proceed. Of Symp. On the Hist. Of High Perf. Polymers at the ACS Meeting in New York, April 1986, (R. B. Seymour and G. S. Kirshenbaum, eds.), Elsevier, New York, 1987.
3. Gangal, S. V., "Polytetrafluoroethylene, Homopolymers of Tetrafluoroethylene," in *Encyclopedia of Polymer Science and Engineering,* 2nd ed., 16:577–600, John Wiley & Sons, New York, 1989.
4. Cottrell, T. L., *The Strength of Chemical Bonds,* 2nd ed., Butterworths, Washington, D. C., 1958.
5. Sheppard, W. A., and Sharts, C. M., *Organic Fluorine Chemistry,* W. A. Benjamin, Inc., New York, 1969.
6. Gangal, S. V., "Polytetrafluoroethylene" in *Encyclopedia of Chemical Technology,* 4th ed., 11:621–644, John Wiley & Sons, New York, 1994.
7. Zisman, W. A., "Surface Properties of Plastics," *Record of Chemical Progress,* 26:1 1965.

2 Fluoropolymers: Properties and Structure

2.1 Introduction

This chapter examines why fluoropolymers exhibit extreme properties. It focuses on the reasons that replacement of hydrogen with fluorine in hydrocarbon macromolecules improves their thermal stability, chemical resistance, electrical properties and coefficient of friction. Understanding of the role of fluorine in determining the properties of a polymer will contribute to a more in depth appreciation of some of the other information in this book. It will also allow the readers to make more informed judgments about fluoropolymers and their applications.

Design engineers routinely encounter questions about the choice of materials in their projects. A thorough understanding of the role of fluorine will aid in the decision to specify fluoropolymer for a given design scenario, and whether a perfluoropolymer or partially fluorinated polymer will meet the requirements of the application.

2.2 Impact of F and C-F Bond on the Properties of PTFE

Fluorine is a highly reactive element with the highest electronegativity of all elements.[1] The change in the properties of compounds where fluorine has replaced hydrogen can be attributed to the differences between C-F and C-H bonds.

A simple way to frame the issues is to explore the differences between linear polyethylene (PE) and polytetrafluoroethylene—their chemical structures are similar except that in the latter, F has replaced H without destructive distortion of the geometry of the former:

Polyethylene

Polytetrafluoroethylene

There are large and significant differences between most properties of PE and PTFE. Four properties are vastly altered in PTFE:

1. PTFE has one of the lowest surface energies among the organic polymers
2. PTFE is the most chemically resistant organic polymer
3. PTFE is one of the most thermally stable among organic polymers
4. PTFE's melting point and specific gravity are more than double PE's

The properties of these two polymers are listed in Table 2.1

Table 2.1. A Comparison of PTFE and PE Properties[2]-[4]

PROPERTY	POLYTETRAFLUOROETHYLENE)	POLYETHYLENE
Density	2.2–2.3	0.92–1
Melting Temperature, °C	342 (first) 327 (second)	105–140
Dielectric Constant (1 kHz)	2.0	2.3
Dynamic Coefficient of Friction	0.04	0.33
Surface Energy, dynes/g	18	33
Resistance to Solvents & Chemicals	Excellent No known solvent	Susceptible to hot hydrocarbons
Thermal Stability[1] $T_{1/2}$, °C k_{350}, %/min E_{act}, kJ/mol	505 0.000002 339	404 0.008 264
Melt Viscosity[2], Poise	10^{10}–10^{12}	-
Refractive Index	1.35	1.51
Chain Branching Propensity	No	Yes

[1] $T_{1/2}$ is the temperature at which 50% of the polymer is lost after 30 minutes heating in vacuum; k_{350} is the rate of volatilization, i.e., weight loss, at 350°C; E_{act} is the activation energy of thermal degradation.

[2] Melt Creep Viscosity for PTFE at 380°C.

Let us compare the C-F and C-H bonds. Table 2.2[1][5] summarizes the key differences arising from the differences in the electronic properties and sizes of F and H. In comparing F and H, several relevant differences are noted:

1. F is the most electronegative of all elements
2. F has unshared electron pairs
3. F is more easily converted to F-
4. Bond strength of C-F is higher than C-H
5. F is larger than H

The electronegativity of carbon (2.5 Paulings) is somewhat higher than that of hydrogen (2.1 Paulings) and significantly lower than fluorine (4.0 Paulings). The consequence of the electronegativity values is that the polarity of the C-F bond is opposite that of the C-H bond, and the C-F bond is more highly polarized (Fig. 2.1). In other words, F has a higher electron density because it pulls the shared pair of electrons closer to itself relative to the center point of the C-F bond. Conversely, in the C-H bond the electron pair is closer to carbon, which has a higher electron density.

Polyethylene melts at 100–140°C, depending on the extent of branching, as compared to PTFE that melts at 327°C (first melting point 342°C). One could expect that weak intermolecular forces in PTFE should result in a lower melting point. On the contrary, its melting point is significantly higher than polyethylene. Why? The nature of the intermolecular forces, which are responsible for its high melting point, in PTFE is not fully understood. The answer may be in the differences between the molecular structure conformation and crystalline structure of polyethylene and polytetrafluoroethylene. Fluorine atoms are much larger than hydrogen resulting in less PTFE chain mobility than polyethylene. Steric repulsion due to the size of the fluorine atoms prevents the polytetrafluoroethylene from forming a polyethylene-like planar zig-zag

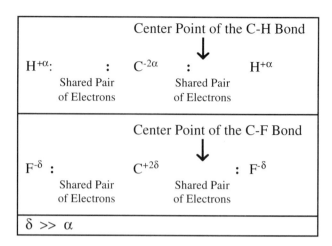

Figure 2.1 Comparative polarization of C-H and C-F bonds.

conformation. Instead, its conformation is helical in which steric repulsion is minimized.

The difference in bond polarity of C-H and C-F affects the relative stability of the conformations of the two polymer chains. Crystallization of polyethylene takes place in a planar and *trans* conformation. PTFE can be forced into such a conformation at extremely high pressure.[6] PTFE, in contrast, below 19°C, crystallizes as an incommensurate helix with approximately 0.169 nm per repeat distance: therefore, it takes 13 C atoms for a 180° turn to be completed. Above 19°C, the repeat distance increases to 0.195 nm which means that 15 carbon atoms will be required for a 180° turn to be completed.[7] At above 19°C the chains are capable of angular displacement, which increases above 30°C until the melting point is reached (327°C).

Substitution of F for H in the C-H bond substantially increases the bond strength from 99.5 kcal/mole for the C-H bond to 116 kcal/mole for the C-F bond. Consequently, thermal stability and chemical resistance of PTFE are much higher than PE because more energy is required to break the C-F bond. Additionally, the size of the F atom and the length of the C-F bond

Table 2.2. Electronic Properties of Hydrogen and Halogens[1][5]

ELEMENT (Preferred Ionic Form)	ELECTRONIC CONFIGURATION	ELECTRO NEGATIVITY, Pauling	IONIZATION ENERGY, kcal/g atom $X^+ + e^- > X$	ELECTRON AFFINITY, kcal/g atom $X + e^- > X^-$	C-X BOND ENERGY IN CX_4, kcal/mole	C-X BOND LENGTH IN CX_4, Å
H (H^+)	$1s^1$	2.1	315.0	17.8	99.5	1.091
F (F^-)	$1s^1$ $2s^2 2p^5$	4.0	403.3	83.5	116	1.317
Cl (Cl^-)	$1s^1$ $2s^2 2p^5$ $3s^2 3p^5 3d^0$	3.0	300.3	87.3	78	1.766

(Table 2.2) are such that the carbon backbone is blanketed with fluorine atoms which render the C-F bond impervious to solvent attack. The polarity and strength of the C-F bond rule out an F atom abstraction mechanism for formation of chain branches in PTFE. In contrast, highly branched polyethylene (>8 branches per 100 carbon atoms) can be synthesized with relative ease.[8] The branching mechanism can be used to adjust the crystallinity of polyethylene to produce polymers with differing properties. This approach is not available in PTFE and instead comonomers with pendent groups have to be polymerized with TFE.

PTFE is insoluble in common solvents. The replacement of H with the highly electronegative F renders PTFE immiscible with protonated material. Conversely, PE can be plasticized and dissolved above its melting point much more easily than PTFE. PTFE absorbs only small amounts of perhalogenated solvents such as perchloroethylene and carbon tetrachloride. The insolubility of PTFE in solvents is one of its most important characteristics for its application in lining equipment for processing corrosive chemicals.

In summary, the characteristics of F and C-F bonds give rise to the high melting point, low solubility, high thermal stability, low friction and low surface energy of PTFE.

2.3 Disturbing the PTFE Structure: Perfluorinated Ethylene-Propylene Copolymer (FEP) and Polychlorotrifluoroethylene

Polytetrafluoroethylene polymerization allows an overwhelming majority of the chains to crystallize, despite their very large molecular weight. This is important to the development of properties such as high modulus, low coefficient of friction, and high heat-deflection temperature (HDT). Crystallinity of virgin PTFE (never melted) is in the range of 92–98%,[9] consistent with an unbranched chain structure. A reasonable question is whether induced branching, or substitution of a different atom for fluorine, can affect the properties of polytetrafluoroethylene.

2.3.1 Disturbing the PTFE Structure: Perfluorinated Ethylene-Propylene Copolymer (FEP)

FEP, a copolymer of TEF and hexafluoropropylene (HFP), contains a tertiary carbon at the branch point bonded to a pendent CF_3. This carbon should have less thermal stability than primary carbons and, to a lesser extent, secondary carbons, that constitute the rest of the backbone of the polymer chain. This is due to a steric effect in which the chain departs from a helix at the branch point. Figure 2.2 shows the results of thermogravimetric analysis (TGA) of PTFE and FEP after one hour of heating in the air. A comparison of degradation curves indicates onset degradation temperatures of 300°C for FEP (0.02% wt. loss) and 425°C for PTFE (0.03% wt. loss).

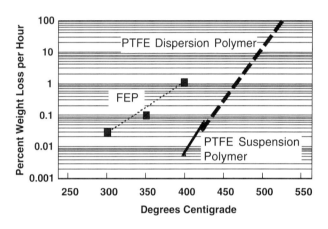

Figure 2.2 A comparison of thermal degradation of FEP and PTFE in air.[10]

Table 2.3 provides a comparison of the properties of FEP and PTFE. Melting point, processing temperature, degradation temperature, and Upper Continuous Use Temperature all have decreased significantly, the most important being the use temperature. The reason for lower thermal stability lies in the greater susceptibility, to oxidation, of the tertiary carbon bonded to the pendent perfluoromethyl group. FEP has about half the crystallinity of PTFE even though its molecular weight is an order of magnitude lower. CF_3 side chains disrupt the crystallization sufficiently to reduce the crystalline content. The melt viscosity of FEP is almost one hundred million times lower than PTFE, which places it among the melt-processible thermoplastic polymers.

Table 2.3. A Comparison of the Properties of FEP and PTFE[10]

PROPERTY	FEP	PTFE
Melting Point, °C	265	327
Processing Temperature, °C	360	400
TGA Loss Temperature of 1%/hr, °C	380	465
Upper Continuous Use Temperature, °C	200	260
MV (380°C), Poise	10^4-10^5	10^{11}-10^{12}
Crystallinity of Virgin Polymer, % wt.	40-50	92-98

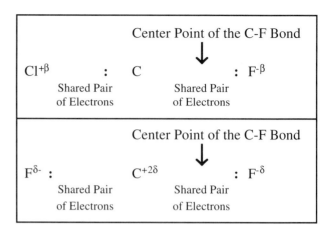

Figure 2.3 Comparative polarization of C-Cl and C-F bonds.

2.3.2 Disturbing the PTFE Structure: Polychlorotrifluoroethylene

Chlorine has a larger atom than fluorine according to the Electronic Configuration data in Table 2.2. It disturbs both the geometric perfection of the chain and its electronic balance because chlorine has a lower electronegativity than fluorine. This means that the symmetry of electron density is disturbed where chlorine has replaced fluorine (Fig. 2.3). The proximity of the shared electron pairs changes in favor of fluorine, therefore rendering the chlorine end of the Cl-C-F less negative. Chlorine in PCTFE happens to be where attack on the molecule commences because of C–Cl bond energy. The various types of attacks include oxidation, branching, chain scission, cross-linking, solvent swelling and partial or complete solubilization.

2.4 Reaction Mechanism

Perfluoroolefins, such as tetrafluoroethylene, are generally much more vulnerable to attack by nucleophiles than electrophiles which is the reverse of hydrocarbon olefins. Nucleophilic attacks occur on the fluoroolefins by the scheme proposed in Figure 2.4. The nucleophile (Nuc) approaches the carbon side of the double bond (**I**) searching for a positive charge leading to the formation of a carbon ion (**II**). For example, if the nucleophilic compound was methoxy sodium, the CH_3-O- side of the molecule would be approaching TFE. The carbon ion (**II**) is unstable and will give off a F^- ion and generate reaction products. Which product is generated depends on the nature of reaction medium. In the example of methoxy sodium, in the absence of a proton donor such as water, F^- would combine with Na^+ to produce NaF and perfluorovinyl methyl ether (**III**).

Generally, PTFE is not susceptible to nucleophilic attack because of the absence of double bonds. It is still susceptible to loss of fluorine by electrophilic attack particularly under heat and over long periods of exposure. Alkali metals, which are highly reactive elements, such as cesium, potassium, sodium, and lithium are among the most likely candidates for abstraction of fluorine from PTFE by an electrophilic mechanism. Certain other metals such as magnesium can attack PTFE if they are highly activated by etching or other means.

$$\underline{\text{Nucleophile}} + CF_2=CF_2 \rightarrow [\underline{\text{Nuc}}]^{\delta+}[CF_2=CF_2]^{\delta-}$$
$$\text{TFE} \qquad \qquad \textbf{I}$$
$$[\text{Nuc}]^{\delta+}[CF_2=CF_2]^{\delta-} \rightarrow \text{Nuc}^+\text{-}CF_2\text{-}CF_2^- \rightarrow F^- \rightarrow \text{Reaction Products}$$
$$\textbf{II}$$
EXAMPLE: Nuc = CH_3-O-Na [No proton donor like water is present]

$$CH_3\text{-O-Na} + CF_2=CF_2 \rightarrow \text{NaF} + CH_3\text{-O-CF}=CF_2$$
$$\textbf{III}$$

Figure 2.4 Proposed reaction scheme for nucleophilic attack on fluoroolefins.[2]

Loss of fluorine destabilizes PTFE's structure. As the fluorine-to-carbon ratio decreases, the color of PTFE changes from white to brown and then to black. The black layer is normally comprised of carbon, some oxygen, and small amounts of other elements.

2.5 Effect of Solvents on Fluoropolymers

Earlier in this chapter, the structure of PTFE was likened to a carbon rod completely blanketed with fluorine atoms which render the C–F bond impervious to solvent attack. This postulate has been proven by testing the effect of almost all common solvents on this polymer. There are no known solvents for PTFE below its melting point. PTFE is attacked only by molten alkali metals, chlorine trifluoride, and gaseous fluorine. Attack by alkali metals results in defluorination and surface oxidation of PTFE parts which is a convenient route to render them adherable.

Small molecules can penetrate the structure of fluoropolymers. Table 2.4 provides a summary of room-temperature sorption of hydrogen-containing and non-hydrogenated solvents into films of PTFE and FEP. Table 2.5 describes the characteristics of the films used in these experiments. Most hydrogen-containing solvents are absorbed into PTFE and FEP at less than 1%. In their case, the extent of swelling does not depend on the solubility parameter. In contrast, halogenated non-hydrogenated solvents penetrate these polymers as a strong function of the solubility parameter. Maximum swelling (11%) takes place at a solubility parameter of 6, and it drops to less than 1% swelling at a solubility parameter of 10.

A useful rule of thumb is that little hydrogen-containing solvent is taken up by perfluoropolymers, irrespective of the solubility parameter. The amount will increase with increasing temperature. For convenience, one can imagine that the solvent molecules are increasingly energized at higher temperatures and the polymer structure becomes more open. Both effects lead to more swelling. For non-hydrogen-containing solvents, swelling decreases when the solubility parameter of the solvent increases. More swelling occurs at higher temperatures, as with the hydrogen-containing solvents. "The more the solvent chemical structure resembles the fluoropolymer structure, the greater the swelling," is the rule of thumb for this group.

Table 2.4. Sorption of Various Compounds by Perfluorocarbon Polymers at Room Temperature[11]

Compound	Solubility parameter (cal/cm^3)$^{1/2}$	Wt gain% PTFE	Wt gain% FEP Resin
Compounds containing hydrogen			
Isooctane	6.85	0.8	0.4
n-Hexane	7.3	0.7	0.5
Diethyl ether	7.4	0.8	0.6
n-Octane	7.55	1.2	0.5
Cyclohexane	8.2	1.1	0.4
Toluene	8.9	0.4	0.3
1,1-Dichloroethane	9.1	1.5	0.6
Benzene	9.15	0.4	0.3
CHCl$_3$	9.3	1.4	1.4
CH$_2$Cl$_2$	9.7	0.5	0.6
1,2-Dichloroethane	9.8	0.8	0.4
CHBr$_3$	10.5	0.5	0.2
Avg		0.8	0.5
Std Dev		0.4	0.3
Compounds without hydrogen			
FC-75^a		10.6	11.0
Perfluorokerosene	6.2	11.2	6.1
Perfluorodimethyl-cyclohexane	6.1	10.1	10.4
C$_6$F$_{12}$^{b}		9.1	8.4
1,2-Br$_2$TFE		6.5	7.2
SiCl$_4$	7.6	5.2	3.6
CCl$_4$	8.6	2.4	1.8
SnCl$_4$	8.7	3.4	2.0
TiCl$_4$	9.0	2.2	1.3
CCl$_2$=CCl$_2$	9.3	1.9	1.4
CS$_2$	10.0	0.4	0.2
Br$_2$	11.5	0.7	0.7

^{a}Structure: CF$_2$–CF$_2$–O–CF$_2$–CF(–CF$_2$–CF$_2$CF$_2$–CF$_3$)

^{b}Cyclic dimer of hexafluoropropylene

Table 2.5. Characteristics of Films in Sorption Studies

	PTFE	FEP
Thickness, μm	50	50
Preparation	Cast from aqueous dispersion	Melt extruded
Crystallinity, %	41	42

2.6 Molecular Interaction of Fluoropolymers: Low Friction and Low Surface Energy

Friction and surface energy (critical surface tension) are very low for fluoropolymers (Table 2.6). Both characteristics are at the root of many applications of these plastics, such as bridge expansion bearings (low friction) and non-sticking cookware (low surface energy). In this section, these properties are related to the intermolecular forces of fluoropolymers and other materials. To help the reader, definitions of the forces are briefly discussed.

Over a century ago (1879), Van der Waals postulated the existence of *attractive* intermolecular forces. His framework for the discussion of these forces was a modified form of the ideal gas law. Other researchers after Van der Waals have classified the intermolecular forces into four components:

- Dispersion (or nonpolar) force
- Dipole-dipole force
- Dipole-induced-dipole (induction) force
- Hydrogen bonding

These forces are referred to as *Van der Waals forces*.[13]-[20] The focus in this section is on short-range forces between two molecules which are fairly close to each other. Van der Waals forces can take place between any pair of molecules. A second class of repulsive forces act in opposition to the Van der Waals forces. The net resultant of two forces is the actual *repulsive* force present between two molecules.

All four forms of attractive energy are proportional to $1/r^6$ therefore allowing the Van der Waals forces to be expressed by Eq. (2.1). Repulsive energy for two neutral molecules which get close to each other are conventionally expressed in Eq. (2.2). Total energy between the two molecules is the sum of the attractive repulsive energies, shown in Eq. (2.3), known as Lennard-Jones potential.[21] We will focus on attractive forces in describing interactions between fluoropolymer molecules or between other molecules and fluoropolymers.

Eq. (2.1) $\quad E_a = A/r^6$

Eq. (2.2) $\quad E_r = B/r^{12}$

Eq. (2.3) $\quad E_t = A/r^6 + B/r^{12}$

where r = distance between two molecules, A, B = constants.

PTFE molecules have little propensity for polarization or ionization, which minimizes the nonpolar energy or force between PTFE molecules and between PTFE and other molecules. There are also no permanent dipoles in its structure, which is not the case for polymers such as polychlorotrifluoroethylene and polyvinylfluoride, minimizing dipole-dipole energy and force. Low polarizability coefficient minimizes dipole-induced-dipole energy. The neutral electronic state of PTFE and its symmetric geometry rule out hydrogen bonding. Consequently, PTFE is very soft and easily abraded. The molecules slip by and slide against each other.[22] Absence of any branches or side chains eliminate any steric hindrance, which could constrain the slipping of PTFE molecules past each other. In PTFE (and fluoropolymers in general) relative to engineering polymers, this characteristics gives rise to properties like:

Table 2.6. Coefficient of Friction and Surface Energy of Unfilled Fluoropolymers

FLUOROPOLYMERS	FORMULA	COEFFICIENT OF FRICTION (DYNAMIC)	CRITICAL SURFACE TENSION,[12] dyne/cm	SURFACE TENSION,[21] (Harmonic-Mean Method) dyne/cm
Polyethylene	$-CH_2-CH_2-$	0.33	31	36.1
Polyvinylfluoride	$-CHF-CH_2-$	0.3	28	38.4
Polyvinylidenefluoride	$-CF_2-CH_2-$	0.3	25	33.2
Polytrifluoroethylene	$-CF_2-CHF-$	0.3	22	-
Polytetrafluoroethylene	$-CF_2-CF_2-$	0.04	18	22.5
Polyvinylchloride	$-CHCl-CH_2-$	0.5	39	41.9
Polyvinylidenechloride	$-CCl_2-CH_2-$	0.9	40	45.4

Low coefficient of friction

Low surface energy

High elongation

Low tensile strength

High cold flow

The electronic balance and neutrality of the molecule of PTFE result in:

High chemical resistance

Low polarizability

Low dielectric constant

Low dissipation factor

High volume and surface resistance

These properties serve as the foundation of the applications for this plastic.

2.7 Conformations and Transitions of Polytetrafluoroethylene

The special size and electronic relationship of fluorine and carbon atoms sets apart the conformational and transitional arrangement of polytetrafluoroethylene from seemingly similar molecules such as polyethylene. Polymerization of tetrafluoroethylene produces a linear molecule without branches or side groups. Branching would require a cleavage of C-F bonds, which does not occur during the polymerization. The linear chain of PTFE does not have a planar zigzag conformation, as is the case with polyethylene. Only under extreme pressure (5,000 atm) does the chain adopt a zigzag conformation.[23]-[25] The chain assumes a helical conformation to accommodate the large atoms of fluorine.

Below 19°C a helix forms with a 13.8° angle of rotation around each carbon-carbon bond. At this angle, a repeat unit of 13 CF_2 are required to complete a 180° twist of the helix. Above 19°C, the number of CF_2 groups to complete a 180° twist increases to 15. The crystalline structure of PTFE changes at 19°C, which is significant due to its closeness to the ambient temperature: the repeat distance is 1.69 nm and separation of chain axes is 0.562 nm.[26] Above 19°C, the repeat distance increases to 1.95 nm and separation of chain axes decreases to 0.555 nm. In the phase III (zigzag) crystal state, at a pressure of 12 kbar, density increases to 2.74 g/cm³ and crystal dimensions are $a = 0.959$ nm, $b = 0.505$ nm, $c = 0.262$ nm, and $\gamma = 105.5°$.[26]

The helical conformation of the linear PTFE molecules causes the chains to resemble rod-like cylinders[22] which are rigid and fully extended. The crystallization of PTFE molecules occurs in a banded structure depicted in Fig. 2.5. The length of the bands is in the range of 10–100 µm while the range of the bandwidth is 0.2–1 µm, depending on the rate of the cooling of the molten polymer.[27] Slowing the cooling rates generates larger crystal bandwidths. There are striations on the bandwidths that correspond to crystalline slices, which are produced by the folding over or stacking of the crystalline segments. These segments are separated by amorphous polymer at the bending point. The thickness of a crystalline slice is 20–30 nm.[28]

PTFE has several first and second order transition temperatures ranging from –110°C to 140°C (Table 2.7).[9] The actual quantity of minor transitions is somewhat dependent on the experimental method used. From a practical standpoint, the two first order

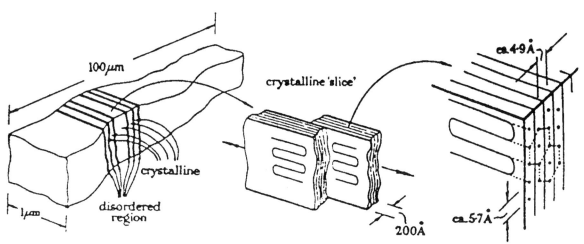

Figure 2.5 Crystalline structure of PTFE.[28]

transitions that occur at 19°C and 30°C are most important. Figure 2.6 shows the phase diagram of polytetrafluoroethylene. It can be seen from this figure that the only phase, which can not be present at the atmospheric pressure, is phase III. It requires elevated pressures under which the polymer molecule assumes a zigzag conformation.

Below 19°C the crystalline system of PTFE is a nearly perfect triclinic. Above 19°C, the unit cell changes to hexagonal. In the range of 19–30°C, the chain segments become increasing disorderly and the preferred crystallographic direction disappears. Between 19°C and 30°C, there is a large expansion in the specific volume of PTFE approaching 1.8%[29] which must be considered in measuring the dimensions of articles made from this plastics.

Table 2.7. Transitions of Polytetrafluoroethylene[9]

Temperature, °C	Order	Region Affected	Type of Disorder
19	1st Order	Crystalline	Angular Displacement
30	1st Order	Crystalline	Crystal Disorder
90	1st Order	Crystalline	
-90	2nd Order	Amorphous	On-set of rotational motion around C-C bond
-30	2nd Order	Amorphous	
130	2nd Order	Amorphous	

2.8 Conformations and Transitions of Polychlorotrifluoroethylene (PCTFE)

The crystal structure of polychlorotrifluoroethylene has been determined by x-ray diffraction and accepted to be pseudohexagonal with lattice parameters of $a = 0.644$ nm and $c = 4.15$ nm.[30][31] The polymer chain is helical with an average of 16.8 monomer units in one turn of the helix. The skeletal angles of the CF_2 and CFCl groups differ by 5–7°, which is accommodated by the helical structure of the polymer chain. Crystal microstructure in PCTFE is spherulitic and the crystallites consist of folded polymer chains.[32]

Density of the polymer depends on its degree of crystallinity. Hoffman and Weeks have estimated the density of completely amorphous and crystalline polychlorotrifluoroethylene.[33] Values of 2.077 g/cm3 and 2.187 g/cm3 for entirely amorphous and crystalline phases can be used to estimate the degree of crystallinity of fabricated parts from Eq. 2.4. The degree of crystallinity of an article is affected by the molecular weight of PCTFE and the cooling rate from melt to solidified state.

Eq. (2.4) $$\theta = \frac{d^{30} - 2.077}{2.187 - 2.077}$$

Note that θ is the degree of crystallinity at 30°C and d^{30} is the density of resin at 30°C.

There is conflicting information about the glass transition temperature of PCTFE in a review article, placing the temperature in the range of 45–60°C.[34] Three transition temperatures at 150, 90 and –37°C have been reported. Another article reported a temperature of 52°C. There is general agreement that glass transition temperature of PCTFE is about 95°C.[35][36] A more recent study has proposed a glass transition temperature 75 ±2°C, based on extensive measurements using dynamic mechanical analysis, thermomechanical analysis and differential scanning calorimetry.[37]

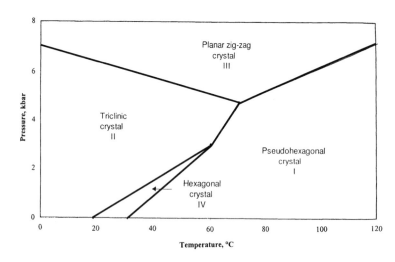

Figure 2.6 PTFE phase diagram.[26]

References

1. Pauling, L., *The Nature of Chemical Bond,* 3rd ed., Cornell Univ. Press, Ithaca, N.Y., 1960.
2. Chambers, R. D., *Fluorine in Organic Chemistry,* 1st ed., John Wiley and Sons, New York, 1973.
3. Hudlicky, M., *Chemistry of Fluorine Compounds,* 1st ed., The McMillan Company, New York, 1962.
4. Van Krevelen, D. W., *Properties of Polymers: Their Estimation and Correlation with Chemical Structure,* 2nd edition, Elsevier, 1976.
5. Patrick, C. R., *Advan. Fluorine Chem.*, 3:(1), 1961.
6. England, D. C., et al., *Proceedings of Robert A. Welch Conference on Chemical Res. XXVI,* R. A. Welch Foundation, pp. 193–243, 1982.
7. Clark, E. S., MUUS, L. T. Z., *Krist,* 117–119, 1962.
8. Raff, R. A. and Doak, K. W., "Crystalline Olefin Polymers," *Interscience*, pp. 678–680, 1965.
9. Gangal, S. V., "Polytetrafluoroethylene, Homopolymers of Tetrafluoroethylene" in *Encyclopedia of Polymer Science and Engineering,* 2nd ed., 17:577–600, John Wiley & Sons, New York, 1989.
10. Baker, B. B., and Kasprzak, D. J. "Thermal Degradation of Commercial Fluoropolymer in Air," *Polymer Degradation and Stability,* 42:181–188, 1994.
11. Starkweather, H. W., Jr., *Macromolecules* 10(5):1161–1163, Sep.–Oct. 1977.
12. Zissman, W. A., "Influence of Const. On Adh.," *Ind. & Eng. Chemistry*, pp. 18–38, Oct. 1963.
13. Margenau, H., and Kestner, N., *Theory of Intermolecular Forces,* 3rd ed., Pergamon Press, London, 1971.
14. Hirschfelder, J. O., ed., *Intermolecular Forces,* Interscience, New York, 1967.
15. "Intermolecular Forces," Discussion, Faraday Society, Vol. 40, 1965.
16. Israelachvilli, J. N., and Tabor, D., *Prog. Surf. Member. Sci.,* 7(1), 1973.
17. Israelachvilli, J. N., *Quart. Rev. Biophys.,* 6(4):341, 1974.
18. Israelachvilli, J. N., in *Yearbook of Science and Technology,* pp. 23–31, McGraw-Hill, New York, 1976.
19. Krupp, H., *Adv. Colloid Interface Sci.,* 1:111, 1967.
20. Hirschfelder, J. O., Curtiss, C. F., and Bird, R. B., *Molecular Theory of Gases and Liquids,* Wiley, New York 1954.
21. Wu, S., *Polymer Interface and Adhesion,* 1st ed., Marcel Dekker, Inc., New York, 1982.
22. Blanchet, T. A., "Polytetrafluoroethyelne," in *Handbook of Thermoplastics,* 1st ed., Marcel Dekker, Inc., New York, 1997.
23. Brown, R. G., "Vibrational Spectra of Polytetrafluoroethylene: Effects of Temperature and Pressure," *Jour. of Chem. Phys.*, 40:2900, 1964.
24. Beecroft, R. I. and Swenson, C. A., "Behavior of Polytetrafluoroethylene (Teflon®) under High Pressures," *Jour. Appl. Phys.*, 30:1793, 1959.
25. Bates, T. W., and Stockmayer, W. H., "Conformational Energies of Perfluoeoalkanes. III. Properties of Polytetrafluoroethylene," *Macromolecules*, 1:17, 1968.
26. Sperati, C. A., and Starkweather, H. W., Jr., *Adv. Polym.. Sci.*, 2:465, 1961.
27. Speerschneider, C. J. and Li, C. H., "Some Observations on the Structure of Polytetrafluoroethylene," *J. Appl. Phys.*, 33:1871, 1962.
28. Makinson, K. R. and Tabor, D., "The Friction and Wear of Polytetrafluoroethylene," *Proc. Roy. Soc.*, 281:49, 1964.

29. McCrum, N. G., "An Internal Friction Study of Polytetrafluoroethylene," *Jour. Polym. Sci.*, 34:355, 1959.
30. Menick, Z., *J. Polym. Sci. Polym. Phys. Ed.*, 11:1585, 1973.
31. Roldan, L. G., and Kaufman, H. S., *Norelco Rep.*, 10(1):11, 1963.
32. Hoffman, J. D., and Weeks, J. J., *J. Chem. Phys.* 37:1723, 1962.
33. Hoffman, J. D., and Weeks, J. J., *J. Res. Nat. Bur. Stand.*, 60:465, 1958.
34. Chandrasekaran, S., "Chlorotrifluoroethylene Polymers", in *Encyclopedia of Polymer Science and Engineering,* 2nd ed., 3:463–480, John Wiley & Sons, New York, 1989.
35. Hoffman, J. D., Williams, G., and Passaglia, E., *J. Polym. Sci.*, 14:173, 1966.
36. Sacher, E. J., *J. Polym. Sci.*, 18:533, 1980.
37. Khanna, Y. P., and Kumar, R., "The Glass Transition and Molecular Motions of Polychlorotrifluoroethylene," *Polymer*, 32(11), 1991.

3 Operational Classification of Fluoropolymers

3.1 Introduction

The discovery of PTFE led to major opportunities to solve industrial material problems. For instance, costly exotic metals could be replaced with an economical plastic. Before these opportunities could be realized, however, a host of processing problems had to be solved. PTFE proved impossible to dissolve in solvents. Its melting point was high (first melting point is 342°C), and after melt, its viscosity was so high (10^{11}–10^{12} poise at 380°C) that it would not flow at all. Tables 3.1–3.5 provide data for a variety of physical, mechanical, thermal, electrical and chemical resistance properties of a number of plastics. A new processing method had to be found to accommodate PTFE's unique properties, which was done by borrowing technology from an unlikely place, the powdered metal processing industry.

Later, PTFE underwent the typical evolutionary process of the other polymer families and was proliferated through copolymerization with a variety of comonomers. The new fluoropolymers could be processed by a variety of techniques such as melt processing, and dispersion and solution coating methods. Per general convention, *melt processing* refers to a polymer developing flow upon melting in normal process equipment such as extruders. This chapter provides a brief summary of major commercial fluoropolymer products and ways in which each is processed. This leads to classification of fluoropolymers according to their processing regime. The normal convention of polymer science will be followed in applying the name *copolymer* to any fluoropolymer containing >1% of a comonomer. A fluoropolymer containing <1% comonomer is categorized as a "modified' homopolymer.

Table 3.1. Physical and Mechanical Properties of Polymers[1][2]

	Specific Gravity	Mold Shrinkage, %	Tensile Strength, MPa	Break Elongation, %	Tensile Modulus, MPa	Flexural Strength, MPa	Flexural Modulus, MPa 23°C	Impact Notched Izod J/m (23°C)	Compressive Strength, MPa
Test Method ASTM	D792	D955	D638	D638	D638	D790	D790	D256	D695
Polystyrene	1.06	0.5	46	2.2	3,172	96	3,103	10.8	96.6
ABS	1.05	0.6	48	8.0	2,069	72	2,621	242	69
SAN	1.08	0.4	72	3.0	3,862	103	3,793	27	103
Polypropylene	0.90	1.5	32	15.0	1,310	41	2,069	27	34.5
Polyethylene	0.96	2.0	30	9.0	1,034	38	1,517	70	27.6
Polyacetal	1.41	1.8	61	60.0	2,827	90	2,552	70	36
Polyester	1.30	2.0	55	200.0	2,758	88	2,345	10.8	90
Polyamide, nylon 6	1.13	1.3	81	200.0	2,758	103	2,759	53.8	90
Polyamide, nylon 6,6	1.14	1.8	79	300.0	1,310	103	1,310	53.8	34
Polycarbonate	1.20	0.6	62	110.0	2,379	93	2,345	161	86
Polysulfone	1.24	0.7	70	75.0	2,482	106	2,690	32	96.6
Test Method ASTM	D792	Measured on Parts	D4894-5 D1708 D638	D4894-5 D1708 D638	D638	D790	D790	D256	
PTFE[a]	2.14-2.22	2-10	20-35	300-550	550	No Break	340-620	188	34.5
PFA[b]	2.15	3.5-6.0	20-26	300	276		551	No Break	
FEP[c]	2.15	3.5-6.0	20-28	300	345	No Break	655	No Break	15.2
ETFE[d]	1.71	1-7	45	150-300	827	38	1,034-1,171	No Break	17.2
PVDF[e]	1.78	0.2-3	31-52	50-250	1,040-2,070	45-74	1,140-2,240	107-427	55-110

Abbreviations used in Tables 3.1–3.5:

[a] PTFE = Polytetrafluoroethylene

[b] PFA = Perfluoroalkoxy resin (copolymer of tetrafluoroethylene and perfluorovinylethers)

[c] FEP = Fluorinated ethylene-propylene copolymer

[d] ETFE = Ethylene-tetrafluoroethylene copolymer

[e] PVDF = Polyvinylidene fluoride

Table 3.2. Mechanical Properties of Polymers[1]-[4]

	Deformation-under-Load (Tensile)	Static Coefficient of Friction
Test Method ASTM	D621 % @23°C (Load)	
Polystyrene	0.5 (13.8 MPa)	-
ABS	0.4 (10 MPa)	-
Polypropylene	2 (6.9 MPa)	-
Polyethylene	3.2 (6.9 MPa)	-
Polyacetal	0.25 (6.9 MPa)	0.38
Polyester	-	0.37
Polyamide, nylon 66	1.5 (10 MPa)	0.50
Polycarbonate	0.45 (20.6 MPa)	0.51
Polysulfone	0.9 (20.6 MPa)	-
Test Method ASTM	D621 (obsolete) @23°C (Load)	
PTFE	3.5 (6.9 MPa)	0.1
PFA	2 (0.69 MPa)	0.2
FEP	1.8 (6.9 MPa)	0.2
ETFE	0.3 (6.9 MPa)	0.4
PVDF [@140°C]	1-2 (0.69 MPa)	-

Table 3.3. Thermal and Electrical Properties of Polymers[1][2]

	Heat Distortion Temperature, °C@0.45 MPa	Coefficient of Thermal Expansion, cm/cm/°C×10^5	Continuous-use Temperature,°C	Flammability	Volume Resistivity, Ω·cm	Dielectric Strength, kV/mm
Test Method ASTM	D648	D696	UL-Sub 94	UL 94	D257	D149
Polystyrene	84.4	6.8	45	HB	10^{16}	
ABS	90.5	7.6	65	HB	10^{16}	16.3
SAN	93.3	6.5	50	HB	10^{16}	17.6
Polypropylene	104.4	9.0	105	HB	10^{16}	26.2
Polyethylene	71.1	10.8	85	HB	10^{16}	
Polyacetal	157.8	8.5	90	HB	10^{14}	19.0
Polyester	154.4	9.5	107	HB	10^{16}	18.2
Polyamide, nylon 6	185.0	8.3	95	V-2	10^{14}	15.9
Polyamide, nylon 66	182.2	8.1	105	V-2	10^{14}	19.8
Polycarbonate	137.8	6.7	121	V-0	10^{16}	15.0
Polysulfone	173.9	5.6	140	V-0	10^{15}	16.9
Test Method ASTM	D648	D696, E831 E228		UL 94	D257	
PTFE	122	12.6-18	260	V-0	$>10^{18}$	19.7
PFA	74	13.7-20.7	260	V-0	$>10^{18}$	19.7
FEP	70	8.3-10.4	204	V-0	$>10^{18}$	19.7
ETFE	81	13.1-25.7	150	V-0	$>10^{17}$	14.6
PVDF	140-168	7-15	150	V-0	$>10^{14}$	63-67

Table 3.4. Water Absorption and the Effect* of Inorganic Chemicals on Polymers[1][2]

Material	Weak Bases and Salts		Strong Bases		Strong Acids		Strong Oxidants		24-hour Water Absorption
	25°C	93°C	25°C	93°C	25°C	93°C	25°C	93°C	Weight Change, %
Polyacetal	1-3	2-5	1-5	2-5	5	5	5	5	0.23
ABS	1	2-4	1	2-4	1-4	5	1-5	5	0.1-0.4
Nylons	1	2	2	3	5	5	5	5	0.2-1.9
Polyester	1	3-4	2	5	3	4-5	2	3-5	0.06-0.09
Polyethylene	1	1	1	1	1	1	1	1	<0.01
Polystyrene	1	5	1	5	4	5	4	5	0.03-0.60
Polysulfone	1	1	1	1	1	1	1	1	0.2-0.3
PVC[f]	1	5	1	5	1	5	2	5	0.04-1.0
PTFE	1	1	1	1	1	1	1	1	0
PFA	1	1	1	1	1	1	1	1	<0.03
FEP	1	1	1	1	1	1	1	1	<0.01
ETFE	1	1	1	1	1	1	1	1	<0.03
PVDF	1	1	1	2	1	2	1	2	0.04
PCTFE[g]	1	1	1	1	1	1	1	1	0.01-0.10

Legend (Tables 3.4 and 3.5):
* Effect is defined by ASTM D543 and D2299
1 = No effect or inert
2 = Slight effect
3 = Mild effect
4 = Softening or swelling
5 = Severe degradation

[f] PVC = Polyvinylchloride
[g] PCTFE = Polychlorotrifluoroethylene

Table 3.5. Effect* of Organic Chemicals on Polymers[1][2]

Material	Aromatic Solvents		Aliphatic Solvents		Chlorinated Solvents		Esters and Ketones	
	25°C	93°C	25°C	93°C	25°C	93°C	25°C	93°C
Polyacetal	1-4	2-4	1	2	1-2	4	1	2-3
ABS	4	5	2	3-5	3-5	5	3-5	5
Nylons	1	1	1	1	1	2	1	1
Polyester	2	5	1	3-5	3	5	2	3-4
Polyethylene	3	4	3	4	3	4	3	4
Polystyrene	4	5	4	5	5	5	4	5
Polysulfone	4	4	1	1	5	5	3	4
PVC	4	5	1	5	5	5	4	5
PTFE	1	1	1	1	1	1	1	1
PFA	1	1	1	1	1	1	1	1
FEP	1	1	1	1	1	1	1	1
ETFE	1	1	1	1	1	1	1	1
PVDF	1	1	1	1	1	1	3	5
PCTFE	1	1	1	1	3	4	1	1

The other major monomers, in addition to TFE, are chlorotrifluoroethylene (CTFE, CClF=CF$_2$), vinylidene fluoride (VDF, CF$_2$=CH$_2$) and vinyl fluoride (VF, CH$_2$=CHF). These monomers are polymerized and copolymerized by different methods to produce commercial fluoropolymers. TFE, CTFE and VDF have associated families of polymers and copolymers. Only one commercial VF polymer is known (Tedlar® by Dupont).

3.2 TFE Homopolymers

Two regimes are commonly used to polymerize tetrafluoroethylene by itself and with other comonomers: *dispersion* and *suspension* processes. The two methods differ in a number of ways but the amount of surfactant added to the polymerization reactor and the shear rate applied during the reaction are the important differences between them. The dispersion method consumes much more surfactant than the suspension (slurry) process and produces small submicron particles (Fig. 3.1) which are the basis for "fine powder" and "dispersion" products. The suspension process generates long stringy particles resembling coconut shreds (Fig. 3.2) which are cut and screened in a finishing step to produce "granular" resins. Both polymerization processes are described in Ch. 5.

Granular PTFE has a high molecular weight (10^6–10^7) and is usually fabricated by an alteration of a metallurgical technique called *compression molding*. In this method a "preform" is made by compressing it in a mold to form the desired shape. This preform is then placed in an oven to be "sintered," that is, melted and densified into the final shape. Other molding techniques are also used to produce preforms. End-use parts are machined from such shapes when directly forming the final design is impractical.

Commercial fine powder PTFE resins are produced in a range of very high molecular weight (10^6–3×10^7). This powder is processed by forming a paste by the addition of an isoparaffin lubricant. This paste is then extruded into tubing, wire insulation, tape, or membrane. Typically, the extruded shape is dried in an oven and the lubricant is removed. In the final step, the dried extrudate is sintered in an oven.

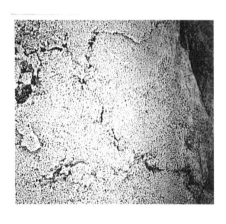

Figure 3.1 Photograph of fine powder particles.

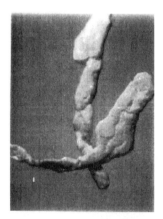

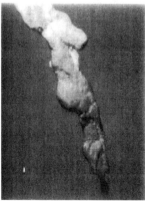

Figure 3.2 Photograph of a granular particle.

The dispersion form of PTFE consists of a fairly high solids content emulsion of PTFE in water. These products are produced by the same polymerization process as fine powder PTFE. Dispersion recovered from the polymerization reactor is usually concentrated to raise the PTFE content. Additional surfactants may be added to prevent coagulation of small particles. Dispersion products are formulated into coatings by the addition of resins, pigments, and other additives. They are applied to substrates by typical latex processing methods such as spin coating, blade drawing, spraying, brushing, and dipping.

3.3 TFE Copolymers

Copolymers of TFE with hexafluoropropylene (FEP), perfluorovinylether (PFA) and ethylene (ETFE) monomers have been made commercially. These polymers can be produced by both suspension and dispersion regimes, aqueously or non-aqueously. The polymers are recovered and finished by extrusion into cubes or pellets. Dispersion forms of TFE copolymers are consumed in industrial and consumer applications. Stabilization steps are usually included in the polymer finishing process to reduce the number of end-groups and chain links because they could initiate degradation during fabrication.

A more recent class of TFE copolymers is one which is soluble in special solvents. These copolymers contain larger fractions of the comonomers, which make them amorphous and thus soluble. For example, FEP contains 10–15% hexafluoropropylene (HFP) and is semicrystalline. Increasing the HFP content beyond 20% generates polymers which are soluble in fluorocarbon solvents. These polymers can be applied as solutions by typical latex processing techniques such as spin coating, blade drawing, spraying, brushing, and dipping.

3.4 CTFE Polymers

Chlorotrifluoroethylene (CTFE) can also be homo- and copolymerized by suspension and dispersion methods in aqueous and nonaqueous media. The copolymer of ethylene and CTFE is ECTFE, a melt processable polymer and a counterpart of ETFE. CTFE is also copolymerized with vinylidene flouride (VDF). Higher VDF contents provide elastomeric polymers. Methods similar to those for PTFE are used to recover the polymer in various forms. Both powder and dispersion products are available. PCTFE can be processed by both non-melt techniques such as compression molding, and melt techniques such as injection molding and extrusion. Thicker parts are usually made by non-melt methods.

3.5 Vinylidene Fluoride Polymers

Vinylidene fluoride (VDF) can be polymerized by a variety of methods such as suspension, dispersion and solution polymerization. It can be copolymerized with a number of fluorinated and non-fluorinated comonomers. Examples of these two groups include perfluoroolefin monomers and acrylic compounds. VDF polymers and copolymers are fabricated by melt processing as well as coating techniques.

3.6 Vinyl Fluoride Polymers

Vinyl fluoride is an exception to almost all major fluorocarbon monomers in that there is only one commercial homopolymer and no copolymers. It is made by suspension polymerization. PVF is fabricated into thin films by melt extrusion of its dispersion in a latent solvent, typically a polar compound such a dimethylformamide.

3.7 Process Classification

The brief description of the commercially important fluoropolymers indicates the techniques by which they can be fabricated. Generally, the processing method is dependent on the rheology of the fluoropolymer in question. Table 3.6 summarizes the structure-rheology-fabrication technique characteristics of various copolymers. Melt viscosity values represent a wide range of shear rate for melt processable polymers in Table 3.6. This volume focuses on all fluoropolymers, which are processed by non-melt methods. A subsequent volume is devoted to the melt processible fluoropolymers.[6]

Table 3.6. Summary of Structure-Rheology-Fabrication Process for Commercial Fluoropolymers

POLYMER	MONOMER UNIT	MELT POINT °C[1]	MELT VISCOSITY, Pa.Sec.[2]	MELT FLOW RATE, dg/min.	FABRICATION TECHNIQUE
Granular PTFE	$-CF_2-CF_2-$	327	$10^{10}-10^{12}$ [3]		Compression Molding
Fine Powder PTFE	$-CF_2-CF_2-$	327	$10^{10}-10^{12}$ [3]		Paste Extrusion
Dispersion PTFE	$-CF_2-CF_2-$	317-337	$10^{10}-10^{12}$ [3]		Coating Methods
FEP	$-CF_2-CF_2-C(CF_3)F-CF_2-$	260-282	10^4-10^5 [4]	0.8-18 [5]	Melt Processing
PFA	$-CF_2-CF_2\ C(O-R_f)F-CF_2-$	302-310	$4\times10^3-3\times10^4$ [4]	1-18 [6]	Melt Processing
ETFE	$-CF_2-CF_2-CH_2-CH_2$	254-279	$0.7-10\times10^3$ [7]	3.7-16 [8]	Melt Processing
PCTFE	$-CClF-CF_2-$	210-215	1-10 [9]		Compression Molding/ Melt Processing
ECTFE	$-CClF-CF_2-CH_2-CH_2-$	240	-	1-50	Melt Processing
PVDF	$-CF_2-CH_2-$	155-192	$0.2-17\times10^3$	5-180 [10]	Melt Processing
PVF	$-CHF-CH_2-$	190	-		Melt Extrusion of Dispersion in Latent Solvent
Soluble Fluoropolymers	$-[CF_2-CF_2]_{n-m}-[C_f]_m$ [11]	305	-	-	Latex Application Methods

[1] Second melt point for PTFE.
[2] From Ref. 5.
[3] Melt creep viscosity by Dynamic Mechanical Analysis at 380°C.
[4] By capillary rheometer ASTM D 2116.
[5] Melt flow rate by ASTM D 2116.
[6] By capillary rheometer ASTM D 3307.
[7] By capillary rheometer ASTM D 3159.
[8] Melt flow rate by ASTM D 3159.
[9] By capillary rheometer at 230°C.
[10] Melt flow rate at 265°C.
[11] C_f is the comonomer m > 0.2 weight fraction.

References

1. *Teflon® PTFE Fluoropolymers Resin,* Properties Handbook No. H-37051-3, Published by DuPont, July 1996.
2. *Handbook. of Plastics Testing Technology,* Shah, V. 1st ed., Wiley-Interscience Pub., John Wiley & Sons, New York, 1984.
3. *The Effect of Creep and Other Time Related Factors on Plastics,* Vol. 1, (A, B), and Vol. 2, *Plastics Design Library*, William Andrew Publishing, New York, 1991.
4. Maier, C., and Calafut, T., *Polypropylene: The Definitive User's Guide and Databook,* 1st ed., *Plastics Design Library*, Norwich, NY, 1998.
5. Sperati, C. A., "Physical Constants of Fluoropolymers," in *Polymer Handbook,* 3rd ed., Vol. 1, *Polymerization and Depolymerization,* John Wiley and Sons, New York, 1989.
6. *Melt Processible Fluoroplastics: The Definitive User's Guide and Data Book, Plastics Design Library,* William Andrew Publishing, New York.

4 Fluoropolymer Monomers

4.1 Introduction

In this chapter, synthesis and properties of major monomers for polymerization of non-melt processible fluoroplastics are discussed, including those used to modify the homopolymers. Tetrafluoroethylene is the primary monomer for polytetrafluoroethylene (PTFE) preparation. Small amounts (<1%) of other monomers are incorporated in the PTFE structure to modify its properties and processing characteristics. These monomers include hexafluoropropylene (HFP), perfluoroalkylvinylethers (PAVE), and chlorotrifluoroethylene (CTFE). CTFE is homopolymerized to produce the polychlorotrifluoroethylene (PCTFE) family of resins. A number of specialty monomers, though less common, are also used to modify the PTFE structure. Examples include perfluoroisopropyl vinyl ether and perfluorobutyl ethylene.

4.2 Synthesis of Tetrafluoroethylene

Tetrafluoroethylene ($CF_2=CF_2$) is the main building block of all perfluorinated polymers, that is, polymers comprised of carbon, fluorine and occasionally a small amount of oxygen. It is difficult to establish exactly the first successful synthesis. Publications in 1890's report a variety of attempts to synthesize TFE by direct reaction of fluorine with carbon, fluorine with chloromethanes and tetrachloroethylene with silver fluoride.[1]-[4] The data presented are insufficient to determine that these efforts actually lead to TFE. Humiston[5] reported the first documented preparation of TFE in 1919 which has been disputed due to erroneous property data.

The first reliable and complete description of synthesis was published in 1933 by Ruff and Bretachneider[6] in which they prepared TFE from decomposition of tetrafluoromethane in an electric arc. Separation of TFE was accomplished by bromination followed by dehalogenation with zinc to purify TFE from the pyrolysis products. Numerous other papers have reported synthesis of tetrafluoroethylene. The works that report commercially significant techniques for TFE preparation list fluorspar (CaF_2), hydrofluoric acid and chloroform as the starting ingredients.[7]-[14] The reaction scheme is shown below:

HF preparation:
$$CaF_2 + H_2SO_4 \rightarrow 2HF + CaSO_4$$

Chloroform preparation:
$$CH_4 + 3\,Cl_2 \rightarrow CHCl_3 + 3\,HCl$$

Chlorodifluoromethane preparation:
$$CHCl_3 + 2\,HF \rightarrow CHClF_2 + 2\,HCl$$
$$(SbF_3\ \text{catalyst})$$

TFE synthesis:
$$2\,CHClF_2 \rightarrow CF_2 \rightarrow CF_2 + 2\,HCl$$
$$(\text{pyrolysis})$$

A few other side compounds are also produced during pyrolysis including hexafluoropropylene, perfluorocyclobutane and octafluoroisobutylene, 1-chloro-1,1,2,2-tetrafluoroethane, 2-chloro-1,1,1,2,3,3-hexafluoropropane and a small amount of highly toxic perfluoroisobutylene.

Sherratt[15] has provided a detailed description of preparation of TFE. The overall yield of TFE production depends on the pyrolysis reaction. It proceeds to yield better than 90% TFE at short contact times, low conversions and subatmospheric pressure in the temperature range of 590–900°C. Results similar to subatmospheric pyrolysis can be achieved if superheated steam is present during the pyrolysis. Tetrafluoroethylene yields approaching 95% can be achieved at 80% chlorodifluoromethane conversion if the molar ratio of steam to $CHClF_2$ is in the range of 7:1 to 10:1.

The products of pyrolysis are cooled, scrubbed with a dilute basic solution to remove HCl and dried. The resulting gas is compressed and distilled to recover the unreacted $CHClF_2$ and to recover high purity TFE.[15] Polymerization of tetrafluoroethylene to high molecular weight requires extreme purity requiring removal of all traces of telogenic hydrogen or chlorine-bearing impurities. Tetrafluoroethylene can autopolymerize if it is not inhibited. Effective TFE autopolymerization inhibitors include a variety of terpenes, such as α-pinene, Terpene B, and d-limonene[16] which appear to act as scavengers of oxygen, a polymerization initiator.

Tetrafluoroethylene is highly flammable and can undergo violent deflagration in the absence of air:

$$C_2F_4 \rightarrow C + CF_4$$

Heat of reaction values between 57–62 kcal/mole (at 25°C and 1 atm) have been reported for TFE deflagration.[17] Similar amounts of heat are released by the explosion of black gun powder.[15] To eliminate transportation concerns, TFE preparation and polymerization are carried out at the same site.

4.3 Properties of Tetrafluoroethylene

Table 4.1 lists the properties of tetrafluoroethylene. It is a colorless, odorless, tasteless, nontoxic gas which boils at -76.3°C and freezes at -142.5°C. Critical temperature and pressure of tetrafluoroethylene are 33.3°C and 39.2 MPa. TFE is stored as a liquid; vapor pressure at -20°C is 1 MPa. Its heat of formation is reported to be -151.9 kcal/mole. Polymerization of tetrafluoroethylene is highly exothermic and generates 41.12 kcal/mole heat. The extent of exothermicity of TFE polymerization can be seen when it is compared with the polymerization of vinyl chloride and styrene which have heats of polymerization of 23–26 kcal/mole and 16.7 kcal/mole, respectively.

A complete description of hazards of tetrafluoroethylene can be found in Ref. 15. Safe storage of TFE requires its oxygen content to be less than 20 ppm. A great deal of research has been devoted to safe handling of tetrafluoroethylene.[18] Temperature and pressure should be controlled during its storage. Increasing temperature, particularly at high pressures, can initiate deflagration in the absence of air. In the presence of air or oxygen, TFE forms explosive mixtures. Detonation of a mixture of tetrafluoroethylene and oxygen can increase the maximum pressure to 100 times the initial pressure.[19]

Table 4.1. Properties of Tetrafluoroethylene[20]

PROPERTY	VALUE
Molecular weight	100.02
Boiling point at 101.3kPa, °C	-76.3
Freezing point, °C	-142.5
Liquid density vs. temperature (°C), g/mL	
$-100 < t < -40$	$1.202 - 0.0041t$
$-40 < t < 8$	$1.1507 - 0.0069t - 0.000037t^2$
$8 < t < 30$	$1.1325 - 0.0029t - 0.00025t^2$
Vapor pressure at T K, kPa	
$196.85 < T < 273.15$	$\log_{10} P_{kPa} = 6.4593 - 875.14/T$
$273.15 < T < 306.45$	$\log_{10} P_{kPa} = 6.4289 - 866.84/T$
Critical temperature, °C	33.3
Critical pressure, MPa	39.2
Critical density, g/mL	0.58
Dielectric constant at 28°C	
at 101.3 kPa	1.0017
at 858 kPa	1.015
Thermal conductivity at 30°C, mW/(m·K)	15.5
Heat of formation for ideal gas at 25°C, ΔH, kJ/mol	-635.5
Heat of polymerization to solid polymer at 25°C, ΔH, kJ/mol	-172.0
Flammability limits in air at 101.3 kPa, vol.%	14-43

Tetrafluoroethylene undergoes free radical addition reactions typical of other olefins. It readily adds Br_2, Cl_2, and I_2, halogen halides IBr and ICl, and nitrosyl halides such as NOCl and NOBr.[21][22] Additional reactions of chlorofluoromethanes and chloromethanes in presence of catalysts like aluminum chloride have been reported.[23] A variety of other compounds such as alcohols, primary amines, and ammonia can be reacted with tetrafluoroethylene to prepare tetrafluoroethers (HCF_2CF_2OR), difluoroacetamide (HCF_2CONHR), and substituted triazines.[24] Oxygen can be added to TFE to produce polymeric peroxide,[25] or tetrafluoroethylene epoxide.[26] In the absence of hydrogen, sodium salts of alcohols will react with TFE to yield trifluorovinylethers ($ROCF=CF_2$) which can be homo- and copolymerized.

4.4 Synthesis of Hexafluoropropylene

Hexafluoropropylene ($CF_3CF=CF_2$) is used as a comonomer in a number of fluoropolymers such as fluorinated ethylene-propylene copolymer. It is also used to "modify" the properties of homofluoropolymers. It was first prepared by Benning et al.[32] by pyrolysis. They identified this compound as hexafluorocyclopropane, erroneously. The full synthesis and identification of HFP was conducted by Henne.[5] A six step reaction scheme beginning with the fluorination of 1,2,3-trichloropropane ($ClCH_2CHClCH_2Cl$) led to 1,2-dichlorohexafluoropropane ($ClCF_2CFClCF_3$). The latter was dehalogenated with zinc in boiling ethanol to yield hexafluoropropylene.

There are a number of ways to prepare HFP. Excellent hexafluoropropylene yields from the thermal degradation of heptafluorobutyrate ($CF_3CF_2CF_2COONa$) have been reported.[27] Cracking of tetrafluoroethylene in a stainless steel tube at 700–800°C under vacuum is an efficient route for the production of HFP. TFE conversions up to 72% and HFP yields of 82% have been reported[28][29] Octafluorocyclobutane (TFE dimer), octafluoroisobutylene and some polymer are the major side products of cracking. The presence of a small amount (3–10%) of chlorodifluoromethane stops the formation of polymer.[30] Thermal decomposition of PTFE under 20 torr vacuum at 860°C yields 58% hexafluoropropylene.[31] HF reaction with 3-chloro-pentafluoro-1-propene ($CF_2=CF-CF_2Cl$) at 200°C, catalyzed by activated carbon, yields HFP.[32] Hexafluoropropylene can be prepared from the catalytic degradation of fluoroform (CHF_3) at 800–1000 °C in a platinum-lined nickel reactor.[15] Another method is copyrolysis of fluoroform and chlorotrifluoroethylene ($CF_2=CFCl$),[16] or chlorodifluoromethane and 1-chloro-1,2,2,2-tetrafluoroethane ($CHClFCF_3$),[32] giving good yields of HFP.

More recently other methods have been reported for the synthesis of hexafluoropropylene. One technique involves the pyrolysis of a mixture of tetrafluoroethylene and carbon dioxide at atmospheric pressure at 700–900°C.[33] Conversions of 20–80% and HFP yields of better than 80% were obtained. The unreacted tetrafluoroethylene and carbon dioxide were distilled from the product and recycled. HFP can be synthesized from hexachloropropylene via a multistep process beginning with fluorination.[34] Later steps convert the initial products to CF_3-$CFCl$-CF_3 which is dehalogenated to HFP. Other techniques report on the synthesis of hexafluoropropylene from the mixture of a variety of linear and cyclic three-carbon hydrocarbons with a partially halogenated three-carbon acyclic hydrocarbon.[34]

4.5 Properties of Hexafluoropropylene

Table 4.2 lists the properties of hexafluoropropylene. It is a colorless, odorless, tasteless, and relatively low toxicity gas, which boils at -29.4°C and freezes at -156.2°C. In a four-hour exposure, a concentration of 3,000 ppm corresponded to LC_{50} in rats.[35] Critical temperature and pressure of hexafluoropropylene are 85°C and 3,254 MPa. Unlike tetrafluoroethylene, HFP is extremely stable with respect to autopolymerization and may be stored in liquid state without the addition of telogen.

Hexafluoropropylene is thermally stable up to 400–500°C. At about 600°C under vacuum, HFP decomposes and produces octafluoro-2-butene ($CF_3CF=CFCF_3$) and octafluoroisobutylene.[26] Under γ-radiation, it reacts with oxygen and produces a 1:1 mole ratio of carbonyl fluoride (COF_2) and trifluoroacetyl fluoride (CF_3COF).[36] Heat of combustion of hexafluoropropylene is 879 kJ/mole.[17] Under basic conditions, hydrogen peroxide reacts with HFP to form hexafluoropropylene epoxide, which is an intermediate in the preparation of perfluoroalkylvinyl ethers.[37][38]

Hexafluoropropylene readily reacts with hydrogen, chlorine, bromine, but not iodine, by an addition

Table 4.2. Properties of Hexafluoropropylene[4]

PROPERTY	VALUE
Molecular weight	150.021
Boiling point at 101.3kPa, °C	-29.4
Freezing point, °C	-156.2
Liquid density vs. temperature (°C), g/mL $-100 < t < -40$ $-40 < t < 8$ $8 < t < 30$	$1.202 - 0.0041t$ $1.1507 - 0.0069t - 0.000037t^2$ $1.1325 - 0.0029t - 0.00025t^2$
Vapor pressure at T K, kPa $196.85 < T < 273.15$	$\log_{10}P_{kPa} = 6.6938 - 1139.156/T$
Critical temperature, °C	85
Critical pressure, MPa	3,254
Critical density, g/mL	0.60
Liquid density, g/mL 60°C 20°C 0°C -20°C	1.105 1.332 1.419 1.498
Heat of formation for ideal gas at 25°C, ΔH, kj/mol	-1078.6
Heat of combustion, kJ/mol	879
Toxicity, LC50(rat), 4h[a], ppm	3,000
Flammability limits in air at 101.3 kPa, vol.%	Nonflammable for all mixtures of air and HFP

[a] Exposure resulting in fatality of 50% of rats in four hours.

reaction similar to other olefins.[18][39]-[41] Similarly HF, HCl, and HBR, but not HI, add to HFP. By reacting hexafluoropropylene with alcohols, mercaptans, and ammonia, hexafluoro ethers (CF_3CFHCF_2OR), hexafluoro sulfides (CF_3CFHCF_2SR), and tetrafluoropropionitrile (CF_3CFHCN) are obtained. Diels-Alder adducts have been identified from the reaction of anthracene, butadiene and cyclopentadiene with HFP.[42] Cyclic dimers of HFP can be prepared at 250–400°C under autogenous pressure.[24][25] Linear dimers and trimers of hexafluoropropylene can be produced catalytically in the presence of alkali metal halides in dimethylacetamide.[21][23]

4.6 Synthesis of Perfluoroalkylvinylethers

Perfluoroalkylvinylethers are synthesized according to the steps shown in Table 4.3. There are also electrochemical processes for the production of perfluoro-2-alkoxy-propionyl fluoride.[46]

Table 4.3. Steps to Synthesize Perfluoroalkylvinylethers

---

1. Hexafluoropropylene is converted hexafluoropropylene epoxy (HFPO) reacting HFP with oxygen under pressure in the presence of an inert diluent at 50–250°C or with an oxidizer such as hydrogen peroxide in a basic solution:[43][44]

$$CF_3CF=CF_2 + H_2O_2 \rightarrow CF_2\text{-}CF\underset{\underset{O}{\diagdown\;\diagup}}{\text{—}}CF_2 + H_2O \quad \text{(Basic Solution)}$$

$$\text{HFPO}$$

---

2. HFPO is reacted with a perfluorinated acyl fluoride to produce perfluoro-2-alkoxy-propionyl fluoride:

$$CF_2\text{-}CF\underset{\underset{O}{\diagdown\;\diagup}}{\text{—}}CF_2 + R_f\text{-}\underset{F}{\overset{}{C}}=O \rightarrow R_fCF_2OCF\text{-}\underset{F}{\overset{}{C}}=O$$
$$\qquad\qquad\qquad\qquad\qquad\qquad CF_3$$

Perfluoro-2-alkoxy-propionyl fluoride

---

3. Perfluoro-2-alkoxy-propionyl fluoride is reacted with the oxygen containing salt of an alkali or alkaline earth metal at an elevated temperature which depends on the type of salt. Examples of the salts include sodium carbonate, lithium carbonate and sodium tetraborate:[45]

$$R_fCF_2OCF\text{-}C=O + Na_2CO_3 \rightarrow R_fCF_2OCF= CF_2 + 2\,CO_2 + 2\,NaF$$
$$\quad\; |\;\; \backslash$$
$$CF_3\; F$$

4.7 Properties of Perfluoroalkylvinylethers

Perfluoroalkylvinylethers (PAVE) forms an important class of monomers in that they are comonomers of choice for the "modification" of the properties of homofluoropolymers in addition to their broad use in the structure of copolymers of TFE. They are capable of suppressing the crystallization of polytetrafluoroethylene efficiently, which imparts useful mechanical properties to lower molecular weight PTFE polymers. The advantage of PAVEs as modifiers over hexafluoropropylene is their remarkable thermal stability. Copolymers of PAVEs and tetrafluoroethylene are as thermally stable as PTFE homopolymers.

Properties of perfluoropropylvinylether (PPVE), a commercially significant example of PAVEs, are listed in Table 4.4. PPVE is an odorless colorless liquid at the room temperature. It is extremely flammable and burns with a colorless flame. It is less toxic than hexafluoropropylene and copolymerizes with tetrafluoroethylene.

Table 4.4. Properties of Perfluoroalkylvinylethers[47]

PROPERTY	VALUE
Molecular weight	266
Boiling point at 101.3 kPa, °C	36
Flash point, °C	-20
Specific gravity at 23 °C	1.53
Vapor density at 75 °C, g/mL	0.2
Vapor pressure at 25 °C, kPa	70.3
Critical temperature, K	423.58
Critical pressure, MPa	1.9
Critical volume, mL/mole	435
Toxicity, average lethal concentration (ACL), ppm	3,000
Flammability limits in air, vol.%	1

4.8 Synthesis of Chlorotrifluoroethylene

This monomer is fairly simple to manufacture compared to the perfluorinated monomers. The commercial process for the synthesis of chlorotrifluoroethylene (CTFE) begins with 1,1,2-trichloro-1,2,2-trifluoroethane (TCTFE). It is dechlorinated by pyrolysis at 500–600°C in vapor phase. An alternative method for preparation of TCTFE is catalytic dechlorination:

$$CCl_3\text{-}CCl_3 + HF \rightarrow CCl_2F\text{-}CClF_2 + 2HCl$$
(catalyst $SbCl_xF_y$)

$$CCl_2F\text{-}CClF_2 + Zn \rightarrow CFCl=CF_2 + ZnCl_2$$
(at 50–100°C in methanol)

A number of compounds including chlorodifluoroethylene, trifluoroethylene, dichlorotrifluoroethane, methyl chloride, dimethyl ether and CTFE dimer are by-products of this reaction.

The reaction stream is put through a number of purification and distillation steps to remove the gaseous and liquid contaminants. Chlorotrifluoroethylene is further purified by the removal of methyl chloride, dimethyl ether and water by passing the gas stream through sulfuric acid. Water and hydrochloric acid are removed by passing the CTFE through an alumina column before condensing it into a liquid. The liquid stream is sent to storage and non-condensable gases are purged.

4.9 Properties of Chlorotrifluoroethylene

Chlorotrifluoroethylene is a colorless gas at room temperature and pressure. It is the monomer for polychlorotrifluoroethylene and is also known as trifluorovinyl chloride and trifluorochloroethylene. Physical properties of CTFE are listed in Table 4.5. It is fairly toxic with an LC50 (rat), 4h of 4,000ppm.[48]

Chlorotrifluoroethylene behaves similarly to other small olefins in its reactions. For example, halogens and hydrohaloacids, such as hydrobromic and hydrochloric acids, add to the double bond:

$$CFCl=CF_2 + HCl \rightarrow CFCl_2\text{-}CHF_2$$

It hydrolyzes slowly in water containing oxygen while it is completely stable in degassed water. The reaction accelerates in the presence of an alkali (e.g., NaOH) at above room temperature.[51] It polymerizes by a free-radical reaction, as well as forming a variety of copolymers with vinylic monomers such as ethylene. Chlorotrifluoroethylene also reacts with amines, alcohols, chloroform, and methylene chloride.

Table 4.5. Properties of Chlorotrifluoroethylene[a]

PROPERTY	VALUE
Molecular weight	116.47
Boiling point at 101.3kPa, °C	-27.9
Freezing point, °C	-157.5
Liquid density at 20°C, g/mL	1.305
Vapor pressure at 21°C, kPa	428
Critical temperature, °C	105.8
Critical pressure, MPa	4.03
Flammability limits, %	16–34[b]
Latent heat of vaporization at -27.9	22.6
Heat of formation at 25°C, kJ/mole	563.2
Heat of combustion at 25°C, kJ/mole	223.8

[a] Ref. 49; [b] Ref. 50

Oxygen and liquid CTFE react and form peroxides at fairly low temperatures. A number of oxygenated products are produced as a result of oxidation of chlorotrifluoroethylene, such as chlorodifluoroacetyl.[52] The same reaction can occur photochemically in the vapor phase. Chlorotrifluoroethylene oxide is a by-product of this reaction. The peroxides act as initiators for the polymerization of CTFE, which can occur violently. This is the reason that all traces of oxygen must be removed for safe storage and shipping without the addition of inhibitors.

CTFE dimers can be produced catalytically and thermally to produce 4,4-dichlorohexafluoro-1-butene. The catalyst of the reaction is sintered porous polytetrafluoroethylene or a mixture of PTFE and sodium or potassium fluoride at 300–450°C.[53] The dimer yield of the reaction is usually very high (>90%). The rest of the CTFE is converted to trimers. Thermally CTFE dimerizes to a mixture of cis- and trans- 1,2-dichlorohexafluorocyclobutanes.[54]

References

1. Chabrie, C., *Compt. Rend.*, 110:279, 1890.
2. Moissan, H., *Compt. Rend.*, 110:276–279, 1890.
3. Moissan, H., *Compt. Rend.*, 110:951–954, 1890.
4. Gangal, S. V., "Fluorine Compounds, Organic (Polymers)," "Perfluorinated Ethylene-Propylene Copolymers," in: *Kirk-Othmer Encyclopedia of Chemical Technology,* 4th ed., 11:644–656, John Wiley & Sons, New York, 1994.
5. Humiston, B., *J. Phys. Chem.*, 23:572–577, 1919.
6. Ruff, O., and Bretschneider, O., *Z. Anorg. Che.*, 210:73, 1933.
7. Park, J. D., et al., *Ind. Eng. Chem.*, 39:354, 1947.
8. Hamilton, J. M., in Stacey, M., Tatlow, J. C., and Sharpe, A. G., eds., *Advances in Fluorine Chemistry*, Vol. 3, Butterworth & Co., Ltd., Kent, U. K., p. 117, 1963.
9. Edwards, J. W., and Small, P. A., *Nature*, 202:1329, 1964.
10. Gozzo, F., and Patrick, C. R., *Nature*, 202:80, 1964.
11. Japanese Patent 60 15,353, Hisazumi, M., and Shingu, H.
12. Scherer, O., et al., US Patent 2,994,723, assigned to Farbewerke Hoechst, Aug. 1, 1961.
13. Edwards, J. W., Sherratt, S., and Small, P. A., Brit. Patent 960,309, assigned to ICI, June 10, 1964.
14. Ukahashi, H., and Hisasne, M., US Patent 3,459,818, assigned to Asahi Glass Co., Aug. 5, 1969.
15. Sherratt, S., in: *Kirk-Othmer Encyclopedia of Chemical Technology 2nd ed.*, (A. Standen. ed.,), 9:805–831, Interscience Publishers, Div. of John Wiley and Sons, New York, 1966.
16. Dietrich, M. A., and Joyce, R. M., US Patent 2,407,405, assigned to DuPont, Sept. 10, 1946.
17. Duus, H. C., "Thermochemical Studies on Fluorocarbons," *Indus. Eng. & Chem.*, 47:1445–1449, 1955.
18. Hanford, W. E. and Joyce, R. M., *J. American Chem. Soc.*, 68:2082–2085, 1946.
19. Teraniski, H., *Rev. Phys. Chem. Jap.*, 28:9–23, 1958.
20. Renfrew, M. M., and Lewis, E. E., *Ind. Eng. Chem.*, 38:870–877, 1946.
21. Haszeldine, R. N., *J. Chem. Soc.*, pp. 2075–2081, 1953.
22. Renn, J. A., et al., *J. of Fluorine Chem.*, 86:113–114, 1997.
23. Coffman, D. D., Cramer, R., and Rigby, G. W., *J. Am. Chem. Soc.*, 71:979–980, 1949.
24. Coffman, D. D., Raasch, M. I., Rigby, G. W., Barrich, P. L., and Hanford, W. E., *J. Org. Chem.*, 14:7470150–753, 1949.
25. Pajaczkowski, A., and Spoors, J. W., *Chem. And Ind.*, London, 16:659, 1964.
26. Gibbs, H. H., and Warnell, J. J., Brit. Patent 931,587, assigned to DuPont, July 17, 1963.
27. Locke, E. G., Brode, W. R., and Henne, A. L., *J. Am. Chem. Soc.,* 56:17260150–1728, 1934.
28. Ruff, O., and Willenberg, W., *Chem Ber.*, 73:724–729, 1940.
29. Hals, L. T., Reid, T. S., and Smith, G. H., *J. Am. Chem. Soc.*, 73:4054, 1951.
30. Hauptschein, M., and Fainberg, A. H., US patent 3,009,966, to Pennsalt Chemicals Corp., Nov. 21, 1961.
31. Lewis, E. E., and Naylor, M. A., *J. Am. Chem. Soc.,* 69:1968–1970, 1947.
32. Downing, F. B., Benning, A. F., and McHarness, R. C., US Patent 2,384,821, assigned to DuPont, Sept. 18, 1945.

33. West, N. E., US Patent 3,873,630, assigned to DuPont, March 25, 1975.
34. Webster, J., et al., US Patent 5,068,472, assigned to DuPont, Nov. 26, 1991.
35. Clayton, J. W., *Occupational Medicine*, 4:262–273, 1962.
36. Lenzi, M., and Mele, A., *Nature*, 205(4976):1104–1105, 1965.
37. Harris, J. F., Jr., and McCane, D. I., US Patent 3,180,895, assigned to DuPont, Apr. 27, 1965.
38. Fritz, G. G., and Selman, S., US Patent 3,291,843, assigned to DuPont, Dec. 13, 1966.
39. Knunyants, I. L., Mysov, E. I., and Krasuskaya, M. P., *Izvezt. Akad. Nauk S. S. S. R., Otdel. Khim. Nauk*, pp. 906–907, 1958.
40. Haszeldine, R. N., and Steele, B. R., *J Chem. Soc.*, pp. 1592–1600, 1953.
41. Miller, W. T., Jr., Bergman, E., and Fainberg, A. H., *J. Am. Chem. Soc.*, 79:4159–4164, 1957.
42. McBee, E. T., Hsu, C. G., Pierce, O. R., and Roberts, C. W., *J. Am. Chem. Soc.*, 77:915–917, 1955.
43. Carlson, D. P., US Patent 3,536,733, assigned to DuPont, Oct. 27, 1970.
44. Eleuterio, H. S., and Meschke, R. W., US Patent 3,358,003, Dec. 12, 1967.
45. Fritz, G. G., and Selman, S., US Patent 3,291,843, assigned to DuPont, Dec. 13, 1966.
46. Brice, T. J., and Pearlson, W. H., US Patent 2,713,593, assigned to Minnesota Mining and Manufacturing Co., July 1955.
47. Gangal, S. V., "Polytetrafluoroethylene," in *Encyclopedia of Chemical Technology*, 4th ed., 11:671–683, John Wiley & Sons, New York, 1994.
48. Carpenter, C. P., Smyth, H. F., and Pozzani, U.C., *J. Industrial Hygiene*, 31:343, 1949.
49. Chandrasekaran, S., "Chlorotrifluoroethylene Polymers," in *Encyclopedia of Polymer Science and Engineering*, 2nd ed., 3:463–480, John Wiley & Sons, New York, 1989.
50. Allied Chem. Corp., Genetron® Chemicals Division, Genetron® Aerosol Propellants Product Data Sheets, PD-6-54, PD-9-54, 1955.
51. Myers, R. L., *Ind. Eng. Chem.*, p. 1783, Aug. 1953.
52. Haszeldine, R. N., and Nyman, F., *J. Chem. Soc. London*, p. 1085, 1959.
53. Teumac, F. N., Harriman, L. W., US Patent 3,214,479, Dow Chemical, Sept. 26, 1965.
54. Atkinson, B., *J. Chem. Soc. London*, pp. 512–519, 1962.

5 Homofluoropolymer Polymerization and Finishing

5.1 Introduction

Tetrafluoroethylene (TFE) is the monomer for polytetrafluoroethylene (PTFE). TFE is polymerized in water in the presence of an initiator, a surfactant, and other additives. Two different regimes of polymerization are common for production of different types of PTFE. *Suspension* polymerization is the route to production of *granular* resins. In this regime TFE is polymerized aqueously in the presence of a very small amount of dispersant accompanied by vigorous agitation. The dispersant is rapidly consumed leading to the precipitation of the polymer. *Emulsion* or *dispersion* polymerization is the method by which *dispersion* and *fine powder* PTFE products are manufactured. Fine powder resins are also called *coagulated dispersion,* which is descriptive of their production method. Mild agitation, ample dispersant, and a waxy substance set the *dispersion* polymerization apart from the *suspension* method. *Dispersion* and *fine powder* products are polymerized by the same method. The finishing steps convert the polymerization product, which is a dispersion, to the two products.

The objective of this chapter is to familiarize the reader with the important types of polytetrafluoroethylene and the methods of producing them. The emphasis is mostly on the review of commercially significant technologies. Although the exact polymerization technologies being practiced by resin manufacturers are closely guarded, the descriptions and discussions of the important public disclosures in patents and other publications should provide an understanding of the subject.

The chapter begins with a brief discussion of the polymerization mechanism by which polytetrafluoroethylene is formed. Next, different PTFE polymers are described which are produced by the polymerization of 100% TFE and the addition of trace amounts of a comonomer to modify the TFE homopolymers. Each polymerization technique and the corresponding finishing steps are discussed next. Characterization methods and the defining properties for each polymer group are described. In separate sections polymerization and characterization of chlorotrifluoroethylene (CTFE) are discussed.

5.2 Polymerization Mechanism

Polymerization of tetrafluoroethylene proceeds by a free radical mechanism. The reaction is initiated by a catalyst or by an initiator, usually, based on the reaction temperature. If polymerization is carried out at low temperatures (<30°C), a redox catalyst is used. (These compounds ionize into charged fragments such as $KMnO_4 \xrightarrow{water} K^+ + MNO_4^-$.) A bisulfite or persulfate is the typical initiator for higher temperature TFE polymerization. The reaction scheme for persulfate initiation is shown in Table 5.1.

There are alternative courses of hydrolysis that can affect the end group at a different stage of the polymerization. The key point is that there is no sulfur when persulfate is the initiator as demonstrated by Bro and Sperati.[1] Bisulfite initiators form sulfonic acid end groups as identified by polymerization of TFE using an iron-bisulfite initiator containing radioactive sulfur ^{35}S.[2] The importance of this study is that it has provided quantitative evidence that the molecular weight of PTFE has similar properties to commercial resins. The estimates for the number average molecular weight ranged from $0.4 \times 10^6 – 10 \times 10^6$ based on the determination of the residual radioactivity of the end groups.

The reactions of free radicals with saturated molecules have appreciable activation energies and negative entropies of activation. This is the reason that termination by disproportionation and transfer to monomer and polymer are not favorable.[3]

Homopolymers of polytetrafluoroethylene are completely linear without detectable branches which is contrary to polyethylene.

5.3 Tetrafluoroethylene Polymers

Tetrafluoroethylene polymerizes linearly without branching which gives rise to a virtually perfect chain structure up to rather high molecular weights. The chains have minimal interactions, as discussed in Ch. 2, and crystallize to form a nearly 100% crystalline structure. Thermoplastics develop good mechanical properties because of the Van der Waals forces arising from inter-chain attractive forces. How can PTFE polymers with useful properties be produced with minimal Van der Waals forces? The answer lies in controlling the crystallinity of the polymer after melting upon recrystallization.

Table 5.1. Reaction Scheme for Persulfate Initiation

Initiator fragments or free radicals are formed because of the degradation of persulfate under heat:

$$K_2S_2O_8 + \text{Heat} \rightarrow 2\overline{SO}_4^\bullet + 2K^+$$

Initiation takes place by formation of new free radicals by reaction of persulfate fragments with tetrafluoroethylene dissolved in the aqueous phase:

$$\overline{SO}_4^\bullet + CF_2=CF_2 \rightarrow \overline{SO}_4(CF_2-CF_2)^\bullet$$

Propagation is the growth of the free radicals of the initiation step by further addition of tetrafluoroethylene:

$$\overline{SO}_4(CF_2-CF_2)^\bullet + n\, CF_2=CF_2 \rightarrow \overline{SO}_4(CF_2-CF_2)_n-(CF_2-CF_2)^\bullet$$

Free radicals undergo hydrolysis where a hydroxyl end group replaces the sulfate:

$$\overline{SO}_4(CF_2-CF_2)_n-(CF_2=CF_2)^\bullet + H_2O \rightarrow HO(CF_2-CF_2)_n-(CF_2-CF_2)^\bullet + H^+ + H\overline{SO}_4$$

$$HO(CF_2=CF_2)_n-(CF_2=CF_2)^\bullet + H_2O \rightarrow COOHCF_2-(CF_2-CF_2)_n^\bullet + 2HF$$

Termination is the last step before the growth of the free radicals

$$COOH-CF_2-(CF_2-CF_2)_n^\bullet + COOH-CF_2-(CF_2-CF_2)_m^\bullet \rightarrow COOH-(CF_2-CF_2)_{m+n}-COOH$$

The only means of controlling the extent of recrystallization after melting in homopolymers of TFE (no other comonomer) is by driving up the molecular weight of the polymer. The extremely long chains of PTFE have a much better probability of chain entanglement in the molten phase and little chance to crystallize to the premelt extent (>90–95%). This is precisely the reason that it is essential to polymerize TFE to 10^6–10^7 for commercial applications. It is speculated that molecular weight may be as high as fifty million.[4] Molecular weight of PTFE can be controlled by means of certain polymerization parameters such as initiator content, telogens and chain transfer agents.

One consequence of very high molecular weight of polytetrafluoroethylene is reflected in its immense melt viscosity. The melt creep viscosity of PTFE is 10 GPa (10^{11} poise) at 380°C.[5] This is more than a million times too viscous for melt processing in extrusion or injection molding, leading to unique processing techniques which are described in this book. PTFE may be a thermoplastic, but it develops no flow upon melting. The closure of voids in articles made from this polymer does not take place with the ease and completeness of the other thermoplastics such as polyolefins. A small fraction of void volume remains in parts made from homopolymers of PTFE due to the difficulty and slow rate of void closure in this polymer. Voids affect permeation and mechanical properties such as flex life and stress crack resistance.

To meet the demands of extreme mechanical properties and resistance to permeation, the residual voids must be eliminated. Solving this problem requires a reduction in the viscosity of PTFE without extensive recrystallization. The remedy has been to polymerize a small amount of a comonomer with tetrafluoroethylene to disrupt the crystalline structure of PTFE. Cardinal et al.[6] have presented one of the early proposals to modify polytetrafluoroethylene in dispersion polymerization whereby the modifier is introduced after some polymerization has taken place. Holmes et al.[7] describes the utilization of perfluoroalkylvinyl ethers such as perfluoropropylvinyl ether as a modifier. Mueller et al.[8] and Doughty et al.[9] have reported modification (copolymerization) of PTFE with hexafluoropropylene and perfluoroalkyvinyl ethers in a suspension polymerization regime to obtain granular powders.

In summary, tetrafluoroethylene can be polymerized by means of two technologies, *suspension* and *dispersion*, to produce PTFE resins. The *suspension* method yields granular polymers, which are processed as molding powders. Homopolymers and modified polymers of PTFE are produced by this technique. *Dispersion* polymerization is the regime by which *fine powder* and *dispersion* PTFE products are manufactured. Fine powder resins are fabricated by paste extrusion where a hydrocarbon is added to the powder as an extrusion aid later removed prior to sintering. Dispersion products are primarily consumed by coating methods and filled co-coagulation applications. Homopolymers and modified polymers are produced by dispersion polymerization. This regime of polymerization allows production of particles, which constitute different polymers at different depth inside the particle.

All three forms of PTFE are produced by batch polymerization under elevated pressure in specially designed reactors. Polymerization media is high purity water, which is virtually devoid of inorganic and organic impurities that impact the reaction by inhibition and retardation of the free radical polymerization. The surfactant of choice in these reactions is anionic and often a perfluorinated carboxylic ammonium salt.

5.4 Preparation of Polytetrafluoroethylene by Suspension Polymerization

Commercially, tetrafluoroethylene is polymerized in an aqueous suspension medium to produce *granular* PTFE resins. In general, the important characteristics of this polymerization regime include little or no dispersing agent and vigorous agitation at elevated temperature and pressure. Disintegrating its particles and drying to obtain a powder, which can be molded by commercial processes, finishes the suspension recovered from the reactor. Finishing processes as illustrated in Fig. 5.1 commonly produce four different types of powder. Polymerization and finishing techniques are reviewed in this section.

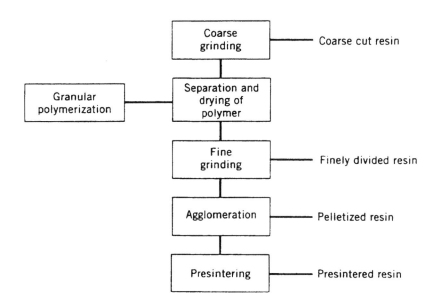

Figure 5.1 Suspension polymerization and finishing processes.[5]

The first reported polymerization of polytetrafluoroethylene was by Plunkett in 1941.[10] Granular PTFE can be prepared by polymerization of TFE alone or in the presence of small amounts of comonomers in the water containing only a free radical catalyst.[11] A great many number of ethylenically unsaturated comonomers can be polymerized with TFE. They include hydrocarbon olefins and their halogen-substituted derivatives. The commercially important comonomers include ethylene, hexafluoropropylene and alkylvinyl ethers and their fluorinated versions.[8][9][12][13] Some of the comonomer-modified granular PTFE resins are commercially significant. See Ch. 6 for information about the manufacturers and grades of these polymers. The polymerization is achieved by introducing the monomer(s) into an aqueous medium containing a polymerization initiator under pressure. A dispersing agent can be added to the aqueous phase to seed the polymerization.

Polymerization of tetrafluoroethylene is done under constant pressure conditions to control the molecular weight and its distribution. It also affects the kinetics of polymerization. Pressure ranges from 0.03–3.5 MPa[14] and is held constant by feeding monomer into the reactor. The initiator and temperature are interrelated. Ionic inorganic initiators such as ammonium persulfate or alkali metal persulfates such as potassium and lithium persulfates would be effective in the range of 40–90°C. Organic peroxides such as bis (β-carboxypropionyl) peroxide also called *disuccinic acid peroxide* can also initiate the polymerization.[15] As the temperature is lowered, at some point the effectiveness of the persulfates is diminished due to insufficient decomposition rate. Redux initiators such as potassium permanganate must replace them.

The amount of the initiator varies in the range of 2–500 ppm, based on the weight of the water. The exact concentration depends on the polymerization conditions. Too much initiator, while all other variables are held constant, reduces the molecular weight. Too little initiator leads to a poor polymerization yield. Water is the carrier for the initiator and the heat transfer medium for the exothermic heat of polymerization (-41 kcal/mole).[14]

Water does not interfere with the reaction but most organic chemicals, even in low concentrations, do. The interference manifests itself as inhibition of the reaction or chain transfer which yields products with undesirable properties. Relatively inert organic compounds such as saturated hydrocarbons inhibit the polymerization reaction unless their solubility in water in very small, thus paraffins up to C_{12} inhibit. Longer paraffins are less inhibitive due to their low aqueous solubility.

A small quantity of an anionic dispersing agent that is predominantly non-telogenic is added to seed the polymerization. The most common dispersants are the ammonium salts of perfluorocarboxylic acids, containing 7–20 carbon atoms.[16] Typical concentration of dispersants is 5–500 ppm, which is insufficient to cause formation of colloidal polymer particles. In the early stages of polymerization, a dispersion forms which becomes unstable as soon as the dispersing agent is consumed. The instability occurs at fairly low solids content at about 0.2% by weight.[14] From here on, most of the polymerization occurs directly onto the larger granular particles, which are, porous water repellent, therefore, float on the water. The reaction tends to continue for some time, even after agitation is stopped, supporting the direct polymerization hypothesis.[14]

Tetrafluoroethylene easily polymerizes at moderate pressures and temperatures. It is necessary to control the rate and to transfer the heat generated by the exothermic polymerization reaction. This is accomplished by circulating a cold fluid through the polymerization reactor jacket and cooling the aqueous phase, which is the heat transfer media. An important concern in the suspension process is the buildup of PTFE on the inner wall of the reactor, which reduces heat transfer. Development of hot spots on the reactor wall can result in deflagration (exothermic and explosive), if it goes unchecked (see Sec. 4.2).

A typical batch begins[14] with the charging of highly purified water (18 MΩ) to a reactor which is equipped with a stirrer, followed by evacuation and pressurization with tetrafluoroethylene. The feed rate of TFE is controlled to maintain a constant pressure in the reactor throughout the polymerization. The content of the reactor is vigorously agitated at 0.0004–0.002 kg·m/sec/ml.[17] Temperature is controlled by adjusting the temperature of the coolant medium in the jacket. Stopping the monomer after a certain feed weight has been reached ends the polymerization. The reaction is allowed to continue in order to consume the majority of the remaining TFE.

The product of suspension polymerization is stringy shaped particles. They have variable size and shape and are elongated as a result of vigorous agitation, resembling shreds of graded coconut (see Fig. 3.2). Two processing steps are required to convert the polymer into usable form. First, water has to be removed and the polymer should be rinsed and dried. Second, the particle size has to be reduced which can be done before or after drying. The manufacturers hold the exact details of particle disintegration technology proprietary.

5.4.1 Fine Cut Granular PTFE

The commercial PTFE powders resulting from size reduction of the suspension polymer particles have a typical average particle size in the range of 20–40 μm (Fig. 5.2). The small particle size of fine cut PTFE imparts the highest physical properties possible to obtain from granular resins to the articles made from them. Fine cut powders have poor "flow" and relatively low apparent density (<500 g/l). Their consistency is similar to wheat flour. They are normally referred to as "fine cut" or "low flow" granular PTFE resin in the trade parlance. They are molded into simple stock shapes, such as cylindrical and rectangular billets, for machining into objects and into films and sheets (see Ch. 7 for fabrication processing information). The small particle size of these powders renders them suitable for compounding with fillers such as glass fiber, carbon black, bronze and others (see Ch. 12 for information about compounding PTFE).

Lower apparent density powders can be prepared by pulverizing the ordinary dry raw suspension.[18] The process is carried out in a mill equipped with a bladed rotor moving at a speed of 3000 meters per minute in vortex of air or another gas. The temperature of the polymer in this process should be above the transition point of PTFE, preferably above 25°C. This process which renders the powder handling more difficult can obtain apparent density below 300 g/l. The great advantage of the "lighter" powders is the improved physical properties of parts made from them (Table 5.2).

5.4.2 Pelletized Granular PTFE

This section offers a review of technologies applied to finely ground powders (ca. 20–40 µm particle size) of PTFE to obtain free flowing agglomerates. Since fillers can be incorporated in the PTFE agglomerate by the same processes, filled and unfilled powders have both been covered.

Fine cut resins have poor flow and low bulk density. These characteristics are disadvantages that render them unsuitable for use in automatic and isostatic molding techniques; both described in Ch. 7. These powders also require fairly large molds because of their low apparent density. The fine cut PTFE resins are converted into soft agglomerates to increase their flow and bulk density. Each agglomerate is comprised of a number of fine cut particles (Fig. 5.3) and has a diameter of the order of a few hundred microns.

Subjecting the fine cut PTFE to an agglomeration process can increase its powder flow and apparent density. The goal of this process is to make the small PTFE particles adhere together. Agglomeration can be achieved by *dry* or *wet* techniques.

The *dry* process entails the use of a water-insoluble organic liquid. An early process[22] consisted of mixing the fine cut PTFE powder with a small amount of an organic fluid which had a low surface tension followed by tumbling of the mixture. After formation of the agglomerates, the organic fluid was removed by heat. Toxicity and flammability are two drawbacks of this method.

In another dry process, organic solvents such as carbon tetrachloride, acetone, p-xylene, and ethanol were the liquid medium for agglomeration of PTFE.[23] Polymer to liquid ratio was 0.6–10 by weight and the process temperature ranged from 20–40°C. The mixture was agitated from 2–140 minutes, after which the agglomerates were removed and dried. Typical average particle size was 500–800 µm while apparent density remained between 400 and 600 g/l. Disadvantages of this process include processing of large quantities of organic liquids and the low apparent density of the product.

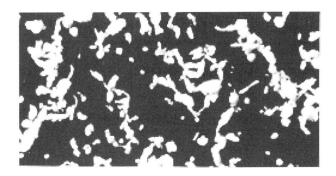

Figure 5.2 Fine-cut PTFE particles. *(Courtesy Daikin America.)*

Figure 5.3 Pelletized PTFE particles. *(Courtesy Daikin America.)*

Table 5.2. Properties of Fine Cut Granular PTFE According to ASTM Method D4894[19][21]

Resin Type	Apparent Density, g/l	Tensile Strength, MPa	Break Elongation, %	Standard Specific Gravity
Teflon® 7A	460	34.5	375	2.16
Teflon® 7C	250	37.9	400	2.16
Polyflon® M-12	290	47	370	2.17
Polyflon® M-14	425	32	350	2.16

The *wet* process requires using mixtures of water and a water-insoluble organic liquid. PTFE powder is mixed with this fluid in a stirred vessel. The mixture is heated to a temperature below the boiling point of the water and agitated to granulate the PTFE. The next steps involve the separation of the agglomerates from the liquid phase and drying. Generally, the wet process is preferable to the dry method because of the easier control of the properties of the agglomerates. There are wet processes in which water is the only agglomeration medium.

Roberts and Anderson[24] reported agglomerating finely ground PTFE powder in water. They added the polymer powder to a tank containing water at 80°C and agitated the mixture for one hour. The resultant slurry was filtered, dried and sieved through a mesh screen having 1,000 μm openings. Seventy percent of the product passed through the sieve and had an apparent density of 540 g/l. The ratio of PTFE powder to water was in the range of 88–149 g/l at a temperature above 40°C. In another experiment, PTFE powder was placed in a cylindrical vessel and rolled for 35 minutes at a speed of 11 rpm without water. The agglomerates obtained in this experiment had significantly worse flow and lower apparent density than those made in the presence of water. Table 5.3 shows a comparison of the properties of the two agglomerates. This process was capable of producing filled agglomerates of PTFE where 10–40 % of the weight of the powder consisted of fillers such as glass fiber, bronze or mica.

A process reported by Browning[25] used a aqueous medium, called *protective colloid*, which contained 500–5000 part per million by weight of a agent such as gelatin which would dissolve in water and form a solution. Other compounds, which could be dissolved, included partially-hydrolyzed polyvinyl acetate, hydroxymethyl cellulose, hydroxyethyl cellulose, polyethylene oxides and starch. Colloid here means materials, which have a long molecular chain with hydrophilic groups, distributed along the chain. This is in contrast to nonionic surfactants such as condensates of ethylene oxide and long chain alcohols, which have a hydrophilic portion at one end of the molecule and a hydrophobic one at the other end. Nonionic surfactants have been used in the agglomeration of granular PTFE powder. [26]

The ratio of the unfilled PTFE powder to the protective colloid solution varied from 2–3 g/ml.[25] The polymer and the solution were mixed at the desired ratio in a conventional V blender in the temperature range of 10–50°C. After tumbling for 60 minutes, the mixture is removed and dried at a temperature in the range of 120–200°C for several hours. For example, after drying for 16 hours at 120°C, the resulting powder had an average particle size of 424 μm and a compaction ratio of 3.21. This powder has an apparent density about 500–700 g/l [Author]. Fillers up to 30% by volume could be incorporated in the PTFE agglomerates by adding them to the mixture in the blender. Blending temperature has to be increased to insure good agglomerates are obtained.

Izumo et al[27] reported a process for agglomeration of finely ground PTFE powders by agitating it in a mixture of water and an organic liquid. This liquid had to have a surface tension <35 dynes/cm and had to be immiscible with water. In a typical process, 200 g/l of PTFE was added to water at 25°C and the mixture was placed in a stirred tank connected to a pipeline homomixer. The agitator speed was 400 rpm and it was operated throughout the process. For the first 3 minutes the mixture was circulated through the homomixer and then tetrachloroethane was added and the agitation continued for another 3 minutes. The agglomerates were recovered from the liquid media and dried. The resulting powder had good flow, an apparent density > 700 g/l and an average particle size in the range of 100–500 μm.

Table 5.3. Properties of PTFE Agglomerates by ASTM Method D4894[24]

Agglomerate Type	Average Particle Size, μm	Powder Flow,(a) g/sec	Apparent Density, g/l	Tensile Strength, MPa	Elongation, %
Made with Water	475	12	540	28.3	325
Made without Water	350	6.3	460	30.3	370

(a) Method to measure flow has been described in US Patent 3,087,921.

In other processes a perhalogenated hydrocarbon was the water-insoluble liquid. Examples of these liquids include trichlorotrifluoroethane, monofluorotrichloromethane and trichloropentafluoropropane. These fluids are nonflammable and noncombustible, are relatively easy to recover from water and have low surface tension (<35 dynes/cm). The boiling point of these liquids ranges from 80–130°C which means the agglomerates are subjected to these or higher temperatures during the recovery process. An increase in the temperature to which the PTFE agglomerates are exposed increases the hardness of the particles, adversely affecting the physical properties of the molded parts. Other drawbacks include high cost of recovery and adverse effect of perhalocarbons on the ozone layer.

To overcome the problems of perhalogenated hydrocarbons, processes have been developed with partially halogenated hydrocarbons,[28] such as 1,1-dichloro-2,2,2-trifluoroethane, 1,1-dichloro-1-fluoroethane and 1,3-dichloro-1,1,2,2,3-pentafluoropropane. Other than ozone safety, these liquids have lower boiling points (40–60°C) and can produce soft agglomerates. It is also possible to add fillers to the PTFE. The surface of hydrophilic fillers can be treated with organosilanes or silicone resin to prevent their separation and migration into the aqueous phase.

More recently, processes have been reported[29] for preparing free flowing PTFE powders in which the finely ground polymer and a filler are wetted with an aqueous phase containing a wetting agent (1–5%) such as ethylene or propylene glycol ethers. PTFE and the filler are dry blended and mixed in a Homoloid Mill (available from Fitzpatrick Company) equipped with a screen. They are then blended with the mixture of the wetting agent and water and mixed by passing through the Homoloid Mill. The wet milled product is next placed on the horizontal rotating pan (Dravo Pan, available from Ferro-Tech, Inc.) to form pellets. The pellets are placed in metal trays and dried in an oven.

For example a powder containing 25% by weight glass, 1% pigment and 74% PTFE was agglomerated by the above process using dipropylene glycol monobutyl ether. The wet agglomerates were dried for four hours at 299°C. The product had good flow and an apparent density of 804 g/l.

The process of agglomeration or granulation can be subdivided into several steps. Practical and mechanistic theories have been described by Hoornaert et. [30] In general, granulation begins by a nucleation step where the primary or the smallest polymer particles begin to assemble or agglomerate under favorable process conditions, described in this section. Figure 5.4 depicts the five processes of granulation or agglomeration mechanism.

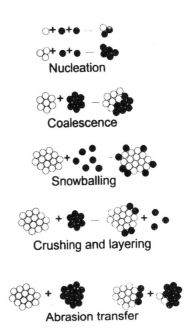

Figure 5.4 Mechanisms of granulation or agglomeration.[30]

A small number of particles form numerous "nuclei." Nucleation lasts a short time and is characterized by rapid growth. It is proceeded by coalescence or growth in the size of the agglomerates. This step is slower and lasts a relatively long time during which most of the agglomerates are formed. This regime endures as long as the agglomerates are deformable and not too large. Snowballing is the third step after the agglomerates have become large and rigid. In this step, individual particles are incorporated into large agglomerates. During the snowballing phase, a dense layer of small particles is deposited on the surface of agglomerates. Crushing and layering and abrasion-transfer are the final two steps of the granulation mechanism, which can also occur simultaneously with the snowballing regime. Crushing and layering is defined as the break up of smaller agglomerates as a result of collision with the equipment surfaces and other agglomerates. Abrasion transfer consists of the transfer of a small part of the surface of one agglomerate being abraded away and transferred to another agglomerate as the two granules contact each other.

5.4.3 Presintered Granular PTFE

Presintered resins are free flowing powders composed of premelted[31] polytetrafluoroethylene, predominantly intended for ram extrusion for the production of thin wall tubes and small diameter solid rods. These resins are prepared by melting as-polymerized polytetrafluoroethylene, cooling the melt and disintegrating the resin back into small particles. The average particle size of these resins is several hundred microns and its melting point is reduced from 342°C to 327°C. The primary advantage of presintered resins is easy feeding in ram extrusion and resistance to fracture caused by excessive back-pressure. See Ch. 7 for a complete description of ram extrusion process.

5.5 Preparation of Polytetrafluoroethylene by Dispersion Polymerization

Commercially, tetrafluoroethylene is polymerized in an aqueous dispersion medium to produce *dispersion and fine powder* PTFE products. In general, the key characteristics of this polymerization regime include ample dispersing agent and mild agitation at elevated temperature and pressure. The dispersion recovered from the reactor is finished by two different series of processes depending on whether a dispersion or a dry powder (fine powder) is the desired final product, as illustrated in Fig. 5.5. Polymerization and finishing techniques are reviewed in this section.

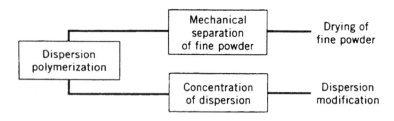

Figure 5.5 Dispersion polymerization and finishing processes.[5]

The polymerization of TFE monomers takes place in the aqueous dispersion medium and the resulting colloidal (Fig. 5.6) polymer remains in a stable emulsion form. An early report of this process was made by Renfrew in 1950[32] who used disuccinic or diglutaric acid peroxide (0.1–0.4% by weight of water) as the polymerization initiator. Gentle agitation was applied to the reactor while held under elevated pressure (0.3–2.4 MPa) at a temperature of 0–95°C. The polymerization product was a stable dispersion of small polymer particles containing 4–6.5 % PTFE. This dispersion easily coagulated after being subjected to agitation. The resulting dispersion still needed the addition of a second dispersant to raise the stability of dispersion while being agitated, a level which would allow transportation and handling.

Figure 5.6 Primary PTFE particles obtained from dispersion polymerization. *(Courtesy DuPont Fluoroproducts.)*

Brinker and Bro[33] reported improvements by the addition of a small amount of methane, ethane, hydrogen or hydrofluoroethanes to the reactor prior to the onset of polymerization. The reactor ingredients included a dispersing agent such as fluoroalkyl carboxylate[34] and an insoluble saturated hydrocarbon as anticoagulant.[35] A typical reaction contained 0.1–3% of a dispersing agent like ammonium perfluorocaprylate. The initiator of choice was a water soluble compound such as ammonium persulfate and disuccinic acid peroxide. Redux initiators, e.g., sodium bisulfite with ferrictriphosphate, could also be used. Initiator concentration depended on the rate and the degree of polymerization, anywhere from 0.01% to 0.5% of the weight of the water. The anticoagulant was a saturated hydrocarbon with more than twelve carbons, also known as *wax*, which is a liquid at the polymerization temperature.

For example, the addition of 0.008% methane (based on TFE) to the reactor at 86°C and 2.8 MPa significantly enhanced stability. The colloidal solids concentration was 36%, well in excess of Renfrew process. Dispersion stability was as defined by the amount of time required for coagulation of PTFE particles when it was agitated at 500 rpm. The Brinker and Bro[33] process nearly tripled the coagulation time to 6–8 minutes.

Up to this point, fairly high molecular weight fine powder PTFE was available which could not be used to create thin parts such as wire coating and tubing. Excessive paste extrusion pressure and flaws developed when high molecular weight resins were converted to parts less than 500 µm in thickness. This was primarily due to the high molecular weight of PTFE. See Ch. 8 for a complete description of the processing of fine powder PTFE. In 1964, Cardinal[6] reported the development of fine powder resins with lower molecular weight as evidenced by their lower melt creep viscosity. This was achieved by the introduction of a modifier to the polymerization kettle. The modifier consisted of a non-polymerizable chain-transfer agent such as hydrogen, methane, propane, carbon tetrachloride, perfluoroalky trifluoroethylene and oxyperfluoroalky trifluoroethylene, the last two containing from three to ten carbon atoms.

The aqueous phase contained water, dispersing agent, initiator and wax. Typical polymerization was conducted at 50–85°C at a pressure of 2.9 MPa. The reactor, also called an *autoclave,* consisted of a horizontal cylinder with length to diameter ratio of 10:1, equipped with a steam/ water jacket and a paddlewheel agitator running the length of the autoclave.[6] The agitator speed was fairly slow compared to the speed of the agitator in the suspension polymerization. The motion of the paddlewheel keeps the aqueous phase saturated with TFE. A typical recipe and some properties are shown in Table 5.4 for hexafluoropropylene and methanol modifiers. Other modifiers such as perfluoropropyl vinyl ether have also been reported.

Table 5.4. An Example of PTFE Emulsion Polymerization Recipe and Some Properties [6]

Reaction Component or Property	Polymer 1	Polymer 2
Modifier Type	hexafluoropropylene	methanol
Modifier content, % by weight of TFE	0.15	0.009
Deionized water	1,500 parts	1,500 parts
TFE	3,000 parts	3,000 parts
Initiator, % by weight of water	0.005 Potassium persulfate	0.006 Ammonium persulfate
Ammonium perfluorononaoate, % by weight of water	0.15	0.15
Wax, % by weight of water	6.3	6.3
Temperature, °C	85	70
Pressure, MPa	2.9	2.9
Agitator speed, rpm	125	125
Solids content of dispersion, %% by weight of water	35	40.5
PTFE particle size, µm	0.17	0.17
Standard Specific Gravity[1]	2.211	2.211
Melt Creep Viscosity at 380°C, Poise	$3\text{-}6 \times 10^9$	-

[1] Determined according to ASTM Method D 4895.

The modifier may be introduced at any time during the polymerization. For example, if it is introduced after 70% of the monomer to be polymerized has been consumed, each PTFE particle will contain a core of high molecular weight PTFE and a shell containing the low molecular weight modified polytetrafluoroethylene. In this example 30% of the outer shell of the particle, by weight, has been modified. The total modifier content of the polymer was extremely small (Table 5.4), yet the impact on the properties was profound. Melt creep viscosity was $3-6 \times 10^{10}$ poise as compared to the polymer made under identical conditions without the modifier which had a melt creep viscosity of 10×10^{10} poise. Paste extrusion pressure was decreased by 20–50% leading to less flaws in tubing and wire insulation made from these resins.

Cardinal et al.'s[6] invention was a significant improvement in the ability to alter and control polymer properties. The importance of perfluoroalkyl vinyl ether comonomers (modifiers) such as perfluoromethyl vinyl ether (PMVE), perfluoroethyl vinyl ether (PEVE), perfluoropropyl vinyl ether (PPVE) and perfluorobutyl vinyl ether (PBVE) in dispersion polymerization was explored by Holmes et al.[7] They developed polymers composed entirely of copolymers of TFE and modifiers such as PPVE with excellent mechanical properties. For example, flex life of eighteen million cycles was obtained after aging the samples at 322°C for 31 days, for PPVE-modified polymer. Standard specific gravity of the PPVE modified polymers was below 2.175 and melt creep viscosity remained below 4×10^{10} Poise at 380°C. Polymerization rates to produce these modified polymers were increased to commercially acceptable rates by incorporation of purified modifiers. This was also accomplished by replacing the disuccinic acid peroxide with a persulfate type initiator such as ammonium persulfate. The latter did not slow down the polymerization reaction.

Polymerization[7] was accomplished in a similar autoclave to that used by Cardinal et al.[6] Length to diameter was about 1.5 to 1. The autoclave was equipped with a four-bladed cage type agitator, rotated at 46 rpm. Typically, the autoclave was first evacuated, charged with wax, water and ammonium perfluorocaprylate (dispersant). The autoclave was heated to 65°C and next ammonium persulfate (initiator) was added while stirring. After heating the autoclave to 72°C, perfluoropropyl vinyl ether was added and the vessel was pressurized with TFE at constant temperature and stirring rate. Temperature was increased to 75°C after reaction had started, as evidenced by a drop in the autoclave pressure. Early in the polymerization, after consumption of about 10% of the total TFE, additional dispersant was added to stabilize the dispersion. After the desired amount of TFE to obtain a 35% polymer concentration in the aqueous phase had been fed, the monomer flow was terminated. After reaching a pressure drop of about 60%, agitation was stopped and the autoclave was vented and its contents discharged. The supernant solids, which were predominantly paraffin wax, were removed. The treatment of the dispersion from this point on depended whether the final product was fine powder or a dispersion. Typical recipe and polymerization data are presented in Table 5.5.

Table 5.5. An Example of PTFE Emulsion Polymerization Recipe and Some Properties[7]

Reaction Component or Property	Polymer 1	Polymer 2
Modifier type	perfluoropropyl vinyl ether	perfluoroethyl vinyl ether
Modifier amount	20.5 ml	3 g
Deionized water, g	21,800	3,600
TFE, g	10,050	1,830
Ammonium persulfate initiator, g	0.33	0.065
Ammonium perfluorocaprylate dispersant, g initial final	 2 26.7	 4.92
Wax, g	855	141
Temperature, °C	65-75	75
Pressure, MPa	2.8	2.8
Agitator speed, rpm	46	105
Solids content of dispersion, % by weight of water	35	33.7
PTFE particle size, μm	0.188	0.10
Standard specific gravity[1]	2.149	2.160
Modifier content of the polymer, % by weight[2]	0.102	0.09
Melt creep viscosity at 380°C, poise	0.9×10^9	2×10^9

[1] Determined according to the ASTM Method D 4895.
[2] The procedure has been described in Ref. 36.

Poirier[36] reported preparation of dispersion polymerized polytetrafluoroethylene with a composite particle structure. The inner portion of the particle (core) contained a higher concentration of the comonomer than its outer portion (shell). The comonomers were from the general family of vinyl ethers such as perfluoroalkyl vinyl ethers. The advantage of these polymers is the possibility of paste extrusion of their fine powders at high reduction ratio without the complication of high extrusion pressure and flaws in the extruded parts such as tubing or wire insulation. The reduction ratio of these polymers can exceed 10,000:1 which is well beyond the conventional commercial range (see Ch. 8 for information about fine powder processing).

The core-shell polymer[36] was made by evacuating the reactor part way through the polymerization and repressurizing the autoclave with tetrafluoroethylene, which resulted in a reduction in the concentration of the comonomer. The core constituted 65–75% of the total weight of the particle. The remaining 25–35% of the polymer formed the shell at a lower comonomer content than the core. Wire insulation was made from these polymers and the number of flaws in the insulation was detected by subjecting the wire to a high voltage (2,000–8,000 volts). The number of flaws were minimized in the core-shell polymer made with a lower concentration of the comonomer in the shell compared to polymer made with TFE alone without the comonomer.

Introduction of the comonomer may improve the paste extrudability of the resin or the properties of the final product. Copolymers tend to improve the transparency of the sintered part such as tubing. Sometimes the modifier improves the extrudability but deteriorates properties of the sintered part. For example, a copolymer of TFE with chlorotrifluoroethylene could be paste extruded at high reduction ratios and low extrusion pressure.[37] Thermal stability of the CTFE copolymer was reduced, substantially. To overcome this problem a core-shell architecture was designed. The core was comprised of a copolymer of a fluoroalkyl vinyl ether and TFE and the shell of a CTFE and TFE copolymer. The thickness of the shell could be fairly low (5% of total weight of the particle) without a loss of good extrusion properties. Table 5.6 shows examples of five polymers with different compositions that show the beneficial effect of core-shell structure.

Table 5.6. An Example of PTFE Emulsion Polymerization and Polymer Properties[37]

Reaction Component or Property	Polymer 1	Polymer 2	Polymer 3	Polymer 4	Polymer 5
Disuccinic acid peroxide, ppm	120	120	60	120	-
Ammonium persulfate initiator, ppm	3.75	3.75	4.1	3.75	10
Temperature, °C	70	70	85	70	70
Solids content of dispersion, % by weight of water	31.9	31.4	32	31.5	31.8
Particle size, μm	0.20	0.20	0.24	0.26	0.18
Core modifier	PPVE	PPVE	CTFE	PPVE	PPVE
Shell modifier	CTFE	CTFE	CTFE	-	-
Chlorotrifluoroethylene content of the polymer, % by weight	0.035	0.280	0.250	-	-
PPVE content of the polymer, % by weight	0.02	0.02	-	0.02	0.10
Standard specific gravity[1]	2.185	2.184	2.183	2.186	2.173
Thermal instability index[1]	10	33	40	1	1
Extrusion pressure at a reduction ratio of 1,500:1, MPa	100	64	52	108	118

[1] Determined according to ASTM Method D 4895.

The discussion of dispersion polymerization should have familiarized the reader with a basic knowledge of the process and the range of conditions and possible products. After the completion of the polymerization, a "raw" dispersion of PTFE particles in water is obtained which must be "finished" before commercial application is possible. In the next two sections the finishing steps to produce *dispersion* and *fine powder* products have been described.

5.5.1 Preparation of Dispersion Polytetrafluoroethylene

Raw dispersions produced by dispersion polymerization contain a wide range of polytetrafluoroethylene solids, often, in the range of <10 to 45% by weight in water. To insure a commercially viable product, the dispersion must be sufficiently stable for transportation, storage and handling. This means that it should not form precipitated PTFE particles, which cannot be reincorporated in the liquid phase by simple mixing. It should also contain the minimum possible water content to decrease the transportation cost. An obvious way to concentrate dispersions is heat to evaporate the excess water. The major drawback to heating is the irreversible coagulation of the polymer, that is, the PTFE coagulates cannot be redispersed.

One approach is to concentrate the dilute PTFE dispersion by the following steps:

1. Adding a surfactant to the dispersion
2. "Breaking the emulsion" by deactivating or insolubilizing the surfactant leading to flocculation the polymer
3. Separating the polymer floc from the bulk of the aqueous phase
4. Forming a concentrated dispersion by peptizing (redispersing) the polymer floc

This method can produce concentrated dispersions in the range of 35–75% PTFE content by weight.[38]

For example, a dispersion containing 3.2% PTFE was concentrated by first neutralizing it with ammonia to a pH of 7.[38] Then a small amount of polyammonium styrene/maleamate or ammonium stearate (ionic surfactant) was added without an apparent change in the dispersion. The next step was acidification of the dispersion with 1% hydrochloric acid, resulting in flocculation of the polymer. The mixture was then centrifuged and the virtually clear supernant liquid was decanted. The PTFE floc was made basic by the addition of ammonium hydroxide and was shaken. A colloidal dispersion containing 40% PTFE was obtained by this method.

Another variation of the above method[38] consisted of the addition of 4 parts (by weight) of 1% aqueous solution of poly-N-vinylcaprolactam (nonionic surfactant) to 25 parts of a 4% polytetrafluoroethylene dispersion. The dispersion remained unchanged at 25°C. Heating it to 30°C brought about a major change by flocculation and settlement of PTFE particles. The supernant liquid was removed by decantation. The floc was cooled to the room temperature and peptized by the addition of 2% of a long chain alkyl sodium sulfate, based on the weight of the polymer. A dispersion containing 40% PTFE was obtained. Using sodium dioctyl sulfosuccinate or a polyglycol ether instead of an alkyl sodium sulfate yielded similar results.

The emulsion can be broken by direct "salt-out" where a water-soluble salt is added to the dilute dispersion. The increase in the ionic strength of the aqueous phase leads to swift flocculation of the polymer. To be effective, the salt should not form an insoluble product with the surfactant. It must also be able to salt out the dispersion at low concentrations. It is highly desirable for the salt to be eliminated from the final article made from the PTFE due to the positive impact on the properties. This is usually possible by the selection of a salt, which degrades below or near the temperature of fabrication processes. Salts such as ammonium carbonate or nitrite decompose below 100°C and meet the other requisites for salting out PTFE dispersions.

Salting out is illustrated by this example where a concentrated dispersion containing almost 60% PTFE was obtained.[38] Sixteen parts (by weight) of a 1% solution of dioctyl ester of sodium sulfosuccinate was added to 100 parts of a 3.2% aqueous dispersion of PTFE. Next, 5.5 parts of ammonium carbonate were dissolved in the dispersion at 25°C. PTFE flocculated completely and settled at the bottom of the container. The supernant liquid was removed and the floc was heated to 80–10°C to drive off most of the ammonium carbonate. After cooling, the polymer was peptized and a 59.5% by weight PTFE dispersion was produced.

To break the emulsion by an acid or a base, some important points must be considered. A dilute acid should be added to anionic type surfactants and a base to the cationic variety. Preferably, the reaction product of the acid or base with the surfactant should be a solid in order to obtain a redispersible floc. For example, ammonium stearate is an anionic surfactant and after acidification forms stearic acid which is a solid at temperatures below 71°C.

The disadvantage of Berry's technique[38] was that it was not easily applicable to higher concentration dispersions. Productivity of dispersion polymerization process was improved and emulsions containing 30–45% colloidal PTFE were produced. Berry's method was not applicable for three reasons:

1. The presence of 0.2–0.4% of ammonium polyfluorocarboxylates (7–10 carbon atoms) in the raw dispersion
2. The strong tendency of the PTFE particles to coagulate, rendering peptization virtually impossible
3. The great sensitivity of the operation to the concentrating procedure made it less practical.

Marks and Whipple[39] proposed a process for concentrating emulsions containing 30–45% colloidal particles of polytetrafluoroethylene.

The process offered by Marks and Whipple was accomplished by the addition of 0.01–1.0% by weight of the dispersion of sodium hydroxide, ammonium hydroxide or ammonium carbonate. Next, 6–12% by weight of dispersion solids of a nonionic surfactant was added with a structure of [R-C_6H_4-(OCH$_2$CH$_2$)$_n$OH] where R is a monovalent hydrocarbon with 8–10 carbon atoms and n = R+1 or R+2 (R = 8–10), or [(tertiary octyl)-C_6H_4-(OCH$_2$CH$_2$)$_{9-10}$OH]. After stirring, the mixture was heated to a temperature of 50–80°C. A cloudy appearance indicated that the nonionic surfactant had begun to insolubilize. PTFE particles settled after a period of time and formed a layer at the bottom of the container. The upper layer, which was relatively clear, was decanted and the lower layer was recovered. The solid content of this dispersion was colloidal PTFE particles at 55–75% content by weight free of coagulated polymer particles.

Marks and Whipple's method does not require peptization which reduces the amount of ionic material in the dispersion. This has a positive impact on the properties of article made from the dispersion, particularly on their electrical properties. The time required for concentration is by one or two orders of magnitude shorter than in Berry's technique. Table 5.7 provides a number of examples for the process. In each case, the process was followed step by step according to the conditions given in this Table. Triton® X-100 was the nonionic surfactant for all the examples and an amount equal to 9% of the PTFE content of the dispersion was added to each mixture. Addition of ammonium carbonate electrolyte yielded the shortest concentration time.

Hoashi,[40] Holmes[41] and Morris[42] have reported improvements in the Marks and Whipple concentration process such as a reduction in the temperature, ease of surfactant removal during fabrication and increased stability.

Another technique reported[43] for concentrating aqueous polytetrafluoroethylene depends on physical dehydration. A water-absorbing organic substance was used which was soluble in water and had a lower density than the dispersion. This material was stratified on top of the PTFE dispersion and the contact between two layers was maintained until sufficient water absorption had taken place. The absorbent material was removed from the floc. Some of the dehydrating agents included glycerine, polyethers such as polyethylene glycol and polypropylene glycol, polysaccharides such as methylcellulose and substituted phenols.

A novel method[44] used a semipermeable membrane to remove water from PTFE by ultrafiltration. In this technique 0.5–12% of a surfactant, relative to the weight of the solid polymer, was added to the dispersion. After mixing the emulsion was circulated over units of semipermeable ultrafiltration membranes at fluid velocity of 2–7 m/sec. The dispersion must be kept from contact with components causing frictional

Table 5.7. Examples of PTFE Emulsion Concentration Process Variable[39]

Starting Dispersion Concentration, Wt.%	Electrolyte Type	Electrolyte Added, Wt. % of Dispersion	Temperature, °C	Concentration Time, min	Final Concentration, Wt.%
47	$CaCl_2$	0.04	80	35	60
47	NH_4Cl	0.04	80	30	63
50.7	NH_4CO_3	0.04	80	12	72
44	NH_4OH	0.36	75	60	68

forces, lest the colloidal polymer particles coagulate. After removal of sufficient quantity of water the concentrated dispersion was removed from circulation. Typically, concentrated products by this technique contained 40–65% polymer solids by weight.

In another procedure for concentration[45] an acrylic polymer containing a large amount of acid groups or its salts were added to PTFE dispersions. The acid content of the acrylic polymer was at least 20% and they had a weight-average molecular weight of 50M–500M; polyacrylic acid was the preferred polymer. Addition of a small amount (0.01–0.5%) of this type of acrylic polymer to the dispersion caused a phase separation to occur. The lower phase contained 50–70% by weight of PTFE and was recovered by decantation. To reduce viscosity and increase stability of the concentrated dispersion, either an ionic or nonionic surfactant could be added before the addition of the acrylic polymer.

Stability is a key requirement of the final product. Any dispersion must have a reasonable shelf life of the order of a few weeks to a few months. It should also be able to withstand transportation and handling during processing. The shear rate inherent in these activities must not lead to coagulation of the polymer particles. For a given dispersion stability is a function of solids content, pH and viscosity. These properties can be adjusted by trial and error to improve the stability.

5.5.2 Preparation of Fine Powder Polytetrafluoroethylene

To produce fine powder from the polymerization dispersion, three processing steps have to take place.

1. Coagulation of the colloidal particles
2. Separation of the agglomerates from the aqueous phase
3. Drying the agglomerates

Diluting the raw dispersion to a polymer concentration of 10–20% by weight and possibly adjusting the pH to neutral or basic[37] carries out coagulation. A coagulating agent such as a water-soluble organic compound or inorganic salt or acid can be added to the dispersion. Examples of the organic compound include methanol and acetone. Inorganic salts such as potassium nitrate and ammonium carbonate and inorganic acids like nitric acid and hydrochloric acid can aid coagulation. The diluted dispersion is then agitated vigorously. Primary PTFE particles form agglomerates which are isolated by skimming or filtration.

Drying of the polytetrafluoroethylene agglomerates is carried out by vacuum, high frequency or heated air such that the wet powder is not excessively fluidized.[37] Friction or contact between the particles, especially at a high temperature, adversely affects the fine powder because of easy fibrillation and loss of its particulate structure leading to poor properties of parts made from this resin. Chapter 8 provides a complete discussion of fibrillation. Drying temperatures range from 100–180°C and have great influence on the paste extrusion of the resin. High drying temperatures result in high extrusion pressures (see Ch. 8).

Fine powder resins must be protected from fibrillation after drying. PTFE does not fibrillate below its transition point at (19°C for TFE homopolymers) during normal handling and transportation. Storage and transportation of the resin after refrigeration below its transition point is the normal commercial practice for handling fine powder polytetrafluoroethylene resins.

5.6 Chlorotrifluoroethylene Polymers and Polymerization

The first documented polymerization of chlorotrifluoroethylene (CTFE) into a solid plastic was reported by Farbenindustrie AG of Main, Germany, in 1937.[46] These polymers had low molecular weight and lacked sufficient mechanical properties. The initial applications of these polymers were limited to lubricants required to perform under severe chemical and thermal environments. High molecular weight polymers of CTFE were synthesized, characterized and utilized during the Manhattan Project[47] in the diffusion process to separate the uranium isotopes. Polychlorotrifluoroethylene homopolymers and low copolymers with vinylidene fluoride are thermoplastics, which are by and large inferior to PTFE with respect to thermal stability and chemical resistance, higher surface energy and coefficient of friction, and less attractive electrical properties. Replacement of one fluorine with chlorine imparts some advantages to CTFE plastics such as increased mechanical strength, lower gas permeability, and improved optical clarity over PTFE. Increasing vinylidene fluoride content of CTFE copolymers to 50% produces an elastomeric gum with excellent resistance to oxidizing acids at elevated temperatures.

High molecular weight homopolymers and copolymers of CTFE were first commercialized by M. W. Kellog Company, under the trade name Kel F®. Allied Signal manufactures a copolymer of CTFE films of which is known as Aclar®. Daikin Industries, Ltd., offers PCTFE resins by the trade name Neoflon® CTFE.

5.6.1 Polymerization of Chlorotrifluoroethylene

The Farberindustrie's pioneering CTFE polymerization was reported to have been conducted under elevated pressure and temperature by solution and aqueous polymerization. Solution polymerization was conducted in two stages, at 40–45°C and 60–65°C, in an alcohol, benzene or petroleum ether. Peroxides such as benzoyl peroxide were the catalyst and an aldehyde the regulator. Aqueous polymerization was also possible in the presence of an emulsifying agent and a peroxide catalyst. Overall, the polymerization resembled that of vinyl chloride.

The synthesis of chlorotrifluoroethylene has been carried out by bulk,[48]–[52] solution,[53]–[55] suspension,[56]–[62] and emulsion[63]–[66] polymerization regimes and high molecular weight polymers have been obtained. Free radical initiators, ultraviolet and gamma ray radiation has been used to initiate the reaction. Emulsion and suspension polymers are more thermally stable than bulk polymerized products. Commercial PCTFE plastic is prepared by the former techniques to maximize continuous use temperature of the polymer. Agitated stainless steel or glass lined vessels are two examples of reactors for PCTFE synthesis. No unusual equipment is required. Typical reaction conditions include pressures of 0.34–1.03 MPa at 21–53°C.

Bulk polymerization is carried out in a small static vessel and is initiated by a halogenated acyl peroxide.[67][68] Ultraviolet light and gamma rays can also initiate the reaction. There are some disadvantages to this method of PCTFE preparation. Temperature control is difficult resulting in poor product reproducibility. Low conversion (<40%), long reaction time (168 hours) and low temperature (-34°C) are the other shortcomings of this technique. The severe temperature gradient in the reaction vessel leads to distinct bimodal molecular weight distribution. After completion of the polymerization the unreacted monomer is distilled off. The polymer is recovered in the form of a porous plug and subsequently pulverized, washed and dried prior to fabrication.

Suspension polymerization is effected in water with organic or inorganic peroxides. The monomer is present in excess simultaneously with aqueous and solid polymer phases and the mixture is stirred. The polymerization was found to be too slow using a persulfate initiator alone.[61] Addition of a reducing agent such as a bisulfite ion along with lowering the pH to 2.5 increased the polymerization rate. Addition of a silver salt further accelerated the reaction rate and improved reproducibility of the yields of a high molecular weight PCTFE.

Polymerization of CTFE by suspension method usually requires a promoting agent to lessen the polymerization time and drive up the molecular weight.[60] These promoting agents fall in three classes: *promoters*, *activators* and *accelerators*. In most cases a promoter must be used which consists of an inorganic compound. Examples include water-soluble persulfates, perborates, perphosphates, percarbonates and hydrogen peroxide. Persulfate and perphosphate salts of sodium, potassium, ammonium and calcium are especially valuable. The required concentration of promoters is in the range of 0.003–0.1 molarity. The exact concentration is dependent on the specific promoter and the desired molecular weight of the polymer. For example to produce PCTFE with a softening point of 200°C, a promoter concentration of 0.003–0.07 is required.[60]

Activators are preferably added to the system in conjunction with the promoters to increase their activity. Some of the more common compounds include sodium bisulfite, sodium thiosulfate, sodium hydrosulfite and trimethylamine. Generally, the activators are water-soluble reducing agents. Equal molar concentrations of the activator and the promoter should be present in suspension polymerization of CTFE.

Regulating agents are added to the aqueous phase to regulate the pH of the mixture. The desirable pH range is 1 to 4 and can be maintained by the inclusion of a buffering agent. Acetic acid, monosodium phosphate and propionic acid are two examples of an effective buffering agent.[60]

Polychlorotrifluoroethylene made by suspension polymerization has generally had an undesirable molecular weight to viscosity relationship, that is, melt viscosity is higher at a given molecular weight.[69] To overcome this deficiency a small amount of vinylidene fluoride is copolymerized with CTFE. Another issue is the tendency of the molecular weight of suspension polymer to be skewed towards the low end, which can have a deleterious effect on part properties.

Emulsion polymerization is carried out in water using an inorganic peroxy initiator and a surfactant. Non-halogenated surfactants are not effective in CTFE polymerization. Halogenated hydrocarbon acids and salts are the more effective surfactants. The most active agents are aliphatic perfluorinated and perfluorochlorinated acids and their corresponding salts with the following general formulas:[70]

$$F(CF_2)_n\text{–COOH} \qquad n = 6\text{–}12$$

$$CCl_3(CF_2\text{–}CFCl)_{n-1}\text{-}CF_2\text{–COOH} \quad n = 3\text{–}6$$

An example of emulsion polymerization is given in a 3M patent[71] using the recipe of Table 5.8 in a batch process. The pH of the reaction mixture was adjusted to 7 by the addition of potassium hydroxide. Perfluorooctanoic acid was the surfactant and a persulfate/bisulfite peroxy package initiated the reaction. Polymerization lasted for 20 hours at 30°C while the reaction vessel was tumbled. At the end of the reaction, the vessel was vented and the latex was recovered.

Table 5.8. Example of CTFE Emulsion Polymerization Recipe[71]

Reaction Component	Amount, parts
Deionized water	300
CTFE	100
Potassium persulfate	2.4
Sodium bisulfite	1.1
Perfluorooctanoic acid	2.4

Freezing helped break the recovered polymer emulsion. PCTFE was filtered out, washed with water and dried at a yield of 46%. The emulsion polymers are the most difficult to recover but they are superior to the resin obtained by other techniques because of higher molecular weight and thermal stability and a greater reproducibility.

5.7 Characterization of Polytetrafluoroethylene

A number of properties have to be measured to characterize and identify each PTFE resin. The basic properties of polytetrafluoroethylene are characterized by standard test methods (Table 5.9) published by American Society for Testing Materials (ASTM). Three major methods specify types and define properties for granular, fine powder and dispersion products. Another set of standards similarly covering the fluoropolymers is published by the International Standards Organization (ISO). This section describes each method and the associated tests.

Table 5.9. Standard Specification Methods for PTFE

PTFE Product Type	ASTM Method	Related ISO Standard
Granular Resins	D 4894 - 97	12086-1 and 12086-2
Fine Powder Resins	D 4895 - 97	12086-1 and 12086-2
Dispersion Products	D 4441 - 96	12086-1 and 12086-2

5.7.1 Granular PTFE Resins

ASTM Method D 4894–97 (last revised in 1997) covers specifications for granular PTFE resins and test methods for the as-produced polymer. Methods for processing granular resins into objects are discussed in Ch. 7 of this book. PTFE resins are thermoplastics in that they can be remelted but they cannot be processed by the normal melt processing technologies due to their extremely high rheology. This polymer does not dissolve in any solvents. These two facts render direct measurement of PTFE molecular weight virtually impossible. An indirect property named *standard specific gravity* is substituted for molecular weight.

Standard specific gravity (SSG) is defined as specific gravity of a sample of PTFE molded and fabricated according to the exact procedures prescribed by ASTM D4894-97. Special emphasis is placed on molding the part identically during each molding. Cooling rate of the molten polymer is closely controlled to maintain a specific rate of crystallization. SSG of samples made with polytetrafluoroethylene with the same structure depends on the crystalline, amorphous and the void content of the sample. Density of crystalline PTFE (2.302 g/cm^3)[73] is significantly higher that of amorphous polymer (2.00 g/cm^3)[73] due to the closer packing of molecules in the crystalline phase. By fixing the molding conditions and cooling rate, SSG value depends on the crystalline/amorphous phase content of the sample. The principle is that, at a specified cooling rate, smaller molecules have higher mobility and pack more easily and quickly than larger molecules which results in a higher extent of crystallinity, therefore, higher SSG.

Properties used in the characterization of granular PTFE are defined in Table 5.10.

A number of operational terms are defined in this method, which have specific meanings. They have been defined in the following paragraphs.

Preforming is the operation in which the PTFE powder is compacted under pressure in a mold. The molded specimen can be handled with care without damage.

Sintering of PTFE is a thermal treatment during which the polymer is melted, coalesced and recrystallized during cooling.

Skiving is machining operation during which a continuous film of PTFE is cut away (peeled) from the outer surface of the cylindrical molding.

Table 5.10. Definition of Basic Properties of Granular Polytetrafluoroethylene According to ASTM D 4894–97

Property	Definition	Reference ASTM Method
Bulk Density[1]	Mass of one liter of resin measured under the test conditions	D 1895
Particle Size[1]	Average particle size and distribution by sieving	E 11
Melting Characteristics[1]	Heat of Fusion and Melting Peak Temperature of resin as determined by Differential Scanning Calorimetry	D 4591
Water Content[1]	Water present in the PTFE resin	
Standard Specific Gravity (SSG)[2]	Specific gravity of a sample of molded and sintered PTFE according to this method	D 792, D 1505
Thermal Instability Index (TII)[2]	A measure of decrease in molecular weight of PTFE material determined by the difference between ESG and SSG: TII = (ESG-SSG) × 1000	
Tensile Properties[2]	Elongation and strength at break of a sample made according to the specified method	D 638
Shrinkage and Growth	The change in the diameter of SSG preform due to sintering	
Extended Specific Gravity (ESG)	The specific gravity of a PTFE specimen molded for SSG after sintering for an extended period of time, compared to the sintering time of SSG	D 792, D 1505
Electrical Properties	Dielectric constant, dissipation factor, dielectric breakdown voltage, dielectric strength	D 149

[1] Properties required for resin specification.
[2] Properties required for on molded specimen.

Reprocessed resin is PTFE powder that has been produced by grinding preformed and sintered polymer.

Reground resin is produced by grinding PTFE that has been preformed but has never been sintered.

ASTM Method D 4894 classifies the different types of granular powder according to the system summarized in Table 5.11.

A special specimen is made to measure the Standard Specific Gravity and Extended Specific Gravity of the resin. The mold has an inside diameter of 28.6 mm and a depth of 76.2 mm. A 12 g quantity of PTFE powder is molded into a preform at a pressure of 34.5 MPa for 2 minutes and sintered according to the specified sintering cycles. The sintered chip is conditioned before measuring its specific gravity according to ASTM D 792 or D 1505 and its outer diameter for shrinkage calculation. SSG can also be measured from a billet made primarily for producing tensile test samples.

Specimens for tensile properties are in the shape of microtensile bars and can be cut from a chip or a billet. Preforming 14.5 g of resin in a mold with an inside diameter of 76.07 mm and a depth of 26 mm makes the chip. An alternative to the chip for Types I, II, II and IV resins is a billet made from a resin charge of 400 g in a mold with an inside 57 mm and a depth of 254 mm. The preform is made at a final pressure of 34.5 MPa for Type I and 17.2 MPa for Types II, III and IV for a period of 2–5 minutes. The preforms are sintered according to specified sintering cycles. The billet must be machined into 0.8 mm thick samples for tensile testing and a chip with suitable thickness for SSG measurement. Electrical properties are measured on a 0.13 mm thick film skived from the billet.

The specifications that differentiate granular polytetrafluoroethylene powders are summarized in Table 5.12. Water content and Melting Peak Temperatures are the other specifications of these resins. Specifications of molded parts are listed in Table 5.13.

Table 5.11. Granular Polytetrafluoroethylene Resin Classification According to ASTM D 4894

ASTM Type	Description
I	General purpose molding and ram extrusion resin
II	Fine cut resin with an average particle size below 100μm
III	Fine-cut or free-flow comonomer-modified resins
IV	Free-flowing resins
V	Pre-sintered resins
VI	Ram extrusion resins, not pre-sintered

Table 5.12. Granular Polytetrafluoroethylene Resin Specifications According to ASTM D 4894

Type	Grade	Bulk Density, g/l	Average Particle Size Diameter, μm
I	1	700 ±100	500 ±150
	2	675 ±50	375 ±75
II			<100
III	1	350 ±75	<100
	2	850 ±50	500 ±100
IV	1	650 ±150	550 ±225
	2	>800	
	3	580 ±80	200 ±75
V		635 ±100	500 ±250
VI		650 ±150	900 ±100

Table 5.13. Molded Granular Polytetrafluoroethylene Specifications According to ASTM D 4894

Type	Grade	Standard Specific Gravity Range	Tensile Strength, MPa	Break Elongation, %
I[1]	1	2.13-2.18	13.8	140
	2	2.13-2.18	17.2	200
II[1]		2.13-2.19	27.6	300
III[1]	1	2.16-2.22	28.0	500
	2	2.14-2.18	20.7	300
IV[1]	1	2.13-2.19	25.5	275
	2	2.13-2.19	27.6	300
	3	2.15-2.18	27.6	200
V[2]	-	-	-	-
VI[2]	-	-	-	-

[1] Thermal Instability Index less than 50.
[2] No molded property specifications.

The actual ASTM Method D 4894-97 must be consulted for a complete description of these procedures.

5.7.2 Fine Powder PTFE Resins

ASTM Method D4895–97 covers specifications for fine powder PTFE resins and test methods for the as-produced polymer. Methods for processing fine powder resins into objects are discussed in Ch. 8 of this book. PTFE resins are thermoplastics in that they can be remelted but they cannot be processed by the normal melt processing technologies due to their extremely high rheology. This polymer does not dissolve in any solvents. These two facts render direct measurement of PTFE molecular weight virtually impossible. An indirect property named *standard specific gravity* is substituted for molecular weight. Section 5.7.1 contains a discussion of the relationship between molecular weight and standard specific gravity.

Properties used in the characterization of fine powder PTFE are defined in Table 5.14.

Specimens for SSG and tensile properties are prepared by similar techniques to those described in Sec. 5.7.1. In this section the properties specific to fine powder characterization are described.

Extrusion pressure is determined in a paste extruder also called a rheometer. It consists of a vertically positioned breech-loaded tubular barrel with an inside diameter of 32 mm. The barrel is approximately 305 mm long which is not critical as long as it can hold enough resin preform to extrude for 5 minutes. The extruder is equipped with a hydraulic system and a ram with an inside diameter of 32 mm, capable of pushing the lubricated PTFE paste out of a small die. There should also be an appropriate pressure-sensing device. The die orifice size is selected to obtain barrel to orifice cross sectional ratios of 100:1, 400:1 and 1600:1 called the *reduction ratio*. The choice of reduction ratio depends on the resin type. The equipment temperature is maintained at 30°C during the extrusion. The agreed upon rate of extrusion is 19 g/min on a dry resin basis.

The lubricant is an isoparaffin also called extrusion aid and should be blended with the resin at a prescribed ratio. The mixture is placed in a jar and blended by rolling. There are alternative techniques to bottle rolling. Chapter 8 provides detailed description of the blending techniques. After blending the jar and its content are stored at 30°C for 2 hours or longer to allow the lubricant to diffuse to the inside surface of the polymer particles. The preform is made by molding the blend in a tube with an inside diameter of 32 mm and a length of 610 mm. The lubricated blend is poured into the tube and is pushed down using a plug in press at a pressure of 0.07 MPa. The preform is loaded in the extruder barrel prior to extrusion during which pressure is recorded.

Table 5.14. Definition of Basic Properties of Fine Powder Polytetrafluoroethylene According to ASTM D 4895-97

Property	Definition	Reference ASTM Method
Bulk Density[1]	Mass of one liter of resin measured under the test conditions	D 1895
Particle Size[1]	Average particle size and distribution by sieving	E 11
Melting Characteristics[1]	Heat of fusion and melting peak temperature of resin as determined by differential scanning calorimetry	D 4591
Water Content[1]	Water present in the PTFE resin	
Standard Specific Gravity (SSG)[2]	Specific gravity of a sample of molded and sintered PTFE according to this method	D 792, D 1505
Thermal Instability Index (TII)[1,2]	A measure of decrease in molecular weight of PTFE material determined by the difference between ESG and SSG: TII = (ESG-SSG) × 1000	
Tensile Properties[1,2]	Elongation and strength at break of a sample made according to the specified method	D 638
Extrusion Pressure[1]	The pressure measured while extruding a paste of fine powder PTFE made with an iso-paraffin under specified conditions	
Stretch Void Index (SVI)[1,2]	A measure of change in specific gravity of a PTFE specimen as a result of being subjected to tensile strain	
Strained Specific Gravity	specific gravity of a PTFE specimen after being subjected to tensile strain	
Untrained Specific Gravity	specific gravity of a PTFE specimen before being subjected to tensile strain	
Shrinkage and Growth	The change in the diameter of SSG preform due to sintering	
Extended Specific Gravity (ESG)	The specific gravity of a PTFE specimen molded for SSG after sintering for an extended period of time, compared to the sintering time of SSG	D 792, D 1505

[1] Properties required for resin specification.
[2] Properties required for on molded specimen.

Stretch void index (SVI) is defined below.

SVI = (Unstrained specific gravity − Strained specific gravity) × 1000

Unstrained specific gravity is measured on a tensile specimen prior to straining it. The strained specific gravity is measured on a sample of PTFE after it has been strained to break at a strain rate of 5.0 mm/min. The break elongation of the PTFE specimen must be greater than 200% or the experiment is repeated. The two specific gravity values are used to calculate SVI.

The specifications that differentiate fine powder polytetrafluoroethylene powders are summarized in Table 5.15. Water content and melting peak temperatures are the other specifications of these resins. Specifications of molded parts are listed in Table 5.16.

The actual ASTM Method D 4895-97 must be consulted for a complete description of these procedures.

Table 5.15. Fine Powder Polytetrafluoroethylene Resin Specifications According to ASTM D 4895

Type	Bulk Density, g/l	Average Particle Size Diameter, μm	Tensile Strength, MPa	Break Elongation, %
I	550±150	500±200	19	200
II	550±150	1050±350	19	200

Table 5.16. Molded Fine Powder Polytetrafluoroethylene Specifications According to ASTM D 4895

Type	Grade	Class	Standard Specific Gravity Range	Extrusion Pressure, MPa	Maximum SVI
I	1	A^1	2.14-2.18	5-15	-
		B^1	2.14-2.18	15-55	-
		C^1	2.14-2.18	15-75	-
I	2	A^1	2.17-2.25	5-15	-
		B^1	2.17-2.25	15-55	-
		C^1	2.17-2.25	15-75	-
I^1	3	C^2	2.15-2.19	15-75	200
		D^2	2.15-2.19	15-65	100
		E^1	2.15-2.19	15-65	200
I^1	4	B^2	2.14-2.16	15-55	50
II^1	-	A^1	2.14-2.25	5-15	NA

[1] Thermal Instability Index less than 50.
[2] Thermal Instability Index less than 15.

5.7.3 Dispersions of PTFE

ASTM Method D 4441-96 covers specifications of dispersions of PTFE and test methods for the as-produced polymer dispersion. Methods for processing dispersions into objects are discussed in Chapter 9 of this book. PTFE resins are thermoplastics in that they can be remelted but they cannot be processed by the normal melt processing technologies due to their extremely high rheology. This polymer does not dissolve in any solvents. These two facts render direct measurement of PTFE molecular weight virtually impossible. An indirect property named *standard specific gravity* is substituted for molecular weight. Section 5.7.1 contains a discussion of the relationship between molecular weight and standard specific gravity.

Properties that characterize dispersions of PTFE are listed in Table 5.17.

ASTM Method D 4441-96 classifies the different types of dispersions according to the system summarized in Table 5.18.

SSG and Melting Characteristics of PTFE in dispersions are measured by the same method as in ASTM D 4895–96. The polymer has to be isolated from the dispersion by coagulation, filtration and drying.

The actual ASTM Method D 4441–96 must be consulted for a complete description of these procedures.

Table 5.17. Definition of Basic Properties of Dispersions of Polytetrafluoroethylene According to ASTM D 4441-96

Property	Definition	Reference ASTM Method
Solids Content[1]	The amount of PTFE in the dispersion as weight %.	D 4441
Surfactant Content[1]	Surfactant added to the dispersion plus the remaining polymerization surfactant.	D 4441
Dispersion Particle Size	Particle size measured in the <u>presence</u> of added surfactant.	D 4441
Raw Dispersion Particle Size	Particle size measured in the <u>absence</u> of added surfactant.	D 4441
Coagulated Polymer	PTFE that has coagulated as a result of handling and processing of dispersion	D 4441
pH	Acidity/alkalinity of the dispersion.	E 70
Standard Specific Gravity (SSG)	Specific gravity of a sample of molded and sintered PTFE isolated from the dispersion according to this method	D 4441, D 792
Melting Characteristics	Heat of Fusion and Melting Peak Temperature of resin as determined by Differential Scanning Calorimetry	D 4441, D 4591

[1] Properties required for dispersion specification.

Table 5.18. Dispersion Polytetrafluoroethylene Classification According to ASTM D 4441-96

ASTM Type	Solids Content, weight %	Surfactant Content, weight %
I	23-27	0.5-1.5
II	25-35	1-5
III	53-57	1-5
IV	58-62	6-10
V	57-63	2-4
VI	58-62	4-8
VII	54-58	6-10
VIII	56-60	5-9
IX[1]	20-45	-

[1] No surfactant is added to stabilize. Addition of a hydrocarbon oil is optional.

5.8 Characterization of Polychlorotrifluoroethylene

ASTM Method D 1430–95 covers polymers of polychlorotrifluoroethylene and test methods for the as-produced polymer. Methods for processing these polymers into objects are discussed in Ch. 10 of this book. PCTFE has very high viscosity and does not dissolve in any solvents. These two facts render direct measurement of PCTFE molecular weight virtually impossible. An indirect factor named *zero strength time* (ZST) is substituted for the molecular weight. ZST is a physical measurement which can be correlated with molecular weigh, melt viscosity or melt index.

Zero strength time is defined as the time required for a compression-molded specimen (50 mm long, 4.8 mm wide and 1.58 mm thick) of PCTFE to lose strength under a 7.5 g weight at 250°C. One end of the bar is fastened in an oven and the weight is hung from the other end.

Properties used in the characterization of PCTFE are defined in Table 5.19. Documents ISO 12086-1 and 12086-2 also specify polychlorotrifluoroethylene resins.

A PCTFE resin is defined to be a powder, if it has an apparent density of less than 800 g/l as determined by Method A of ASTM D 1895. Preforming is defined as compaction of the polymer powder under pressure in a mold

Different types of PCTFE have been classified in Table 5.20. The specifications that differentiate polychlorotrifluoroethylene polymers are summarized in Table 5.21.

Table 5.19. Definition of Basic Properties of Polychlorotrifluoroethylene According to ASTM D 1430-95

Property	Definition	Reference ASTM Method
Specific Gravity[1]	Specific gravity of a sample of PCTFE compression-molded according to this method	D 792
Zero Strength Time[1]	Time required for a compression-molded specimen of polymer to lose strength at 250°C	D 1430
Deformation under Load[1]	Viscoelastic flow of polymer under a constant load	D 621[2]
Melting Point[1]		D 2117
Dielectric Constant[1]	A measure of effectiveness of material as a dielectric	D 150
Dissipation Factor[1]	A measure of conversion of electrical energy to heat in a dielectric (insulation) material	D 150
Apparent Density	Mass of unit volume (one liter) of polymer measured under the test conditions	D 1895

[1] Properties required for resin specification.
[2] Obsolete.

Table 5.20. Polychlorotrifluoroethylene Resin Classification According to ASTM D 1430

ASTM Type	Description
I	Homopolymer – No Comonomer
II	Modified Homopolymer - < 1% by weight Comonomer content
III	Copolymer – 10% or less Comonomer

Table 5.21. Molded Polychlorotrifluoroethylene Specifications According to ASTM D 1430

Type	Grade	Standard Specific Gravity Range, 23°C/23°C	Zero Strength Time, sec	Deformation Under Load[1], %	Melting Point, °C	Max Dielectric Constant in the frequency range		Max Dissipation Factor in the frequency range	
						kHz	MHz	kHz	MHz
I	1	2.10-2.12	100-199	10	210-220	2.70	2.50	0.030	0.012
	2	2.10-2.12	200-299	10	210-220	2.70	2.50	0.030	0.012
	3	2.10-2.12	300-450	10	210-220	2.70	2.50	0.030	0.012
II	1	2.10-2.12	100-199	15	200-210	2.70	2.50	0.030	0.012
	2	2.10-2.12	200-299	15	200-210	2.70	2.50	0.030	0.012
III	1	2.08-2.10	100-199	20	190-200	2.70	2.50	0.035	0.015
	2	2.07-2.10	200-299		190-200	2.70	2.50	0.035	0.015

[1] At 1112 N, 27 h, 70°C
[2] Thermal Instability Index less than 15.

References

1. Bro, M. I. and Sperati, C. A., "End Groups in Tetrafluoroethylene Polymers," *J. Polymer Science*, 38:289–305, 1959.
2. Doban, R. C., Knight, A. C., Peterson, J. H., and Sperati, C. A., Paper Presented at 130th Meeting Am. Chem. Soc., Atlantic City, Sep., 1956.
3. Sperati, C. A., and Starkweather, H. W., "Fluorine-Containing Polymers. II Polytetrafluoroethylene," Chapter from *Fortschr. Hochpoly.*, Forsch., 2465–495 1961.
4. "Fluorocarbon Resins from the Original PTFE to the Latest Melt Processable Copolymers," Technical Paper, Reg. Technical Conf. SPE, Mid Ohio Valley Bicentennial Conf. On Plastics, Nov. 30–Dec. 1, 1976.
5. Gangal, S. V., "Polytetrafluoroethylene, Homopolymers of Tetrafluoroethylene," in *Encyclopedia of Polymer Science and Engineering*, 2nd ed., 17:577–600, John Wiley & Sons, New York, 1989.
6. Cardinal, A. J, Edens W. L., and Van Dyk, J. W., US Patent 3,142,665, assigned to DuPont, July 28, 1964.
7. Holmes D. A., and Fasig, E. W., US Patent 3,819,594, assigned to DuPont, June 25, 1974.
8. Mueller, M. B., Salatiello, P. P., and Kaufman, H. S., US Patent 3,655,611, assigned to Allied, Apr. 11, 1972.
9. Doughty, T. R., Jr, Sperati, C. A., and Un, H. W., US Patent 3,855,191, assigned to DuPont, Dec. 17, 1974.
10. Plunkett, R. J., US Patent 2,230,654, assigned to DuPont, Feb. 4, 1941.
11. Brubaker, M. M., US Patent 2,393,967, assigned to DuPont, Feb. 5, 1946.
12. Kuhls, J., Steininger, A., and Fitz, H., US Patent 4,078,134, assigned to Hoechst Aktiengesellschaft, March 7, 1978.
13. Kometani, Y., Tatemoto, M., Takasuki-shi, and Fumoto, S., US Patent 3,331,822, assigned to Thiokol Chemical Corp., July 18, 1967.
14. Sherratt, S., in *Kirk-Othmer Encyclopedia of Chemical Technology* 2nd ed., (A. Standen, ed.), 9:805–831, Interscience Publishers, Div. of John Wiley and Sons, New York, 1966.
15. Renfrew, M. M., US Patent 2,534,058, assigned to DuPont, Dec. 12, 1950.
16. Gangal, S. V., US Patent 4,189,551, assigned to DuPont, Feb. 12, 1980.
17. R. F., Edens, W. L., and Larsen, H. A., US Patent 3,245,972, Anderson, assigned to DuPont, April 12, 1966.
18. Thomas, P. E., and Wallace, C. C., Jr., US Patent 2,936,301, assigned to DuPont, April 10, 1960.
19. Product information, Teflon® 7A, No. E-89757-2, DuPont Co., May, 1997.
20. Product information, Teflon® 7C, No. E-89758-2, DuPont Co., May, 1997.
21. Daikin Industries of Fluoroplastics, Daikin Industries, Ltd., Orangeburg, New York, March, 1998.
22. British Patent No. 1,076,642, assigned to Pennsalt Chemical Corp., Sep. 19, 1963.
23. Kometani, Y., et al., US Patent 3,597,405, assigned to Daikin Industries, Ltd., Aug. 3, 1971.
24. Roberts, R., and Anderson, R. F. US Patent 3,766,133, assigned to DuPont, Oct. 16, 1973.
25. Browning, H. E., US Patent 3,983,200, assigned to Imperial Chemical Industries, Ltd., Sep. 28, 1976.
26. Banham, J. and Browning, H. E., US Patent 3,882,217, assigned to Imperial Chemical Industries, Ltd., Apr. 6, 1975.
27. Izumo, M., Nomura, S., and Tanigawa, S., US Patent 4,241,137, assigned to Daikin Industries, Ltd., Dec. 23, 1980.

28. Honda, N., Sawada, K., Idemori, K., and Yukawa, H., US Patent 5,189,143, assigned to Daikin Industries, Ltd., Feb. 23, 1993.
29. Harvey, L. W., and Martin, E. N., US Patent 5,502,161, assigned to ICI America's, Inc., Mar. 26, 1996.
30. Hoornaert, F., Wauters, P. A. L., Meesters, G. M. H., Pratsinis, S. E., and Scarlett, B., "Agglomeration Behavior of Powders in Lodige Mixer Granulator," *Powder Technology,* 96:116–128 1998.
31. Product information, Teflon® 9B, No. E-89762-3, DuPont Co., May, 1997.
32. Renfrew, M. M., US Patent 2,434,058, assigned to DuPont, Dec. 12, 1950.
33. Brinker, K. C., and Bro, M. I., US Patent 2,965,595, assigned to DuPont, Dec. 20, 1960.
34. Berry, K. L., US Patent 2,559,752, assigned to DuPont, July 10, 1951.
35. Bankoff, S. G., US Patent 2,612,484, assigned to DuPont, Sep. 30, 1952.
36. Poirier, R. V., US Patent 4,036,802, assigned to DuPont, July 19, 1977.
37. Shimizu, T., and Hosokawa, K., US Patent 4,840,998, assigned to Daikin Industries Ltd., June 20, 1989.
38. Berry, K. L., US Patent 2,478,229, assigned to DuPont, Aug. 9, 1949.
39. Marks, B. M., and Whipple, G. H., US Patent 2,037,953, assigned to DuPont, June 5, 1962
40. Hoashi, J., US Patent 3,301,807, assigned to Thiokol Chemical Corp., Jan. 31, 1967.
41. Holmes, D. A., US Patent 3,704,272, assigned to DuPont, Nov. 28, 1972.
42. Morris, P. S., and Hutzler, R. H., US Patent 3,778,391, assigned to Allied Chemical Corp., Dec. 11, 1973.
43. British Patent No. 1,189,483, assigned to Montecatini Edison S. P. A. of Milan, Italy, Apr. 29, 1970.
44. Kuhls, J., and Weiss, E., US Patent 4,369,266, assigned to Hoechst Aktiengesellschaft, Jan. 13, 1983.
45. Jones, C. W., US Patent 5,272,186, assigned to DuPont, Dec. 21, 1993.
46. British Patent No. 465,520, assigned to Farbenindustrie, I. G., May 3, 1937.
47. Miller, W. T., US Patent No. 2,564,024, assigned to US Atomic Energy Commission, Aug. 14, 1951.
48. Miller, W. T., Dittman, A. L., and Reed, S. K., US Patent 2,586,550, assigned to USAEC, Feb. 19, 1952.
49. Miller, W. T., US Patent 2,792,377, assigned Minnesota Mining and Manufacturing Co., May 14, 1957.
50. Dittman, A. L., and Wrightson, J. M., US Patent 2,636,908, assigned to M. W. Kellog Co., Apr. 28, 1953.
51. Brit. Patent 729,010, assigned to Farbenfabriken Bayer AG, Apr. 27, 1955.
52. French Patent 1,419,741, assigned to Kureha Chemical Co., Dec. 3, 1965.
53. Young, D. M., and Thompson, B., US Patent 2,700,662, assigned to Union Carbide Co., Jan. 25, 1955.
54. Hanford, W. F., US Patent 2,820,027, assigned to Minnesota Mining and Manufacturing Co., Jan. 14, 1958.
55. Lazar, M., *J. Polymer Sci.*, 29:573, 1958.
56. Roedel, G. F., US Patent 2,613,202, assigned to General Electric Co., Oct. 7, 1952.
57. Caird, D. W., US Patent 2,600,202, assigned to General Electric Co., June 10, 1952.
58. French Patent 1,155,143, assigned to Society d'Ugine, Apr. 23, 1958.
59. Herbst, R. L., and Landrum, B. F., US Patent 2,842,528, assigned to Minnesota Mining and Manufacturing Co., July 8, 1958.
60. Dittman, A. L, Passino, H. J., and Wrightson, J. M., assigned to M. W. Kellog, US Patent 2,689,241, Co., Sep. 14, 1954.
61. Hamilton, J. M., *Ind. Eng. Chem.*, 45:1347, 1953.

62. Passino, H. J., Dittman A. L., and Wrightson, J. M., US Patent No. 2,820,026, assigned to 3M Co., Jan. 14, 1958.
63. Hamilton, J. M., US Patent 2,569,524, assigned to DuPont, Oct. 21, 1951.
64. Passino, H. J., et al., US Patent 2,744,751, assigned to M. W. Kellog Co., Dec. 19, 1956.
65. Brit. Patent 840,735, F. Fahnoe and B. F. Landrum, assigned to Minnesota Mining and Manufacturing Co., July 6, 1960.
66. Benning, A. F., US Patent 2,559,749, assigned to DuPont, July 10, 1951.
67. Miller, W. T., US Patent 2,579,437, assigned to M. W. Kellog Co., Dec. 18, 1951.
68. Rearich, J. S., US Patent 2,600,804, assigned to M. W. Kellog Co., June 17, 1952.
69. Bringer, R. P., "Chlorotrifluoroethylene Polymers," in *Encyclopedia of Polymer Science and Technology*, 1st ed., 7:204–219, John Wiley and Sons, New York, 1967.
70. Muntell, R. M., and Hoyt, J. M., US Patent No. 3,043,823, assigned to 3M Co., Jul. 10, 1962.
71. British Patent No. 805,103, Nov. 26, 1958.
73. Sperati, C. A., "Physical Constants of Fluoropolymers," in *Polymer Handbook*, 3rd ed., Vol. 1, *Polymerization and Depolymerization,* John Wiley and Sons, New York, 1989.

6 Commercial Grades of Homofluoropolymers

6.1 Introduction

This chapter presents information published by resin manufacturers about the commercially available grades of fluoroplastic homopolymers. The commercial grades have been grouped in two ways. They have been classified according to the near equivalency to each other. To accomplish this, key properties of the polymers have been selected to classify each product line. Granular resins have been classified according to their apparent density. Fine powder polymers have been classified, based on their reduction ratios. Dispersion polymers have grouped by their similarity to each other.

The second series of tables present the properties of commercial grades according to the manufacturers publications. Polymer properties have been classified by resin suppliers in separate tables. The reader can easily find the similar products in each grade and find their properties in these tables.

6.2 Granular PTFE

Granular PTFE grades have been divided into homopolymer and modified resins. Each group has been further divided into fine cut or *low flow* and pelletized resins or *free flow*. Apparent density is the key property by which the groups have been organized. *Modified* and *ram extrusion* powders are classified in separate tables. Tables 6.1 through 6.4 provide equivalency information for a number of granular polytetrafluoroethylene resins. It is important that no two resins are ever perfectly "equivalent," but when they are close they have been grouped together. Granular PTFE resins are specified by ASTM Method D 4894. It also provides procedures or references to other ASTM methods for the measurement of resin properties.

6.3 PTFE Dispersions

Polytetrafluoroethylene dispersions are produced by dispersion (emulsion) polymerization, followed by surfactant stabilization and usually a form of concentration. These dispersions can be further formulated by the addition of various organic and inorganic additives and converted into coatings. A number of techniques are available for the application of PTFE based coatings to metallic and nonmetallic substrates (see Ch. 9). As a result of the overwhelming impact of other formulation ingredients, it is difficult to find one common variable to classify the different dispersions, like the apparent density for granular polytetrafluoroethylene and the reduction ratio for fine powders. Normally, both the solids content and surfactant content are used to classify simple dispersions. Commercial PTFE dispersions offered by major manufacturers have been listed in the Tables 6.10–6.14. ASTM Method D 4441 specifies PTFE dispersions. Viscosity of the dispersion is measured by ASTM Method D 2196. It also provides procedures or references to other ASTM methods for the measurement of resin properties.

6.4 Fine Powder PTFE

Fine powder PTFE is produced by dispersion (emulsion) polymerization followed by coagulation, isolation and drying. Fine powder or coagulated dispersion powders are classified according to processing characteristics. A crucial factor in the selection of a powder for paste extrusion is its *reduction ratio* range, which is often specified by the manufacturer. Chapter 8 can be consulted for a complete understanding of reduction ratio. To help the reader, it is briefly discussed here. Reduction ratio is the cross section of the paste in the extruder barrel to that of the extrudate. The resin selected to make a particular part must be reducible to the extrudate at a reasonable pressure in the extruder. This means that it should generate sufficient pressure for fibrillation of the resin, yet the pressure must not exceed the normal range of commercial equipment. Reduction ratio is primarily related to the molecular weight of the polymer; the lower the molecular weight, the higher the reduction ratio.

Most manufacturers specify the recommended range of reduction ratio or its upper limit. It is difficult to find the exact match of any specific product different resin producers product line. Table 6.15 shows a broad categorization of five manufacturers' powders according to reduction ratios. Tables 6.16

through 6.20 should be consulted to obtain the exact reduction ratio and other properties of each resin, as published by the manufacturer. A review of these data reveals that certain resins have a broader range of reduction ratio than the others, which makes them more versatile with respect to processing. The variation in the properties shown in Tables 6.16–6.20 is due to the difference in the type of properties that are published by the fine powder manufacturers. ASTM Method D 4895 specifies fine powder PTFE resins. It also provides procedures or references to other ASTM methods for the measurement of resin properties.

Dispersions of copolymers of tetrafluoroethylene are also coated on surfaces. Fluoropolymer manufacturers should be contacted for information.

6.5 PCTFE Dispersions

Polychlorotrifluoroethylene (PCTFE) consumption is much smaller than PTFE. Few manufacturers produce it. Polychlorotrifluoroethylene dispersions are produced by dispersion (emulsion) polymerization. The polymer is a copolymer and contains vinylidene fluoride (VDF). VDF content of the copolymer determines its properties, which can range from stiff to soft flexible coatings. These dispersions can be further formulated by the addition of various organic and inorganic additives and converted into coating. A number of techniques are available for the application of PCTFE based coating to metallic and nonmetallic substrates (see Ch. 9). Commercial PCTFE dispersions offered by AlliedSignal Corporation have been listed in the Tables 6.21.

6.6 Polychlorotrifluoroethylene Polymers

Polychlorotrifluoroethylene (PCTFE) consumption is much smaller than PTFE. Few manufacturers pro-

duce it. The important factor by which any PCTFE is classified is *zero strength time* (ZST) which is an indicator of molecular weight. This material is offered in powder and pellet forms at different ZST's (molecular weights). The powders are fabricated by compression molding and extrusion processes. Injection molding requires pellets. Tables 6.22 and 6.23 provide information for a few commercial grades of PCTFE. These resins are specified by ASTM Method D 1430 that also provides procedures or references to other ASTM methods for the measurement of resin properties.

The other manufacturer of PCTFE is AlliedSignal Corporation, which does not normally sell resin but a film. Oriented melt extruded films of polychlorotrifluoroethylene are offered under the trademark of Aclar® by AlliedSignal. The primary applications of PCTFE films are in packaging where moisture and chemical barrier properties are required.

6.7 Fluoropolymer Manufacturers

There are several major manufacturers that offer one or more of the fluoropolymers. The business domain of these companies varies widely. Similarly, their manufacturing bases range from domestic to international. Table 6.24 provides information about the major manufacturers of fluoropolymers. DuPont, Imperial Chemical Industries, Daikin, and Allied have production facilities in the United States and overseas. Dyneon, Asahi Glass and Ausimont produce polymer in their home countries and export them.

Polytetrafluoroethylene is also manufactured by companies in India (Hiflon®) Russia (Fluoroplast®), People's Republic of China, and Poland. Most of the Russian production is low quality granular and fine powder resin and is exported. The other countries consume the majority of their production locally. Little reliable information is available about the types and ASTM or ISO properties of these resins.

Information on the following trademarked products is included in some tables in this chapter.

Teflon® is a trademark of DuPont Company.
Fluon® is a trademark of Imperial Chemical Industries.
Hostaflon® is a trademark of Dyneon Company.
Polyflon® is a trademark of Daikin Company.
Algoflon® is a trademark of Ausimont Company.

Table 6.1. Grouping of Granular Fine Cut (Low Flow) Polytetrafluoroethylene Resins by Apparent Density

Apparent Density	DuPont Teflon®[1]	Imperial Chemical Industries Fluon®[2]	Dyneon Hostaflon®[3]	Daikin Polyflon®[4]	Ausimont Algoflon®[5]
Low (<300 g/l)	7C, 7J,			M-12	
Medium (<400 g/l)		G-585	TF 1702, TF 1740, TF 1750		F5, F6, F-31
High (<500 g/l)	7A, 701N, 707N, 7AJ, 710J	G-140, G-163, G-170, G-190, G-570, G-580		M-15 M-14	F2, F5/S, F7
Very High (>500 g/l)	703N	G-171			F3

Table 6.2. Grouping of Granular Pelletized (Free Flow) Polytetrafluoroethylene Resins by Apparent Density

Apparent Density	DuPont Teflon®	Imperial Chemical Industries Fluon®	Dyneon Hostaflon®	Daikin Polyflon®	Ausimont Algoflon®
Medium (<800 g/l)	8, 8A, 8B, 801N, 820J	G-307, G-320, G-311	TF 1665	M-31	A-20, A-25, S 111
High (>800 g/l)	850A, 807N, 809N, 800J, 810J	G-350	TF 1620, TF 1641, TF 1645	M-32, M-33	A-27, A-30, S-121, S-131

Table 6.3. Homopolymer Polytetrafluoroethylene Ram Extrusion Granular Resins

Apparent Density	DuPont Teflon®	Imperial Chemical Industries Fluon®	Dyneon Hostaflon®	Daikin Polyflon®	Ausimont Algoflon®
Medium (<800 g/l)	9B, 901N	G-201	TF 1101		E2, E2BP
High (>800 g/l)				M24, M-25	

Table 6.4. Modified Polytetrafluoroethylene Granular Resins

Type	DuPont Teflon®	Dyneon Hostaflon®	Daikin Polyflon®
Fine Cut (*low flow*)	NXT 70, NXT 75, 70J	TFM 1700, TFM 1705	M-111, M-112
Pelletized (*free flow*)	TE-6462, TE-6472, 170J	TFM-1600	M-137
Ram Extrusion Grade		TF-1502, TFM-1105	

Table 6.5. Properties of DuPont Granular Resins[1]-[24]

Teflon®	Bulk Density, g/l	Specific Gravity	Average Particle Size, μm	Tensile Strength, MPa	Elongation, %	Shrinkage, %	Compression Ratio
Fine Cut							
7A	460	2.16	34	34.5	375	3.4	3.2:1
701N	430	2.16	20	36	400	4.9	-
703N	580	2.16	225	34	340	3.9	-
707N	490	2.16	20	36	360	3.6	-
7AJ	450	2.16	25	37.9	350	34.3	4.9:1
7C	250	2.16	28	37.9	400	6	5.5:1
7J	280	2.17	35	33	300	6	-
Free Flow							
8	725	2.16	600	27.6	300	2.8	3.2:1
8A	690	2.16	480	27.6	300	2.8	3.3:1
8B	705	2.16	450	28	300	2.8	3.2:1
850A	900	2.16	400	28	300	2.7	2.4:1
801N	730	2.16	500	32	290	2.9	-
807N	1,005	2.16	420	32	330	2.2	2.0:1
809N	950	2.16	1,150	37	320	-	-
820J	700	2.15	650	30	300	3.2	3.1:1
800J	850	2.16	330	32	300	2.7	2.5:1
Ram Extrusion							
9B	575	-	550	-	-	-	-
901N	600	2.15[1]	625	28[1]	290[1]	-	-
Modified							
NXT 70	440	2.17	33	31.5	450	4.7	-
NXT 75	440	2.17	33	31.5	500	4.7	-
TE-6462	780	2.17	400	28	400	4.7	2.4:1
TE-6472	780	2.17	400	28	400	4.7	2.4:1
70J	340	2.18	25	32	490	3.6	-
170J	620	2.18	515	28	400	5	-

[1] Properties of a 40 mm solid rod.

Table 6.6. Properties of Imperial Chemical Industry Granular Resins[25]

Fluon®	Bulk Density, g/l	Specific Gravity	Average Particle Size, μm	Tensile Strength, MPa	Elongation, %	Shrinkage, %	Compression Ratio
Fine Cut							
G-140	430	2.17	18	31	390	3.8	
G-163	420	2.16	33	35	325	2.8	
G-170	440	2.16	30	34	330	4.8	4.5:1
G-171	550	2.15	57	37	300	2.5	
G-190	435	2.17	24	44	450	3.3	
G-570	485	2.18	30			4.0	
G-580	430	2.17	30	34	300	4.7	4.6:1
G-585	360	2.17	25	34	300	4.9	4.8:1
Free Flow							
G-307	760	2.16	740	31	275	2.8	2.7:1
G-311	760	2.17	600	31	275	2.8	2.7:1
G-320	695	2.17	430	31	330	2.9	
G-350	930	2.16	380	29	300	2.3	
Ram Extrusion							
G-201	600	2.16	575	17	350	7.5	

Table 6.7. Properties of Dyneon Granular Resin[26][27]

Hostaflon®	Bulk Density, g/l	Specific Gravity	Average Particle Size, μm	Tensile Strength, MPa	Elongation, %	Shrinkage, %
Fine Cut						
TF 1702	350	2.16	20	40	580	6.2
TF 1740	380	2.17	40	32	500	3.7
TF 1750	370	2.16	20	42	450	4.3
Free Flow						
TF 1620	870	2.15	220	34	400	3.0
TF 1641	830	2.15	425	33	400	2.7
TF 1645	830	2.15	425	32	350	2.6
TF 1665	800	2.16	650	25	400	3.4
Ram Extrusion						
TF 1502	670	2.16	800	30	400	10.5
TF 1101	650	2.15	600	21	350	8.5
Modified						
TFM 1700	400	2.16	20	32	650	6
TFM 1705	420	2.16	25	31	580	5.4
TFM 1600	850	2.16	425	32	650	3.5
Modified Ram Extrusion						
TF 1105	845	2.16	760	21	250	8.5
TF 1615	750	2.16	800	25	400	10.5

Table 6.8. Properties of Daikin Granular Resins[28][29]

Polyflon®	Bulk Density, g/l	Specific Gravity	Average Particle Size, µm	Tensile Strength, MPa	Elongation, %	Shrinkage, %
Fine Cut						
M-12	290	2.17	25	47	370	
M-14	425	2.16	150	32	350	
M-15	380	2.17	30	46	360	
Free Flow						
M-25	840	2.16	640	20	220	
M-31	670	2.17	700	46	350	
M-32	790	2.17	700	43	350	
M-33	830	2.17	250	40	350	
Ram Extrusion						
M-24	600	2.15	400	20	250	
Modified						
M-111		2.17		26	525	
M-112	460	2.15				3.5
M-137	840	2.16				3.3

Table 6.9. Properties of Ausimont Granular Resins[30]

Algoflon®	Bulk Density, g/l	Specific Gravity	Average Particle Size, µm	Tensile Strength, MPa	Elongation, %	Shrinkage, %	Compression Ratio
Fine Cut							
F2	440	2.175	20	37	350	2.8	4.9:1
F5	390	2.175	15	39	370	3.0	-
F6	370	2.175	15	40	380	3.1	5.9:1
F5/S	430	2.175	15	40	400	3.4	5:1
F7	430	2.175	15	41	400	3.4	5:1
F31	360	2.180	20	35	500	3.9	6:1
Free Flow							
A25	650	2.170	650	35	310	2.7	3.3:1
A20	780	2.165	550	35	310	2.6	2.8:1
A27	880	2.170	650	33	310	2.6	2.5:1
A30	910	2.170	750	33	300	2.6	2.4:1
S111	680	2.170	650	37	350	2.8	3.2:1
S121	800	2.170	550	37	350	2.9	2.7:1
S131	910	2.170	650	35	340	2.9	2.4:1
Ram Extrusion							
E2	640	2.150	550	23	340	-	3.3:1
E2BP	670	2.150	700	25	350	-	3.2:1

Table 6.10. Properties of DuPont's Dispersions[31]-[41]

Teflon®	30	30B	35	B	304A	305A	306A	307A	308A	313A	33
Solids Content, % by Weight	60	60	32	61	45	60	60	60	25	60	3.5
Non-ionic Surfactant Content, % by Weight			2.5 (anionic)		6	8	6	6	0.9	7	0.22
Dispersion Particle Size, µm	0.22	0.22	0.05-0.5	0.22	0.22	0.22	0.22	0.16	0.2	0.22	1.22
Specific Gravity at 20°C	1.5	1.5	1.22	1.44	1.33	1.5	1.5	1.5	1.16	1.5	
Brookfield Viscosity at 25°C, mPa·S	20	20			1,700 (at 35°C)	20	20	20	6	20	
pH (minimum)	9.5	9.5	4	8.5 (max.)	9.5	9.5	9.5	9.5	9.5	9.5 (min.)	9.5 (min.)
Melting Point of Polymer, °C	337/327	337/327		337/327	337/327	337/327	337/327	337/327	337/327	337/327	337/327

Table 6.11. Properties of Imperial Chemical Industry's PTFE Dispersions[42]

Fluon®	AD-1	AD-1HT	AD-1L	AD-1S	AD-2	AD-057	AD-059	AD-584
Solids Content, % by Weight	60	59-62	59-62	44-46	55	25-30	25-30	32-34
Non-ionic Surfactant Content, % by Weight	7	7-8	5-6	3-4	2.5	NONE	NONE	2-3
Dispersion Particle Size, µm	0.2				0.1-0.2			
Brookfield Viscosity at 20°C, mPa·S	19							
pH	9		9 (minimum)	9 (minimum)		4 (maximum)	4 (maximum)	4 (maximum)

(Cont'd.)

Table 6.11. *(Cont'd.)*

Fluon®	AD-133	AD-502	AD-639	AD-704	AD-730
Solids Content, % by Weight	31-34	25-30	58	23-27	59-62
Non-ionic Surfactant Content, % by Weight	3-4	0.5-1.5	10	0.5-1	6.5-7.5
Dispersion Particle Size, µm			0.3		
pH	9 (minimum)	4 (maximum)		4 (maximum)	9 (minimum)

Table 6.12. Properties of Dyneon's Dispersions[43]

Hostaflon®	TF 5032	TF 5033	TF 5034	TF 5035	TF 5039	TF 5135	TF 5136	TF 5137	TF 5040	TF 5041	TF 5235	PFA 6900
Polymer	PTFE	PTFE	PTFE	PTFE	PTFE	PTFE	PTFE	PTFE	PTFE	PTFE	PTFE	PFA
Solids Content, % by Weight	60	35	61	58	55	58	59	58	55	60	60	60
Non-ionic Surfactant Content, % by Weight	5	4	3	5 (ionic)	10	5	4	7	11	8	6	5
Dispersion Particle Size, µm	0.18	0.18	0.14	0.25	0.25	0.3	0.24	0.2	0.15	0.3	0.23	0.25
Density, g/cm^3	1.5	1.25		1.5	1.4							
Brookfield Viscosity at 20°C, mPa.S			16			10	22	85	55	500	35	5
pH	8.5	10		8.5	8.5	11	11	10	9	9.5	8	2.5

Table 6.13. Properties of Daikin PTFE Dispersions[44]

Polyflon®	D-1	D-2	D-2C	D-3
Solids Content, % by Weight	60	60	60	60
Non-ionic Surfactant Content, % by Weight	7	6	6	6
Dispersion Particle Size, µm	0.2–0.4	0.2–0.4	0.2–0.4	0.2
Specific Gravity at 25°C	1.5	1.5	1.5	-
Viscosity at 25°C, mPa·S	25	25	25	-
pH	9–10	9–10	9–10	-
Melting Point of Polymer, °C	335	335	335	-

Table 6.14. Properties of Ausimont's PTFE Dispersions[30]

Algoflon®	D60/G	D60/G6	D60FX	D60 EXP1	D3310	D60/A	D60/27	D60/EXP96	D3310
Solids Content, % by Weight	60	57.5	60	61	58	60	60	59	59
Non-ionic Surfactant Content, % by Weight	3.75	5.6	3	3.8	3.5	3	2.75	3	3
Dispersion Particle Size, µm	0.24	0.24	0.24	0.24	0.24	0.24	0.24	0.24	0.24
Specific Gravity at 20°C	1.51	1.49	1.52	1.53	1.50	1.52	1.52	1.50	1.50
Brookfield Viscosity at 20°C, mPa·S	26		25	30	30		25	16	15
pH (minimum)	9	9	9	9	9	9	9	9	9
Conductivity, µS/cm	700	600	600	750	1,100	700	650	1,300	1,300

Table 6.15. Fine Powder Polytetrafluoroethylene Grouping by Reduction Ratio

Reduction Ratio	DuPont Teflon®	Imperial Chemical Industries Fluon®	Dyneon Hostaflon®	Daikin Polyflon®	Ausimont Algoflon®
<100	603J, 604J, 601A, 602A, 613A	CD-147	TF-2029	F-103	DF 200
<300	637N, 669N, 65N,	CD-141, CD-126	TF 2021, TF 2027	F-301	DF 210 DF 280X
<800	60A, 67A	CD-123	TF-2025, TF-2026	F-104	DF 1
<1,600	6C, 62, 636N, 600A	CD-086, CD 1, CD 014	TFM-2001, TF-2053, TF-2071	F-302, F303	DF 380, DF 381X
<3,000	6C, 640J	CD 509			
<4400	6C, 610A, CFP-6000, 614A. 60A, 600A	CD-076, CD-090, CD 506	TF-2072	F-201, F203	DFC

Table 6.16. Properties of DuPont Fine Powder PTFE[45][61]

Teflon®	Reduction Ratio Range	Specific Gravity	Average Particle Size, µm	Thermal Instability Index (TII)	Stretch Void Index (SVI)	Extrusion[1] Pressure, MPa (Reduction Ratio)	Bulk Density, g/l
601A	25-150	2.15	550	-		44.8 (400:1)	-
602A	25-150	2.16	550	-		37.3 (400:1)	
613A	25-150	2.15	500	<15		41.4 (400:1)	-
612A		2.16	450	<15	<75	24 (400:1)	-
65N	30-300	2.16-2.18	525-825	-	-	6-9 (100:1)	
669N	10-300	2.18	470	-	-	7.6 (100:1)	500
637N	10-500	2.175	500	-	-	9.5 (100:1)	-
638RFF-N	10-500	2.170	650	<20	-	7.4 (100:1)	530
67A	10-500	2.22	500	-	-	11 (100:1)	400
62	100-1200	2.16	500	-	-	-	475
636N	100-1200	2.21	460	-	-	6.9 (100:1)	480
60A	250-4400	2.20	425	-	-	--	475
6C	100-4400	2.17	450	-	-	-	495
600A	100-1200	2.20	550	-	-	9 (100:1)	500
610A	250-2500	2.17	450	<15	<200	48.3 (1,600:1)	
640J	250-2500	2.17	400	-	-	34.5 (1,600:1)	
CFP 6000	250-4400	2.175	450	<15	-	41.4 (1,600:1)	510
614A	250-4400	2.175	450	<15	-	34.5 (1,600:1)	510

[1] This pressure is measured according to the ASTM Method D 4895.

Table 6.17. Properties of Imperial Chemical Industry's Fine Powder PTFE[62]

Fluon®	Reduction Ratio Range	Specific Gravity	Average Particle Size, μm	Thermal Instability Index (TII)	Stretch Void Index	Extrusion[1] Pressure, MPa (Reduction Ratio=900:1)	Bulk Density, g/l
CD 147	25-150						
CD 141	30-300						
CD 126	10-300	2.19	575			138	475
CD 123	10-800	2.17	500			114	500
CD 1	30-1,000	2.205	550			90	500
CD 086	200-1500	2.15	500	5	57	82.8	475
CD 014	30-1500		550			76	500
CD 509	800-3,000	2.19	500			60	500
CD 506	800-4,000		500			41.4	500
CD 076	250-4400	2.19				46.9[2]	490
CD 090	800-4,000	2.19	500		177	48.3	475

[1] This pressure is measured by paste extrusion through a 20° angle die at 16% by weight Isopar® H lubricant.
[2] At a reduction ratio of 1,600:1, by rheometry at 18.4% by weight Isopar® G lubricant.

Table 6.18. Properties of Dyneon's Fine Powder PTFE[63]

Hostafluon®	Reduction Ratio Range	Specific Gravity	Average Particle Size, μm	Tensile Strength, MPa	Elongation at Break, %	Extrusion[1] Pressure, MPa (Reduction Ratio)	Bulk Density, g/l
TF 2029	5-100	2.15	500	30	380	50 (400:1)	500
TF 2021	20-300	2.15	500	30	380	30 (400:1)	500
TF 2027	30-300						
TF 2025	20-600	2.15	500	22	380	38 (400:1)	470
TF 2026	10-800	2.195	480	-	-	13 (100:1)	460
TF 2053	100-1200	2.15	450	24	390	65 (1,600:1)	-
TFM X2001	100-1,200						
TF 2071	20-1,600	2.16	450	24	390	60 (1,600:1)	450
TF 2072	400-4,400	2.17	450	24	390	40 (1,600:1)	450

[1] This pressure is measured according to the ASTM Method D 4895.

Table 6.19. Properties of Daikin's Fine Powder PTFE[64]

Polyflon®	Reduction Ratio Range	Specific Gravity	Average Particle Size, μm	Tensile Strength, MPa	Elongation at Break, %	Extrusion[1] Pressure, MPa (Reduction Ratio)	Bulk Density, g/l
F-103	<100		450				600
F-301	<300						
F-104	<1,000		500				450
F-302	<1,200						
F-303	<1,200						
F-203	<4,000	2.17				34.5 (1600:1)	
F-201	<4,000	2.18	450			40.5 (1600:1)	450

[1] This pressure is measured according to the ASTM Method D4895.

Table 6.20. Properties of Ausimont's Fine Powder PTFE[30]

Algoflon®	Reduction Ratio Range	Specific Gravity	Average Particle Size, μm	Tensile Strength, MPa	Elongation at Break, %	Extrusion[1] Pressure, MPa (Reduction Ratio)	Bulk Density, g/l
DF200	25-150						
DF1			500				540
DF210	30-300	2.16	500	30	300	8 (100:1)	500
DF230X	30-300	2.16	500	30	300	9 (100:1)	500
DF280X	30-450	2.16	350	30	300	6 (100:1)	500
DF380	30-1100	2.16	350	30	375	16 (400:1)	450
DF381X	80-1100	2.15	350	30	375	24 (400:1)	450
DFC	250-4,400						

[1] Aclon® is a trademark of AlliedSignal Company.

Table 6.21. Properties of AlliedSignal's Aclon®[1] PCTFE Dispersions[65]

Property	404	400A	400LT
Appearance	Milky White	Milky White	Milky White
Solids Content, % by Weight	46-50	46-50	46-50
Surfactant Type	Nonionic	Nonionic	Nonionic
Surface Tension, Dynes/cm	30-40	30-40	30-40
Dispersion Particle Size, µm	0.1-0.2	0.1-0.2	0.1-0.2
Density at 25°C, g/ml	2.1-2.16	2.1-2.16	2.1-2.16
Viscosity at 25°C, Centipoise	5-10	5-10	5-10
pH	9-10	9-10	9-10
Coating Process Temperature, °C (1-2 Minutes)	210-250	100-200	180-200
Coating Characteristic	Stiffest	Flexible	Stiff
Chemical Resistance	Most Resistant	Least Resistant	Resistant

[1] Aclon® is a trademark of AlliedSignal Company.

Table 6.22. Properties of Daikin's Neoflon®[1] PCTFE[66]

Product	ZST, sec	Apparent Density, g/l	Flow Value × 10^3, ml/sec	Form
M-300	200-300	600	1-3	Powder (10-60 mesh)
M-300H	200-300	950	1-3	Granular Powder
M-300P	200-300	1,100	1-3	Pellet
M-400H	301-450	950	0.3-0.8	Granular Powder
M-400P	301-450	1,100	0.3-0.8	Pellet

[1] Neoflon® is a trademark of Daikin Company.

Table 6.23. Physical Properties of Daikin's Neoflon® PCTFE[66]

Property	Test Method (ASTM)	M-300H	M-400H
Zero Strength Time, sec	D 1430	200-300	301-450
Specific Gravity	D 792	2.10-2.17	2.10-2.17
Tensile Strength, MPa	D 638	32-38	34-40
Elongation, %	D 638	50-200	100-250
Tensile Modulus of Elasticity, MPa	D 638	1,300-1,500	1,200-1,400
Compressive Strength at 1% strain, MPa	D 695	40-45	37-42
Compressive Modulus of Elasticity, MPa	D 695	1,400-1,600	1,200-1,400
Flexural Strength, MPa	D 790	69-74	67-72
Flexural Modulus of Elasticity, MPa	D 790	1,600-1,900	1,400-1,700
Impact Strength, ft-lb/in	D 256	2.5-3.5	2.5-3.5
Hardness, shore D		75-85	75-85
Deformation under Load, % 24 hrs at 7 MPa 25°C 80°C 100°C	D 621	<0.2 1.7-1.9 7.0-9.0	<0.2 1.4-1.6 4.5-6.5

Table 6.24. Major Manufacturers of PTFE and PCTFE Fluoropolymers

Company	Home Country	Fluoropolymers	Manufacturing Sites	Comments
DuPont	USA	All Forms of PTFE	USA, The Netherlands, Japan	
Imperial Chemical Industries	United Kingdom	All Forms of PTFE	USA, United Kingdom	
Dyneon	USA/Germany	All Forms of PTFE	Germany	Joint Venture of 3M and Hoechst
Daikin	Japan	All Forms of PTFE and PCTFE	USA, Japan	
Ausimont	Italy	All Forms of PTFE	Italy	Imports resin and finishes it in the USA
AlliedSignal	USA	PCTFE film and Dispersion	USA	
Asahi Glass	Japan	All Forms of PTFE	Japan	

References

1. Product information, Teflon® PTFE 7A, Granular Compression Molding Resin, DuPont Co., Wilmington, Delaware, May, 1997.
2. Product information, Teflon® PTFE 7C, Granular Compression Molding Resin, DuPont Co., Wilmington, Delaware, May, 1997.
3. Ref. Product information, Teflon® PTFE 8, Granular Compression Molding Resin, DuPont Co., Wilmington, Delaware, May, 1997.
4. Product information, Teflon® PTFE 8A, Granular Compression Molding Resin, DuPont Co., Wilmington, Delaware, May, 1997.
5. Product information, Teflon® PTFE 8B, Granular Compression Molding Resin, DuPont Co., Wilmington, Delaware, May, 1997.
6. Product information, Teflon® PTFE 850A, Granular Compression Molding Resin, DuPont Co., Wilmington, Delaware, July, 1997.
7. Product information, Teflon® PTFE 9B, Granular Ram Extrusion Resin, No., DuPont Co., Wilmington, Delaware, May, 1997.
8. Product information, Teflon® NXT 70, Modified PTFE Granular Molding Resin, DuPont Co., Wilmington, Delaware, April, 1998.
9. Product information, Teflon® NXT 75, Modified PTFE Granular Molding Resin, DuPont Co., Wilmington, Delaware, April, 1998.
10. Product information, Teflon® TE-6462, Chemically Modified Free-Flow Granular Molding Resin, DuPont Co., Wilmington, Delaware, May, 1997.
11. Product information, Teflon® TE-6472, Chemically Modified Free-Flow Granular Molding Resin, DuPont Co., Wilmington, Delaware, May, 1997.
12. Product information, Teflon® PTFE 701N, Granular Compression Molding Resin, DuPont International, S. A. Switzerland, Geneva, Switzerland, 1990.
13. Product information, Teflon® PTFE 703N, Granular Compression Molding Resin, DuPont International, S. A. Switzerland, Geneva, Switzerland, 1990.
14. Product information, Teflon® PTFE 707N, Granular Compression Molding Resin, DuPont International, S. A. Switzerland, Geneva, Switzerland, 1990.
15. Product information, Teflon® PTFE 801N, Granular Compression Molding Resin, DuPont International, S. A. Switzerland, Geneva, Switzerland, 1990.
16. Product information, Teflon® PTFE 807N, Granular Compression Molding Resin, DuPont International, S. A. Switzerland, Geneva, Switzerland, 1990.
17. Product information, Teflon® 809N, DuPont International, S. A. Switzerland, Geneva, Switzerland, 1990.
18. Product information, Teflon® PTFE 901N, Granular Ram Extrusion Resin, DuPont International, S. A. Switzerland, Geneva, Switzerland, 1990.
19. Product information, Teflon® PTFE 7AJ, Granular Compression Molding Resin, Mitsui DuPont Fluorochemical Co., Tokyo, Jpn., 1990.
20. Product information, Teflon® PTFE 7J, Granular Compression Molding Resin, Mitsui DuPont Fluorochemical Co., Tokyo, Jpn., 1990.
21. Product information, Teflon® PTFE 800J, Granular Compression Molding Resin, Mitsui DuPont Fluorochemical Co., Tokyo, Jpn., 1990.

22. Product information, Teflon® PTFE 820J, Granular Compression Molding Resin, Mitsui DuPont Fluorochemical Co., Tokyo, Jpn., 1990.
23. Product information, Teflon® 70J, Modified PTFE Granular Molding Resin, Mitsui DuPont Fluorochemical Co., Tokyo, Jpn., 1990.
24. Product information, Teflon® 170J, Chemically Modified Free-Flow Granular Molding Resin, Mitsui DuPont Fluorochemical Co., Tokyo, Jpn., 1990.
25. Fluon® Polytetrafluoroethylene, Granular Molding Powders, ICI Fluoropolymers, Business Unit of ICI Americas, Inc., Wilmington, Delaware.
26. Hostaflon® Polytetrafluoroethylene, Product Comparison Guide, DYNEON, A 3M-Hoechst Enterprise, Minneapolis, Minnesota, Feb., 1998.
27. Hostaflon® Polytetrafluoroethylene Compression Molding Powders, DYNEON, A 3M-Hoechst Enterprise, Minneapolis, Minnesota, Sep., 1998.
28. Daikin Industries of Fluoroplastics, Daikin Industries, Ltd., Daikin Americas, Orangeburg, NY, March, 1998.
29. Fluorocarbon Polymers of Daikin Industries, Daikin-Polyflon® TFE Molding Powders, Daikin Industries, Ltd., Osaka, Jpn., March, 1992.
30. ALGOFLON® Polytetrafluoroethylene, Properties and Application Selection Guide, Ausimont Corp., Morristown, NJ.
31. Product information, Teflon® PTFE 30, Fluoropolymer Resin, Aqueous Dispersion, No. H-03234-5, DuPont Co., Wilmington, Delaware, Dec., 1996.
32. Product information, Teflon® PTFE 30B, Fluoropolymer Resin, Aqueous Dispersion, DuPont Co., Wilmington, Delaware, Dec., 1996.
33. Product information, Teflon® PTFE 35, Fluoropolymer Resin, Aqueous Dispersion, DuPont Co., Wilmington, Delaware, Dec., 1996.
34. Product information, Teflon® PTFE B, Fluoropolymer Resin, Aqueous Dispersion, DuPont Co. Wilmington, Delaware, Dec., 1996.
35. Product information, Teflon® PTFE 304A, Fluoropolymer Resin, Aqueous Dispersion, DuPont Co., Wilmington, Delaware, Dec., 1996.
36. Product information, Teflon® PTFE 305A, Fluoropolymer Resin, Aqueous Dispersion, DuPont Co., Wilmington, Delaware, Dec., 1996.
37. Product information, Teflon® PTFE 306A, Fluoropolymer Resin, Aqueous Dispersion, DuPont Co., Wilmington, Delaware, Dec., 1996.
38. Product information, Teflon® PTFE 307A, Fluoropolymer Resin, Aqueous Dispersion, DuPont Co., Wilmington, Delaware, Dec., 1996.
39. Product information, Teflon® PTFE 308A, Fluoropolymer Resin, Aqueous Dispersion, DuPont Co., Wilmington, Delaware, Dec. 1996.
40. Product information, Teflon® PTFE 313A, Fluoropolymer Resin, Aqueous Dispersion, DuPont Co., Wilmington, Delaware, Dec., 1996.
41. Product information, DuPont PTFE K-20, Fluoropolymer Resin, Aqueous Dispersion, DuPont Co., Wilmington, Delaware, Dec., 1996.
42. Fluon® PTFE, Aqueous Dispersion, ICI Fluoropolymers, Business Unit of ICI Americas, Inc., Wilmington, Delaware.
43. Hostaflon® TF Dispersions, DYNEON, A 3M-Hoechst Enterprise, Minneapolis, Minnesota, Sep., 1998.
44. Daikin Fluorocarbon Polymers, Daikin PTFE Dispersions, Daikin Kogyo, Ltd., Osaka, Jpn., 1986.

45. Product information, Teflon® PTFE 600A, Fluoropolymer Resin, Fine Powder Lubricated Extrusion Resin, DuPont Co., Wilmington, Delaware, Feb., 1999.

46. Product information, Teflon® PTFE 601A, Fluoropolymer Resin, Fine Powder Lubricated Extrusion Resin, DuPont Co., Wilmington, Delaware, Feb., 1999.

47. Product information, Teflon® PTFE 602A, Fluoropolymer Resin, Fine Powder Lubricated Extrusion Resin, DuPont Co., Wilmington, Delaware, Feb., 1999.

48. Product information, Teflon® PTFE 610A, Fluoropolymer Resin, Fine Powder Lubricated Extrusion Resin, DuPont Co., Wilmington, Delaware, Feb., 1999.

49. Product information, Teflon® PTFE 613A, Fluoropolymer Resin, Fine Powder Lubricated Extrusion Resin, DuPont Co., Wilmington, Delaware, Feb., 1999.

50. Product information, Teflon® PTFE 614A, Fluoropolymer Resin, Fine Powder Lubricated Extrusion Resin, DuPont Co., Wilmington, Delaware, Feb., 1999.

51. Product information, Teflon® PTFE 6C, Fluoropolymer Resin, Fine Powder Lubricated Extrusion Resin, DuPont Co., Wilmington, Delaware, Feb., 1999.

52. Product information, Teflon® PTFE 60A, Fluoropolymer Resin, Fine Powder Lubricated Extrusion Resin, DuPont Co., Wilmington, Delaware, Feb., 1999.

53. Product information, Teflon® PTFE 62, Fluoropolymer Resin, Fine Powder Lubricated Extrusion Resin, DuPont Co., Wilmington, Delaware, Feb., 1999.

54. Product information, Teflon® PTFE 67, Fluoropolymer Resin, Fine Powder Lubricated Extrusion Resin, unpublished, DuPont Co. Wilmington, Delaware, May, 1999.

55. Product information, Teflon® PTFE 640, Fluoropolymer Resin, Fine Powder Lubricated Extrusion Resin, DuPont Co., Wilmington, Delaware, Feb., 1999.

56. Product information, Teflon® CFP 6000, Fluoropolymer Resin Extrusion Powder, DuPont Co., Wilmington, Delaware, Feb., 1999.

57. Product information, Teflon® 65N, DuPont International, S. A., Geneva, Switzerland, 1990.

58. Product information, Teflon® 669N, DuPont International, S. A., Geneva, Switzerland, 1990.

59. Product information, Teflon® 636N, DuPont International, S. A., Geneva, Switzerland, 1990.

60. Product information, Teflon® 637N, DuPont International, S. A., Geneva, Switzerland, 1990.

61. Product information, Teflon® 638RFFN, DuPont International, S. A., Geneva, Switzerland, 1990.

62. Fluon® PTFE, Coagulated Dispersion, ICI Fluoropolymers, Business Unit of ICI Americas, Inc., Wilmington, Delaware.

63. Hostaflon® TF Fine Powder, DYNEON, A 3M-Hoechst Enterprise, Minneapolis, Minnesota, September, 1998.

64. Fluorocarbon Polymers of Daikin Industries, Daikin-Polyflon® TFE Fine Powder, Daikin Industries, Ltd., Osaka, Jpn., July, 1986.

65. Ref. *Allied Signal Publication*, "Keep Your Cool with ALCON® PCTFE Dispersions," Morristown, NJ, 1998.

66. Ref. *Daikin PCTFE*, Daikin Fluorocarbon Polymers, Daikin Kogyo Co., Feb., 1981

PART II
7 Fabrication and Processing of Granular Polytetrafluoroethylene

7.1 Introduction

This chapter describes the fabrication of suspension polymerized or *granular* polytetrafluoroethylene into shapes and articles for conversion to parts for end-use applications. This type of PTFE is fabricated by a modified metallurgy technology named *compression molding,* where the PTFE powder is compressed into a "preform" at ambient temperature. The preform has sufficient strength to be handled, roughly equivalent to blackboard chalk. After removal from the mold, the preform is heated in an oven above its melting point and is sintered. The consolidation of particles during sintering is referred to as *coalescence,* which produces a homogenous and strong structure (Sec. 7.3.3.2). Varying the cooling rate by which the part reaches room temperature controls crystallinity of a part for a given polymer.

There are four basic molding techniques for processing granular resin. All four rely on the principles of compression molding of PTFE. These procedures are applied to convert granular resins into parts ranging in weight from a few grams to several hundred kilograms (Table 7.1).

The only continuous process for manufacturing parts from granular PTFE is called *ram extrusion* (see Sec. 7.6)

7.2 Resin Selection

Selection of resin depends on the desired properties of the PTFE part in the application and the manufacturing method used to produce it. Electrical insulation, reactor liners, and most gaskets are typically made using fine cut resins to achieve the best properties. Mechanical parts such as bridge and heavy equipment bearings do not rely on these properties to the same extent, and can therefore be made from free flow (pelletized) resins.

Resin flow is a function of the apparent density of the resin. Resins with an apparent density of >500 g/l are usually obtained by pelletizing fine cut resins and are known as *free flow* (Fig. 7.1a). The consistency of these resins is similar to granulated sugar in contrast to fine cut resins (with apparent density of <500 g/l) which have a consistency close to that of flour.

Resin flow and property improvement are inversely related (Fig. 7.1b). A free flow resin produces a part that has lower elongation, tensile strength, specific gravity, and dielectric breakdown strength than the same part produced with a fine cut PTFE powder (Figs. 7.2 and 7.3). Improvement in resin flow raises the efficiency of mold filling. Good resin flow is a requirement for automatic molding, isostatic molding, and ram extrusion processes.

Table 7.1. Selection of Granular Fabrication Process Based on Part Geometry

FABRICATION PROCESS	SHAPE	PART DIMENSION	PART WEIGHT
Billet/Block Molding	Rectangular, cylindrical	One centimeter in diameter or height to 5 meters in diameter or 1.5 meters in height	Ten grams to several hundred kilograms
Sheet Molding	Flat sheet	One centimeter to 1 meter in width and thickness of 3–75 millimeters	A few hundred grams to a few tens of kilograms
Automatic Molding	Small round	A few millimeters to a few centimeters in diameter	A few grams to a few hundred grams
Isostatic Molding	Complex geometry	A few centimeters to 0.5 meters in the major dimension	A few tens of grams to a few tens of kilograms
Ram Extrusion	Rod or tube	2–400 millimeter in diameter	Continuous process

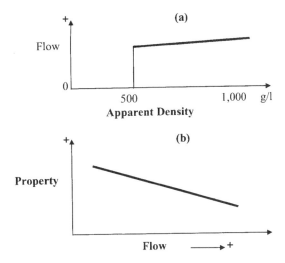

Figure 7.1 The relationships of property and apparent density vs. flow.

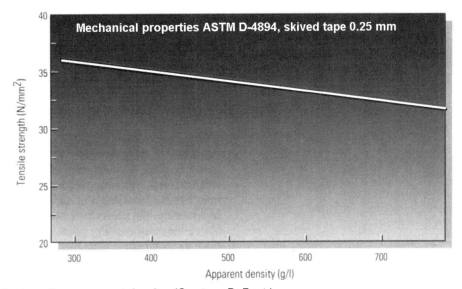

Figure 7.2 Tensile strength vs. apparent density. *(Courtesy DuPont.)*

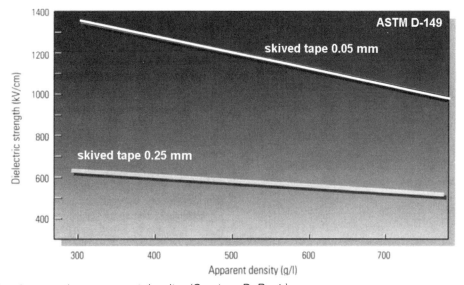

Figure 7.3 Dielectric strength vs. apparent density. *(Courtesy DuPont.)*

Small particles and bimodal or multimodal particle size distribution of resin powder yield the highest packing density[1] and consequently the lowest void content, leading to improvement in part properties. The frictional interaction among the particles also affects the degree of compaction.

7.3 Compression Molding

Compression molding is the method by which massive (700 kg or 1,500 lbs), cylindrical (billet), rectangular, and sheet shapes of PTFE are produced. The blocks and cylinders can be solid or annular and are by far the heaviest objects produced from any fluoropolymer. The height of a cylinder can exceed 1.5 m (60 in). These billets are cut (skived) into wide thin films (<0.5 mm thick) or sheets (7 mm thick). Sheets, blocks, and cylinders are utilized as stock shapes for machining more complex shapes. The same principles are applied to mold any shape. This section will focus on the more common cylindrical shape, the billet.

PTFE's specific gravity is high compared to other plastics. A solid billet with a wall thickness of 130 mm and 300 mm tall may weigh about 50 kg. Table 7.2 presents sizes of common billets. The selection of the size of the billet depends on the properties required in the application. For example, PTFE has low thermal conductivity and a thermal gradient forms across the wall thickness during sintering. Dielectric strength is more influenced than tensile strength by this thermal gradient. This is the reason thin (0.05–0.125 mm) electric-grade tapes are skived from billets with a wall thickness of 75–100 mm. Mechanical grade sheets are skived from heavier wall billets (125–175 mm). Billet height is determined by the desired width of the film or sheet. Electrical tapes are commonly made from 300 mm billets. Sheets for mechanical applications and for lining chemical processing equipment are sometimes made from 1.5 m tall billets.

The large quantity of polymer and the length of time required to produce a shape require careful attention to issues that affect productivity such as the handling and storage of resin. High temperature storage of granular PTFE can lead to compaction during handling. Polymer should be conditioned at temperatures of 21–25°C before molding to reduce clumping and ease handling. Dew point conditions should be avoided to prevent moisture from condensing on the cold powder which will expand during sintering and crack the molding. Molding below 20°C should be avoided because PTFE will undergo a 1% volumetric change at a transition temperature of 19°C. Preforms molded below 20°C can crack during sintering.

Table 7.2. Approximate OD-ID-Height-Weight Relationships of Typical Billet Moldings

OD (mm)	ID (mm)	Wall Thickness (mm)	Weight/Height (kg/m)
500	150	175	386
500	200	150	356
400	100	150	255
480	150	150	305
180	200	140	323
300	50	125	148
300	100	100	136
250	100	75	89
200	50	75	64
150	25	62.5	37
100	20	40	16
75	35	20	7

Molding areas should be equipped with positive pressure to keep out dust and airborne contaminants. Parts intended for the semiconductor industry are preferably molded in cleanrooms. Dust, oil, and particles of an organic nature must be prevented from contaminating the resin because during sintering they will carbonize into dark specks.

7.3.1 Equipment

Relatively simple equipment is used for billet molding. It consists of a stainless steel mold and hydraulic press for fabrication of the preform, and an oven for sintering. A lathe and skiving blades are required for preparation of film and sheet.

7.3.1.1 Mold Design

PTFE resins are molded in molds similar to those utilized for thermosetting resins or metal powders. A complete mold consists of a cylindrical or rectangular die and upper and lower end plates and a mandrel for annular parts. These parts are normally made of tool steel to allow machining, and plated with chromium or nickel to protect them from corrosion. Occasionally, the end plates are made of brass or plastics such as nylon. A small diametrical clearance is designed in the end plates to allow easy assembly and air escape. Figures 7.4 and 7.5 show examples of mold assemblies for a range of billet sizes.

Molds should be designed carefully to avoid distortion under the preform pressure. The minimum wall thickness can be determined from Eq. 7.1.

Eq. (7.1) $$t_m = F_S \cdot \frac{P_m \cdot d_i}{2 \delta_m}$$

where: t_m = Minimum wall thickness (mm)
F_s = Safety factor, a value of 2.5 is commonly used
P_m = Maximum internal pressure = 0.7 maximum preform pressure (N/mm²)
d_i = Inside diameter (mm).
δ_m = Allowable metal yield stress (N/mm²)

The internal dimensions of the mold depend on the properties of the resin. The height of the mold is a function of the apparent density or compression ratio defined as:

Eq. (7.2) $$C.R. = \frac{H_F}{H_P}$$

$C.R.$ = Compression Ratio
H_F = Filled height, mm
H_P = Preform height, mm

Figure 7.4 Typical mold assembly for small to medium size billets. *(Courtesy DuPont.)*

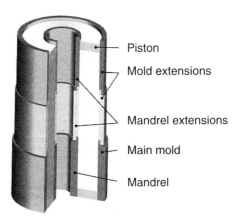

Figure 7.5 Typical mold assembly for larger billets. *(Courtesy DuPont.)*

The apparent density of the resin determines the maximum value of compression ratio and therefore the maximum mold length. Equation 7.3 shows the theoretical value of compression ratio as the ratio of theoretical height of the resin in the mold and the preform height. PTFE powder is compacted during handling and charging of the mold and, therefore, its apparent density increases. Height H_F is smaller than H_T because a given weight of the resin assumes a lower height in the mold than it would have if it had not been compacted under the force of its own weight.

Eq. (7.3) $$(C.R.)_T = \frac{H_T}{H_P}$$

H_T = Filled height without compaction, mm

Equations 7.2 and 7.3 can be used to calculate the actual filled height and theoretical maximum required height of the mold, if the preform height and the compaction ratio of the resin are known. The actual mold is designed to a height between the H_F and H_T.

Equation 7.3 can be rearranged to obtain Eq. 7.4, in which S is the surface area of the cross-section of the mold and W is the weight of the resin being molded. The numerator and denominator of the right side of Eq. 7.4 represent the definition of preform density and apparent density, respectively.

Eq. (7.4) $$(C.R.)_T = \frac{\frac{W}{H_P S}}{\frac{W}{H_t S}}$$

Eq. (7.5) $$(C.R.)_T = \frac{\rho_P}{\rho_T}$$

where: ρ_P = Preform density, g/cm³

ρ_T = Apparent powder density, g/cm³

Preform density is typically about 1.9 g/cm³ for a fine cut PTFE powder with an apparent density of 450 g/liter (0.45 g/cm³). A typical value for compression ratio for this resin is 3.2. The maximum value for this resin is given by substituting for values in Eq. 7.5.

$$(C.R.)_T = \frac{1.9}{0.45} = 4.2$$

The mold length has to be 3.2–4.2 times the height of the tallest billet expected to be made; a 1.5 meter tall billet would require a mold that is 4.8–6.3 m long. Extensions can be added to the mold to obtain the desired height as long as the joint between the extension and the main body of the mold are smooth; otherwise stress concentration may lead to cracking of the billet. Mold diameter is determined by resin shrinkage, which means that each mold is designed for a specific resin since each has a specific shrinkage value. The word *shrinkage* refers to the shrinkage of the part after sintering has been completed. Usually the initial reference is the corresponding mold dimension in the calculation of shrinkage.

7.3.1.2 Presses

Hydraulic presses are recommended for preform production. Important elements of a press are smooth pressure application, maximum opening ("daylight"), ram stroke, flatness and levelness of the platens, and tonnage. A programmable press allows smooth application and removal of pressure, which is critical to producing a good part. Jerky and uneven motion of the ram will result in nonuniform application of pressure to the resin resulting in cracking during sintering. Figure 7.6 shows a typical preforming press. The tonnage of the press determines the maximum diameter of the preform. The typical maximum required preform pressure is 60 MPa for unfilled resin and 100 MPa for filled resin.

Figure 7.6 An automatic press. *(Courtesy DuPont.)*

7.3.1.3 Ovens

PTFE is an excellent thermal insulator. Its thermal conductivity (0.25 W/m·K), roughly 2,000 times less than copper, impacts the rate of sintering of a preform. The most common way of delivering heat to a preform is by circulation of hot air. A large volume of air has to be recirculated because of its low thermal capacity. Ideally, the sintering oven is electrically heated for use up to 425°C and should be equipped with override controllers to prevent overheating.

Good temperature control is critical to achieving uniform and reproducible part dimensions and properties. The interior of the oven should be designed to maximize air circulation and temperature uniformity, and prevent the formation of "hot spots." A highly rated insulation will minimize heat loss, which is particularly important during the sintering of a full oven load. Temperature monitoring at various locations in the oven reveals hot and cold zones, which should be corrected.

Leakage usually occurs in the door area and can be prevented by maintaining the door gasket and hanging additional insulation outside the oven door. Heat loss can lead to a variety of problems with the parts such as incomplete sintering, distortion, cracking, and poor physical properties.

Controlled cooling is accomplished by fresh air intake during the cool-down portion of the cycle. Very little air enters the oven during the heating part of sintering, only an amount sufficient for the removal of the off-gases. The exhaust should be directly from the oven to the atmosphere. A hood should be placed over the oven door, where leaks are most likely to remove PTFE fumes. Adequate ventilation of the sintering area is very important. Fumes and off-gases must not be inhaled because of health hazards. See Ch. 18 for detailed discussion of safety and health issues.

7.3.2 Densification and Sintering Mechanism

Resin powder particles are separated by air which is removed during preforming and sintering. Figure 7.7 depicts the changes that PTFE undergoes from the powder stage to a sintered part. The powder is charged to the mold and compressed and held for a dwell period. After the preform is made, it is removed from the mold and allowed to rest for stress relaxation and degassing.

The preform expands due to relaxation and recovery. The pressure placed on the molded resin exerts three types of changes in the particles of resin. Resin particles undergo plastic deformation and are intermeshed together leading to the development of cohesive or *green* strength. Particles also deform elastically and experience cold flow under pressure. The air trapped in the space between the resin particles is compressed. Removal of pressure allows the recovery of elastic deformation, which creates a quick "snap back" of the preform. Over time, stress relaxation partly reverses the cold flow, and the preform expands.

The trapped air in the preform is under high pressure, theoretically equal to the preforming pressure. The air requires time to leave the preform because it is mostly contained in the void areas surrounded by enmeshed particles. Immediate sintering would lead to a rise in the already high pressure of the air and catastrophic cracking of the part as the PTFE melts and the mechanical strength declines. The preform should be allowed to degas which equalizes the internal air pressure to atmospheric pressure.

Sintering of the preform takes place in an oven where massive volumes of heated air are circulated. Initial heating of the preform leads to thermal expansion (Fig. 7.8) of the part. After PTFE melts, relaxation of the residual stresses occurs (stored because of the application of pressure to the polymer) where additional recovery takes place and the part grows. The remaining air begins to diffuse out of the

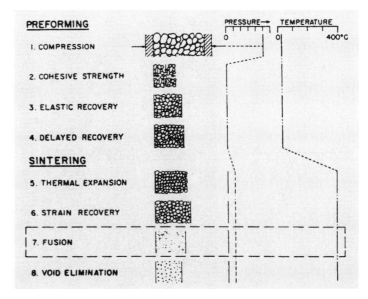

Figure 7.7 Schematic diagram of preforming and sintering sequence with polytetrafluoroethylene.[2]

preform after heating starts. The adjacent molten particles begin to coalesce slowly; usually hours are required because of the massive size of PTFE molecules (molecular weight is 10^6–10^7). Fusion of the particles is followed by elimination of the voids, where almost no air is left. It is noteworthy to remember that the elimination of all the voids in PTFE is quite difficult because of limited mobility of the large polymer molecules.

7.3.3 Billet Molding

In this section the three processes for producing a billet from granular PTFE: preforming, sintering and cooling, are described.

7.3.3.1 Preforming

Preforming consists of charging the mold with the powder and compaction by the application of pressure to prepare a green part with sufficient strength to allow handling. Demolding or removal from the mold and placement in the oven are the steps that require green strength. Occasionally, a preform may be machined which increases the importance of green strength.

A key variable is resin temperature during molding. The powder is harder and has better flow below the transition temperature of 19°C (Fig. 7.8), but it does not respond well to pressure. The preforms produced below the transition temperature have low green strength and are more likely to crack during sintering. To avoid these problems, the resin should be conditioned at 21–25°C for 24 hours. The molding area should ideally be maintained at >21°C.

Adequate and uniform application of pressure is the determining factor of properties of the final part in the molding step. PTFE becomes softer and exhibits higher plastic flow as the temperature increases and can thus be molded at lower pressures. An increase of temperature from 21°C to >31°C is roughly equivalent to 2 MPa of molding pressure. The effect of temperature is helpful, to a moderate extent, when press capacity is limited. In the summer weather, a decrease in preforming pressure may be necessary to eliminate cracking problems. Economics of raised temperature molding such as lengthy heat-up time to condition the resin versus the cost of additional press capacity should be calculated before a decision is made. Figures 7.9 and 7.10 show the relationship of preform temperature and pressure to the key properties of the finished part.

Filling the mold must be done uniformly because uneven filling leads to nonuniform density in the preform and cracking. Charging the mold is much simpler with a free flow resin than a fine cut powder. Free flow resins more or less assume the shape of the mold and require little distribution. Conversely, significant effort has to be expended to distribute the fine cut resin evenly in the mold. All lumps should be gently broken up by a scoop or a mesh screen. A key consideration is to completely fill the mold prior to

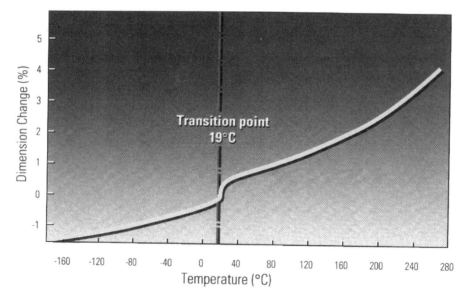

Figure 7.8 Transition point and linear thermal expansion of PTFE. *(Courtesy DuPont.)*

any pressure application. "Tamping" the resin in the mold, even with the slightest force, to make room for more powder will lead to the formation of a charge line at the interface and possible cracking during sintering. This is the step where introduction of contamination in the part is likely to happen. The molding areas should be isolated from the rest of the process such as machining where oil and dust are present. Increasingly, processors are molding in cleanrooms (usually portable) to meet the more stringent quality specifications of customers. Wiping and cleaning the exterior of resin drums prior to opening and resealing of the unused portion of the powder are among the helpful practices for prevention of contamination.

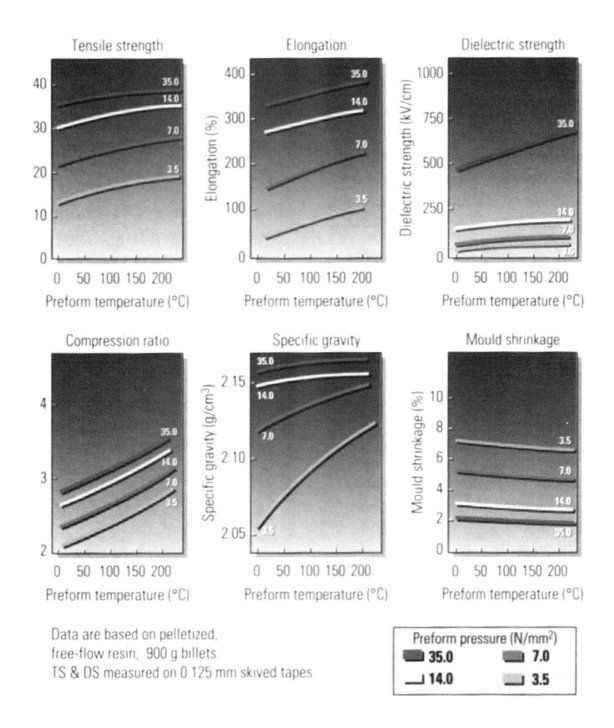

Figure 7.9 Properties vs. preform temperature and pressure. *(Courtesy DuPont.)*

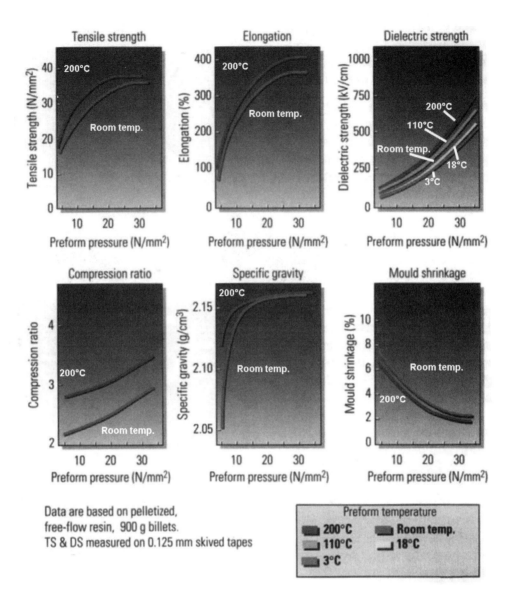

Figure 7.10 Properties vs. preform pressure and temperature. *(Courtesy DuPont.)*

The compression of PTFE particles involves elastic and plastic deformations. At the early stages of pressurization, the powder particles pack together by rolling, and slipping, thus eliminating interparticle void space and removal of air. This process increases the number of points where direct particle to particle contact takes place. The increase in pressure sets off plastic deformation by the enlargement of contact areas and enmeshing of the particles together. Intraparticle voids are eliminated by the flow of PTFE under pressure. At some point the plastic regime diminishes and elastic compression of the preform takes over which recovers rapidly after the removal of pressure. Time must be allowed for transmission of pressure through the resin because of its compressible nature. Excessive pressure leads to a phenomenon known as "plane slippage" due to severe plastic deformation resulting in the formation of cracks in the preform.

Pressurization rate or ram closing speed depends on the size and shape of the billet and the type of resin. The apparent density of the powder determines the air-filled void space, which must be eliminated. The slower the ram speed, the more completely the air will leave the preform, but process productivity suffers at a low closing rate. Very fast ram speeds lead to entrapment of air, resulting in high porosity and low density areas, even billet cracking. Table 7.3

provides ram speeds that offer compromises between productivity and part quality. The pressurization profile of a typical preform is shown in Fig. 7.11. Note that the dwell time after reaching maximum pressure is almost as long as the time to reach that pressure.

Table 7.3. Press Closing Rates (mm/min) *(Courtesy DuPont)*

Billet size (mm)	% Compaction		
	~ 60	Next ~ 20	Last ~ 20
Small to medium (~100)	500–150	250–25	50–5
Medium to tall (100-500)	150–25	50–5	10–2

Maximum pressure during the preform molding has a direct bearing on void-closure and the final part properties. Dielectric breakdown strength (DBV) and shrinkage are strongly affected by the pressure. DBV rapidly deteriorates with increasing void content. Resin type is the determining factor in the selection of a molding pressure which minimizes voids and avoids plane slippage. Figure 7.12 presents the effect of pressure on DBV and tensile properties, clearly suggesting that there is a range for optimal preforming pressure. Figures 7.13 through 7.18 show the quantitative impact of preform pressure on DBV and tensile properties of typical commercial resins; fine cut resins have superior properties in all cases.

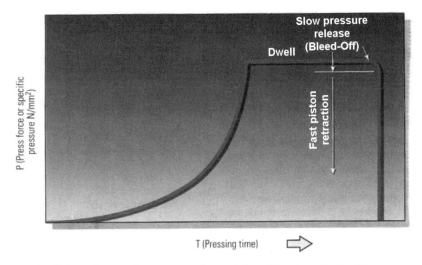

Figure 7.11 Press force vs. dwell time diagram for compression molding. *(Courtesy DuPont.)*

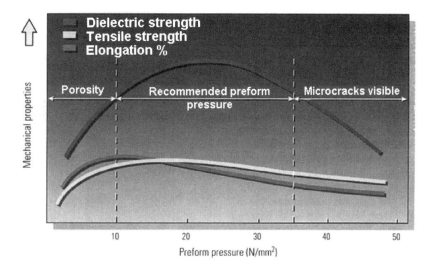

Figure 7.12 Mechanical properties vs. preform pressure. *(Courtesy DuPont.)*

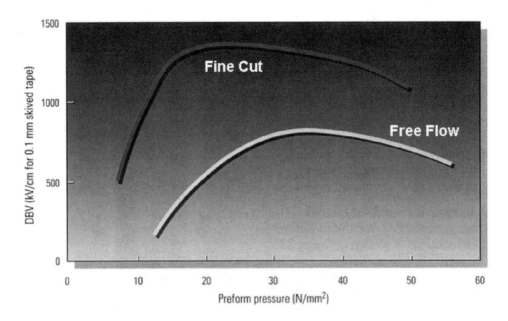

Figure 7.13 Effect of preform pressure on dielectric breakdown voltage. *(Courtesy DuPont.)*

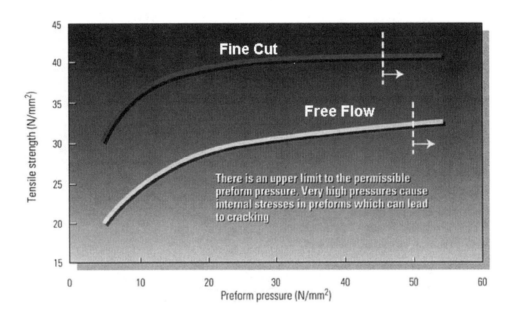

Figure 7.14 Effect of preform pressure on tensile strength. *(Courtesy DuPont.)*

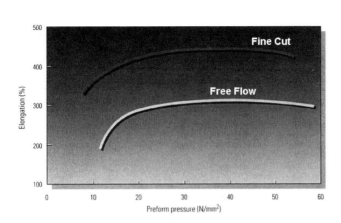

Figure 7.15 P/T diagram for compression molding. *(Courtesy DuPont.)*

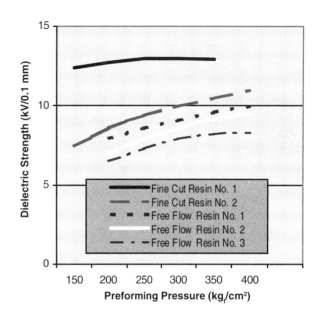

Figure 7.16 Relationship of molding pressure and dielectric strength.[3]

Molding Conditions:		
Dimensions:	100 (O.D.) × 50 (I.D.) × 50 (H) mm	
Preforming pressure:	150 ~ 400 kg$_f$/cm²	
Dwell period:	5 min	
Heating rate:	200 ~ 300 °C, 33.3 °C/h	
	300 ~ 370 °C, 17.5 °C/h	
Sintering period:	5 hrs at 370 °C	
Cooling rate:	370 ~ 300 °C, 17.5 °C/h	
	300 ~ 200 °C, 20 °C/h	

Test specimens for tensile properties were punched according to the No. 3 Dumbbell type of test method JIS K 6301 from skived film 0.5 mm in thickness.

Test specimens for dielectric strength were skived to films of 0.1 mm.

1 MPa = 10 kg$_f$/cm² = 1N/mm²

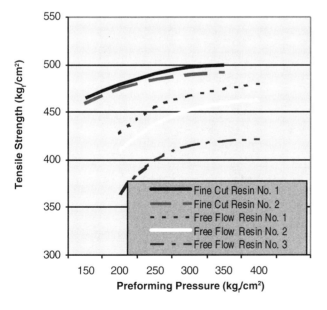

Figure 7.17 Relationship of molding pressure and tensile properties.[3]

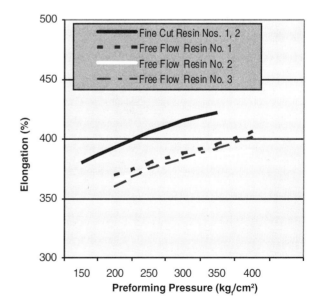

Figure 7.18 Relationship of molding pressure and elongation properties.[3]

The preform shrinks in the radial (cross direction or CD) during sintering and tends to grow in height or machine direction (MD) as shown in Fig. 7.19. Raising the preform pressure reduces the shrinkage and increases the growth. Typical numerical values for shrinkage and growth of a few commercial resins can be found in Fig. 7.20 and 7.21. There is no general relationship between the shrinkage of fine cut and free flow powders except that they are fairly close.

Dwell time is necessary to obtain even compaction of the resin in the preform. Insufficient dwell leads to density gradients in the billet, "hourglassing" which refers to the shape of the billet after sintering, property variation, air entrapment and microcracks. Dwell time depends on the type of resin, rate of pressurization and the size and shape of the preform. The rule of thumb is 2–5 minutes of dwell per 10 mm of final height for billets <100 mm in diameter, and 1–1.5 minute for 10 mm of height for large billets (>100 mm diameter)

Pressure release at the end of the dwell should begin very slowly until the initial expansion or elastic recovery has taken place. Sudden depressurization results in the formation of visible microcracks because of rapid expansion of the pressurized entrapped air.

Pressure decay during the molding can be a serious problem in billets that are taller than 100 mm. This can be corrected by pressurizing the resin from both sides if a double action press is available. Otherwise, an arrangement similar to that shown in Fig. 7.22 can be used. Separate spacers around the shell and under the end cap temporarily support the mold until the first 30–60% of the compression has been reached. The shell spacers are removed at this point which allows pressurization from the opposite end of the ram. The mold can also be turned over to compress the billet from the opposite side in a single action pressure. Repressurization must be done very slowly until the maximum pressure is reached.

At the completion of depressurization, the preform has to be removed from the mold. After expansion, the preform tends to remain engaged with the mold and should be removed carefully. Excellent surface finish of the mold contact surfaces facilitates demolding. The arrangements sketched in Fig. 7.23 can be utilized to release the preform if a double acting press is not available. Complex shapes may be made in more elaborate telescopic molds (Fig. 7.24).

Degassing is the last step in preparation of the preform prior to sintering. Air and residual stress remains entrapped in the preform and should be relieved prior to sintering. All the air does not exit the resin during the compression cycle and a small remaining volume is pressurized. This air needs time to escape from the preform; otherwise it will increase substantially during the heat-up segment of the sintering cycle and crack the billet. Stresses remaining in the preform can be equally potent and lead to billet cracking during the heat-up period. The higher the heating rate, the more magnified the effect of residual stresses. A time interval is required to relax stresses and allow air to escape. Figure 7.25 presents typical times sufficient to prevent cracking.

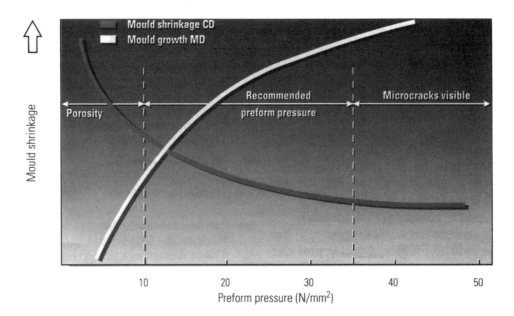

Figure 7.19 Mold shrinkage vs. preform pressure. *(Courtesy DuPont.)*

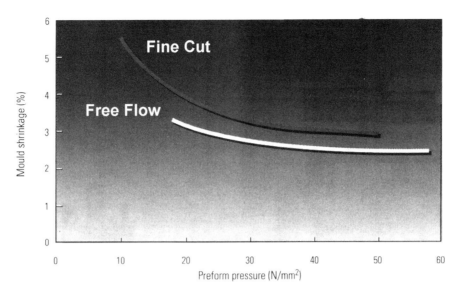

Figure 7.20 Effect of preform pressure on mold shrinkage. *(Courtesy DuPont.)*

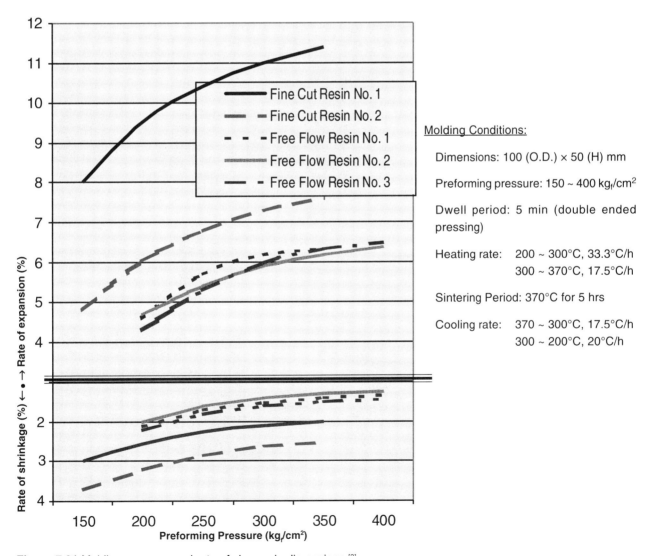

Figure 7.21 Molding pressure and rate of change in dimensions.[3]

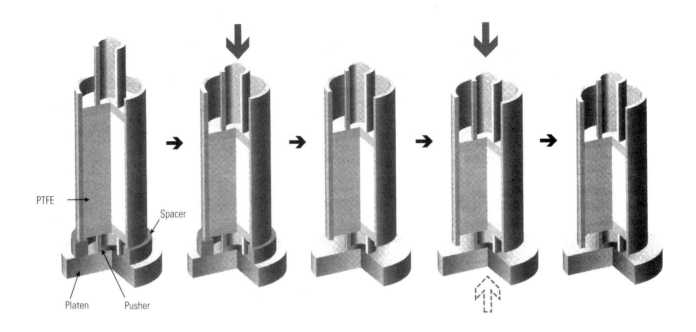

Figure 7.22 Double acting compression molding. *(Courtesy DuPont.)*

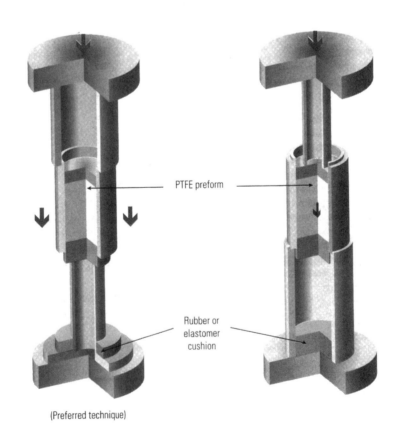

Figure 7.23 Systems for preform removal. *(Courtesy DuPont.)*

Fabrication and Processing of Granular Polytetrafluoroethylene

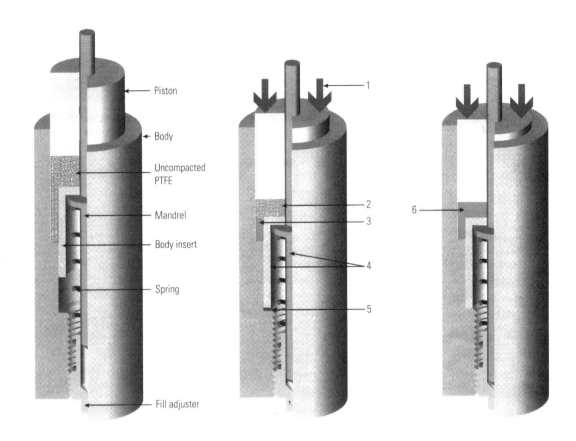

Figure 7.24 Telescopic mold. *1.* Pressure applied to piston. *2.* Cup seal top is partially compacted. *3.* Cup seal skirt is completely compacted. *4.* Body insert and mandrel move together, compressing spring. *5.* Movement of mandrel and body insert stops when insert reaches shoulder. *6.* Piston continues to compact cup seal top after insert stops. Piston movement stops when cup seal top is fully compacted. *(Courtesy DuPont.)*

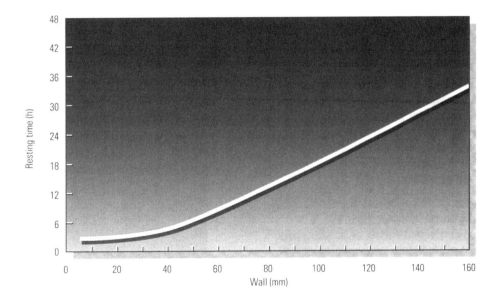

Figure 7.25 Resting time for degassing after preforming. *(Courtesy DuPont.)*

7.3.3.2 Sintering

A preform has limited cohesive strength and is essentially useless; sintering allows coalescence of the resin particles, which provides strength and void reduction. Sintering cycle profiles of time and temperature affect the final properties of the billet. Sintering temperatures exceed the melting point of PTFE (342°C) and range from 360°C to 380°C.

Figure 7.26 shows the various steps of the sintering process. At first the preform completes its elastic recovery and begins to thermally expand past the PTFE melting point, 342°C. The expansion can reach up to 25–30% by volume depending on the type of resin, powder, preforming pressure and temperature. Above 342°C, PTFE is a transparent gel due to the absence of a crystalline phase. At the sintering temperature, adjacent melted PTFE particles fuse together and coalesce. After two particles have completely coalesced, they would be indistinguishable from a larger particle and voids are eliminated under the driving force of surface tension.[4] Smaller particle resins and higher preform pressures improve coalescence.

Coalescence and void elimination require time because of the limited mobility of PTFE molecules. Melt creep viscosity of PTFE is in the range of 10^{11}–10^{12} poise at 380°C which severely inhibits any flow similar to that known for thermoplastics. The sintering temperature is held for a period of time to allow fusion, coalescence and void elimination to proceed and maximize properties in the part. A time is reached beyond which the part properties no longer improve and degradation begins. Property development should be balanced against cost in selecting a sintering cycle. Figure 7.27 shows the effect of sintering temperature on the specific gravity and tensile strength of the billet. Specific gravity increases while tensile strength decreases. Degradation of PTFE above 360°C leads to a lowering of molecular weight material, which crystallizes more easily and has decreased tensile strength.

The PTFE preform should be heated slowly because of its low thermal conductivity. This means that large volumes of turbulently heated air must be recycled through the oven to heat up the part. A thermal gradient develops between the exterior part of the preform and its inside. This gradient is required for heating the interior of the preform. Heating helps relax residual preforming stresses, which increase as the maximum pressure and mold closing rate increase. Ideally, the slowest possible rate is best because the thermal gradient also induces mechanical stresses in the billet which, along with the residual molding stress, can surpass the cohesive strength of the preform and lead to its cracking. Economics of sintering favor the fastest rate. The compromise value is the highest heating rate, which allows relaxation of stresses in the part yet does not result in cracking of the billet. It depends on oven temperature, airflow and billet wall thickness. Maximum heating rate should be determined experimentally using the following guidelines.

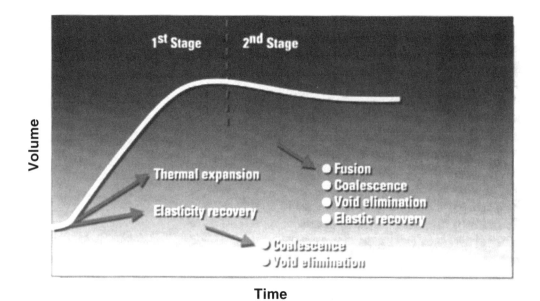

Figure 7.26 Mechanism of sintering Teflon®. *(Courtesy DuPont.)*

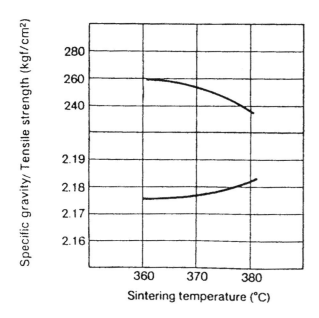

Figure 7.27 Seven-hour sintering: sintering temperature and product quality.[3]

The maximum heating rate for small parts is 80–100°C/hr. A typical heating rate is no more than 50°C/hr up to 150°C, 30°C/hr up to 300°C and 6–10°C/hr at higher temperatures. Only very small preforms can be subjected to faster rates. Table 7.4 provides recommended heating rates for various size preforms. Notice the decreasing heating rate at higher temperatures for massive billets (150 and 300 kg); they can withstand a smaller thermal gradient at higher temperatures.

A helpful strategy to maximize heating rate, particularly, at the early stages of heating is to introduce a number of hold periods (Fig. 7.28) to allow heating of the interior sections of the part. This helps reduce the thermal gradient, thus decreasing the probability of cracking. The hold period near the melting point (>300°C) is especially important because of the relatively large volume increase (about 10%) of PTFE during melting.

Table 7.4. Examples of TFE Sintering Conditions[3]

Preform size		Sintering Cycle		
Size (mm) (dia × length) (O.D./I.D.) × (L)	Weight (kgf)	Heating rate	Sintering	Cooling rate
50 × 50	0.2	50°C/h	5 h at 370°C	50°C/h
100 × 100	1.7	30°C/h	10 h at 370°C	30°C/h
174/52 × 130	6.0	30°C/h	12 h at 370°C	30°C/h
420/150 × 600	150	50°C/h (25°C → 150°C) 3 h at 150°C 25°C/h (150°C → 250°C) 3 h at 250°C 15°C/h (250°C → 315°C) 5 h at 315°C 10°C/h (315°C → 365°C)	20 h at 365°C	10°C/h (365°C → 315°C) 10 h at 315°C 10°C/h (315°C → 250°C) 25°C/h (250°C → 100°C)
420/150 × 1200	300	50°C/h (25°C → 150°C) 5 h at 150°C 25°C/h (150°C → 250°C) 5 h at 250°C 15°C/h (250°C → 315°C) 5 h at 315°C 10°C/h (315°C → 365°C)	30 h at 365°C	10°C/h (365°C → 315°C) 10 h at 315°C 10°C/h (315°C → 250°C) 25°C/h (250°C → 100°C)

Notes: * Preforming pressure: 150 kg$_f$/cm^2 (dual press)
* Compression speed: 40 to 60 mm/min (pressure applied in 4 stages)
* Dwell time: 30 min or more

** Preforming pressure: 150 kg$_f$/cm^2 (dual press)
** Compression speed: 40 to 60 mm/min (pressure applied in 4 or 5 stages)
** Dwell time: 45 min or more

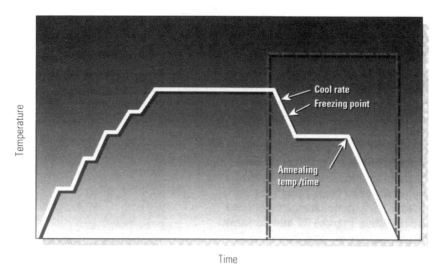

Figure 7.28 A general sintering cycle. *(Courtesy DuPont.)*

Resin particles coalesce and voids are eliminated in the *gel state*. The rate of sintering near the melting point is low and therefore the maximum temperature is selected above the PTFE melting point. The maximum hold temperature for unfilled resin is 385°C and for filled PTFE is 370°C. Thermal degradation accelerates significantly above these temperatures. The typical hold temperature range is 365–380°C.

The optimum hold time at the maximum temperature is determined through repeated trials. Figures 7.29 and 7.30 show the effect of sintering time and temperature on tensile strength, elongation and dielectric strength of skived film. It is clear in this example that 12 hours hold time at 377–382°C results in the best properties. Beyond this point properties generally decline. Why are these long hold times needed? Once the gel state is reached, PTFE requires about 2 hours for the completion of sintering. The rest of the hold time is for the interior of the preform to reach temperature. Prolonged sintering times are beneficial as long as degradation can be minimized.

The rule of thumb for determining the sintering time is 1 hour per centimeter of wall thickness of solid billets and 1.5 hours per centimeter of thickness for billets with a hole in the middle. Small parts need 0.8 hours per centimeter of time sintering temperature.

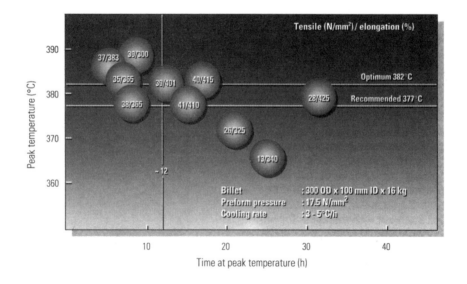

Figure 7.29 Peak temperature/time effect on properties (TS/EL). *(Courtesy DuPont.)*

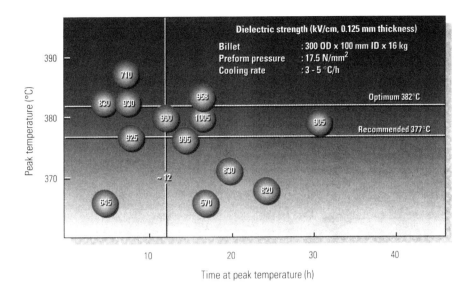

Figure 7.30 Peak temperature/time effect on properties (DBV). *(Courtesy DuPont.)*

7.3.3.3 Cooling

Figure 7.28 highlights a general cooling cycle, which immediately begins at the end of the sintering time. It plays two important roles: crystallization and annealing of the sintered billet. Many of the properties of PTFE (similar to other semicrystalline polymers) are governed by the crystalline phase content of the part. Crystallinity is determined by the cooling rate. At 320–325°C, the molten resin reaches the freeze point and crystallization begins to take place. Polymer chains, which were randomly distributed in the molten state, begin to pack in an orderly manner during the crystallization process. The slower the cool down, the higher the number of crystalline structures will be. This means that controlling the cooling rate can control the properties of the part. Table 7.5 shows the crystallinity of the granular parts as a function of cooling rate. It is interesting that the minimum attainable crystallinity by quenching in ice water is 45%, because of the low thermal conductivity of PTFE.

The actual cooling, similar to heating, is a strong function of the thermal conductivity of PTFE. Slow cooling, especially for thick parts, is necessary to avoid large thermal gradients, which can cause cracking of the part. This is especially important during the freezing transition because of the large volume decrease that the polymer experiences while going from melt to solid phase. Large stresses are generated in the part, which can fracture the melt if the cooling rate is not sufficiently slow. The cooling rate depends on the melt strength and wall thickness. The melt strength of a polymer increases with increasing molecular weight and it can withstand a higher thermal gradient.

Generally, large billets (150 kg and 300 kg in Table 7.4) should be cooled at rates between 8–15°C/hr down to 250°C. This slow cooling rate allows the middle of the wall of the part to reach the freeze point before faster cooling is commenced. Between 250°C and 100°C, cooling rate can be increased to 25°C/hr and below 100°C the oven doors can be opened. Smaller parts can be cooled at higher rates below 300°C.

Annealing refers to removal of residual stresses in the billet by holding it for a period of time between 290°C and 325°C during the cooling cycle. It also minimizes thermal gradients in the billet by allowing the wall interior to catch up with the exterior surface. The crystallinity of the part depends on the annealing temperature. A part which is annealed below the crystallization temperature range (<300°C) will only undergo stress relief. Annealing at a temperature in the crystallization range (300–325°C) results in higher crystallinity (higher specific gravity and opacity) in addition to stress relief. Figure 7.31 shows the effect of annealing and sintering temperatures on the tensile strength, elongation, dielectric strength and specific gravity of the part. The highest crystallinity part, as evidenced by specific gravity and opacity, is obtained by sintering and annealing the part at the highest temperature.

Table 7.5. Effect of cooling rate on crystallinity, typical for granular molding powders. *(Courtesy DuPont.)*

Cooling rate °C/min	% Crystallinity
Quenched in ice water	45
5	54
1	56
0,5	58
0,1	62

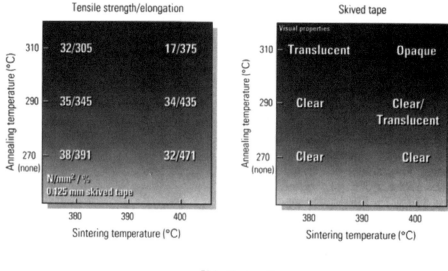

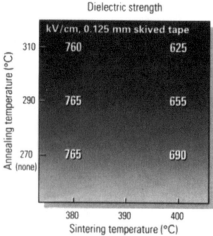

Figure 7.31 Effect of annealing on properties (typical for a high property resin). *(Courtesy DuPont.)*

7.3.4 Sintering Cycles

Table 7.4 and Fig. 7.32 provide examples of sintering cycles for a variety of cylindrical shapes and dimensions. These examples should be used as a conservative starting cycle, which allow a margin for shortcomings in the molding and degassing operations. They can be further refined and optimized.

7.3.5 Hot Compression-Molding

PTFE can be compression-molded under heat–that is, pressure and heat can be applied simultaneously followed by cooling. Sintering and cooling have to be performed in the mold. This process can be used to produce parts from unfilled and filled resins, almost free from porosity and with unusual properties. It has been reported[5][6] that increased resistance to cold flow (creep) and impact can be obtained by hot compression molding. Addition of a perfluorinated paraffin (e.g., $C_{25}F_{52}$) can further improve these properties. The fluorinated paraffin is to be mixed with the resin powder before charging the mold. The product of this is reported to have little tendency to flow under pressure. (Table 7.6)

The mold is placed inside the oven after the preform has been made. Pressure is then applied while sintering and cooling the part. It is significantly more expensive to equip an oven for hot compression molding. Additional heat is also required because the melting point of PTFE increases with pressure (Fig. 7.33). Filled PTFE parts are more likely to be made by this process.

7.4 Automatic Molding

Automatic molding is a process for automatic charging of resin into the mold followed by compression. It is usually utilized for mass production of small parts with a fairly simple geometry. The main requirement of the resin is good flow for easy and complete filling of the mold and part-to-part uniformity. Consistent resin shrinkage is mandatory to obtain consistent size parts. The combination of high productivity and low labor requirements of automatic molding render this process highly desirable for production of rings, seals, spacers, valve seats etc. where large numbers of relatively inexpensive parts are needed.

Figure 7.34 presents a schematic diagram of the four stages of automatic molding. In the first step, the free flowing resin is charged into the mold cavity formed by the lower ram (punch) and the outer mold. Pressure is actuated during step 2 and the upper ram compresses the resin for a few seconds. In step 3 the upper ram is retrieved. Finally, the lower punch pushes the preform up out of the cavity during step 4, also known as *de-molding*. These operations take place automatically according to preset conditions.

PTFE powders get increasingly softer and stickier above their transition temperature and tend to form aggregates. This can cause bridging in the feed section of the mold, which leads to uneven filling of the cavity or non-uniformity in each charge. The molding area should be maintained at 23–25°C.

A higher pressure than ordinary compression molding is required for automatic molding because of the short duration of the compression cycle. The effect of pressure on the specific gravity of the preform has been shown in Fig. 7.35 which shows that the specific gravity increases with increased pressure independent of the dwell time. The specific gravity begins to level off as a function of pressure above 400 kg_f/cm^2 (40 MPa). The combinations of pressure and dwell period have to be optimized to arrive at the highest productivity of parts with acceptable properties. Figure 7.36–7.38 show the impact of dwell time and molding pressure on the shrinkage and tensile properties of parts. The optimal molding pressure and dwell period are 400 kg_f/cm^2 and 10 seconds to obtain the maximum tensile strength and elongation at break for the polymer in this example. Both properties decline at above 400 kg_f/cm^2 molding pressure.

7.5 Isostatic Molding

This technology was originally invented for ceramic and powder metal processing early in the twentieth century. It has been adopted to produce parts from granular PTFE powders. Isostatic molding (sometimes called *hydrostatic molding*) is another technique for producing PTFE preforms by the application of hydrostatic pressure to the powder. The powder is loaded in a closed flexible mold. Compaction of the powder into a preform takes place by pressure applied through the flexible part (bladder) of the mold. The bladder is usually made of an elastomeric material such as polyurethane. This method allows molding of complex shapes by the placement of mandrels inside the flexible bladder.

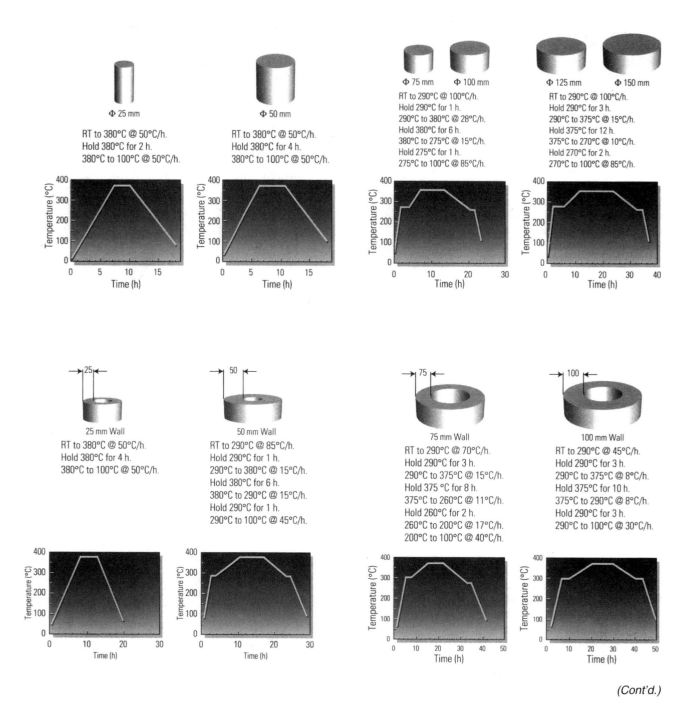

Figure 7.32 Examples of sintering cycles for rods and billets of various shapes and sizes. *(Courtesy DuPont.)*

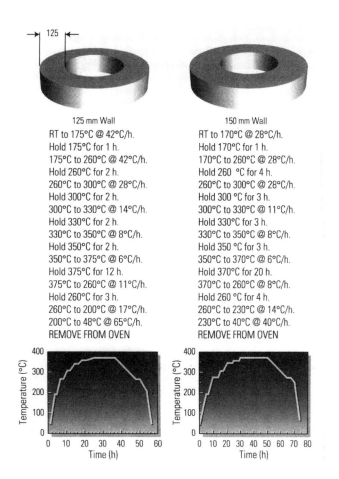

Figure 7.32 *(Cont'd.)*

Table 7.6. Effect of Compression Molding Process on Properties *(Courtesy DuPont.)*

Process Type	Deformation Under load, % (ASTM D621)	Notched Izod, ft.lb/in (ASTM D256)	Elongation at Break, % (ASTM D638)	Tensile Strength, psi (ASTM D638)
Hot Compression Molding without $C_{25}F_{52}$	0.98	3.1	349	11,200
Hot Compression Molding with $C_{25}F_{52}$	0.87	8.3	169	2810
Standard Compression Molding	2.35	3	480	13,800

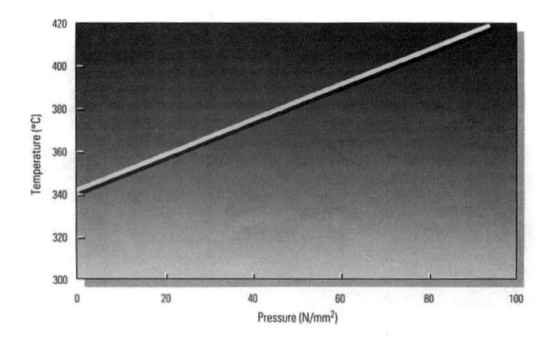

Figure 7.33 Dependence of the melting point of Teflon on pressure. *(Courtesy DuPont.)*

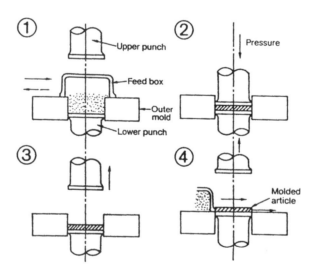

Figure 7.34 Automatic compression molding diagram.[3] *(Courtesy DuPont.)*

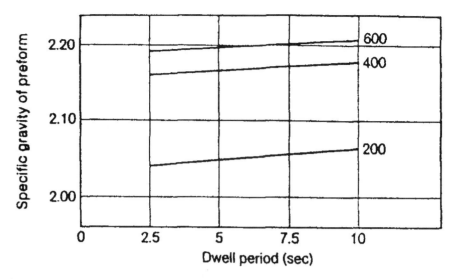

Note: The numbers in the graph indicate the molding pressure (kgf/cm^2).
Dimensions of the molded article: 64 (O.D.) × 52 (I.D.) × 15 (Length) mm

Figure 7.35 Relationship of molding pressure/dwell period and specific gravity of preform.[3]

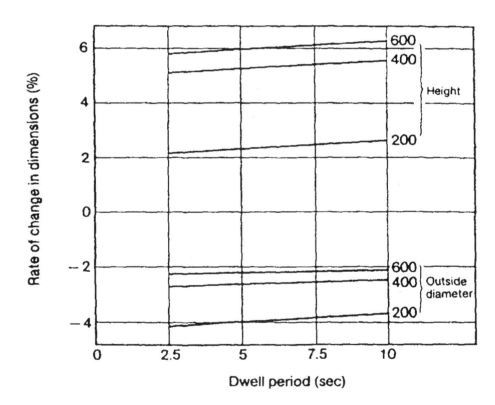

Note: The numbers in the graph indicate the molding pressure (kgf/cm^2).
Dimensions of the molded article:
64 (O.D.) × 52 (I.D.) × 15 (H) mm
Sintering conditions: Heating rate 100°C/hr, 370°C (698°F)
maintained for 3 hr
Cooling rate 100°C/h

Figure 7.36 Relationship of molding pressure/dwell period and rate of change in dimensions.[3]

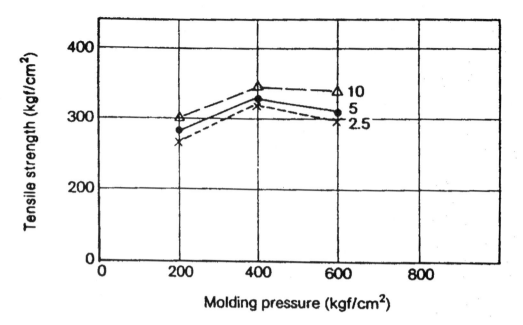

Figure 7.37 Relationship of molding pressure/dwell period and tensile strength.[3]

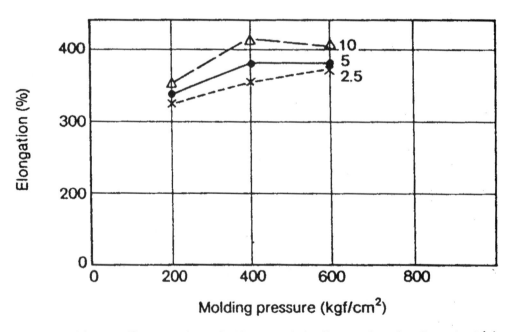

Figure 7.38 Relationship of molding pressure/dwell period and elongation.[3]

7.5.1 Introduction to Isostatic Molding

Isostatic molding is a suitable alternative to compression and automatic molding techniques for the production of PTFE parts with complex shapes in a wide range of sizes. Compression molding can supply a stock shape which can be machined to obtain the desired shape. The drawback to this option is the extensive machining and material cost which can drive up the cost of the object. Isostatic molding requires relatively low-cost tooling and allows significant savings in machining and material costs. Complicated parts in the exact or nearly exact shape and size, requiring some finishing, can be molded and sintered by this method. A bellows is an example of a part which can be directly molded by isostatic molding, while extensive machining is required to achieve the curved contour of the bellows. Isostatic molding is the method by which all shapes of preforms (Fig. 7.39) can be made.

Figure 7.39 Isostatically pressed samples of various materials by both wet- and dry-bag methods. *(Courtesty Halogen Joint-Stock Co.)*

Figure 7.40 shows the principal steps for isostatically molding a simple solid cylinder. The mold cavity is formed inside an elastomeric membrane shaped like a hollow cylinder. In this case, it does not include any mandrels and is completely filled with the powder. The elastomeric bag is closed, sealed and placed inside a pressure vessel. The vessel containing a fluid is sealed, pressurized and held for a dwell period during which the powder is compacted by the action of a pressurized fluid. At the end of the dwell time, the vessel is depressurized and the mold is removed and disassembled for the removal of the preform.

The flexible nature of the bag renders the definition of its volume difficult when compared to a metallic mold. The shape of the bag may also change during the filling step unless it is supported while being charged. The change in the shape of the bag depends on several factors.

- The original mold shape.
- Fill uniformity.
- The geometry and wall-thickness of the flexible segments.
- Elastic properties of the bag.
- The fastening of the rigid and elastic sections of the mold.
- The extent and the rate of powder compaction.
- The residual stress in the bag.

Exertion of pressure on the bag is multidirectional which conforms the resin powder to all patterns and nonuniformities in the bag. Consequently, the surfaces adjacent to the bag are less smooth than those adjacent to the smooth metallic surfaces. The importance of surface formation is one of the considerations that determines the selection of the type of molding process.

Isostatic molding is ideal for manufacturing thin long objects from small tubes (5 mm diameter) or very large diameter thin wall tubes (30 cm diameter). Examples include pipe liners, liners for valves and fittings, flanged parts, closed end articles, and a host of other shapes which would require extensive machining. Table 7.7 offers examples of isostatically-molded parts and their applications.

7.5.2 Comparison of Isostatic with Other Fabrication Techniques

Very little pressure decay occurs during isostatic compaction because of the absence of die wall friction. The absence of pressure decay permits making a great range of shapes, complexity and sizes. This process yields a stress-free homogenous preform that exhibits uniform shrinkage during sintering. It results in a component free of distortion with uniform physical properties. The even and constant application of pressure throughout the cycle time results in a preform with lower void content. This is why shrinkage

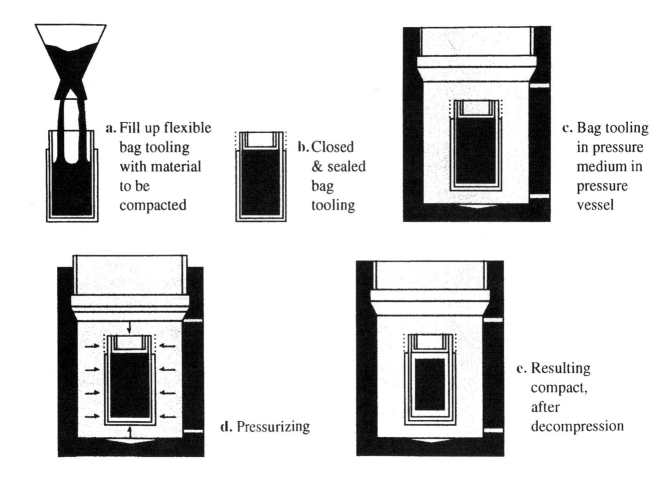

Figure 7.40 Principle steps of isostatic molding.[7]

is lower, and specific gravity and physical strength are higher than the other technique based on uniaxial compaction. The high extent of compaction allows the use of lower pressures in isostatic molding.

A drawback of isostatic molding is that the bag cannot produce sharp and perpendicular corners. Furthermore, the bag follows the contours of the resin particles and produces a less smooth surface than compression or automatic molding methods. Some machining will be required to achieve sharp corners and a smooth surface. The lower cost of isostatic molds, labor and material render the finishing machining affordable without upsetting the economics of this process. Simple parts such as straight tubes can be made with tight tolerances, better than ±2%.[8]

Isostatic cycle times as short as 12 seconds are possible for small simple parts. The length of the cycle increases with the complexity and size of the article. In such instances, this technique is often the only available method for the fabrication of those parts. An example is in-situ formation of a PTFE liner inside a fitting such as a T-piece.

Table 7.7. Example of Isostatically Molded Parts

ISOSTATICALLY MOLDED PART	APPLICATIONS
Thin Wall Objects	Pipe liners: 2.5-30 cm diameter, 2.5-50mm wall thickness up to 6 m length
	Pump, valve and fitting liner
Flanged Articles	T pieces, spacers, elbows
Closed End Articles	Cups, nose cones, radomes, bottle and caps, test tubes, crucibles and laboratory ware
Encapsulation	Magnetic stirrers, butterfly valve flaps and thermocouples
Textured Objects	Beehive insulators, embossed articles, valve gate covers vessel covers and threaded parts
Solid Parts	Pyramids, balls and stopcocks,
Parts with Embedments	Reinforced mesh or sheet, bolt head or stud, conductor

7.5.3 Basic Isostatic Process

Solid parts such as rods and balls can be produced by the basic process described in the previous section as shown in Fig. 7.41. The only part required is a flexible bag with end plug that can be clamped. The simplest form of isostatic molding process entails filling a flexible mold, i.e., the bag, with granular PTFE and inserting the filled bag in a pressure vessel containing a neutral fluid, closing the vessel and pressurizing the fluid. The pressure in the vessel is held for a period of time followed by decompression, removal of the mold from the vessel and de-molding of the preform. The preform will assume the shape of the mold and its size will depend on the extent of compaction of the PTFE powder. The part is sintered similarly to the preforms made by the other molding techniques.

7.5.4 Complex Isostatic Molding

To make more complex hollow or closed end parts such as tubes, liners, test tubes and beakers, multipart molds are required. Figure 7.42 depicts three methods for producing a tube shape PTFE preform.

1. Resin is loaded between two flexible bags, which after exposure to pressure compact from the outward and inward radial directions.

2. The outside bag is replaced with a rigid cylinder and the resin is loaded between them. The direction of compaction is outward radially and the resin is squeezed against the surface of the outer rigid cylinder.

3. The inner bag is replaced with a rigid cylinder and the resin is loaded between them. The direction of compaction is inward radially and the resin is squeezed against the surface of the inner rigid cylinder.

A mold for the production of a beaker is shown in Fig. 7.43.[8] In this case the rigid part is inside the bag. This rigid part, which is usually made of a polished metal, aids in the formation of a shape, and is often referred to as *mandrel*. The adjacent surface to mandrel assumes the surface formation of the mandrel and, in the case of a polished metal, can be quite smooth. Pipe liners are normally manufactured with an outside flexible bag and inside polished metal mandrel to produce a smooth surface for exposure to processing fluids. A smooth surface minimizes material trapping and build up in the pores of the rough areas. The rigid mandrel prevents true isostatic compaction because of a friction effect, but this phenomenon is relatively small in the majority of the molds.

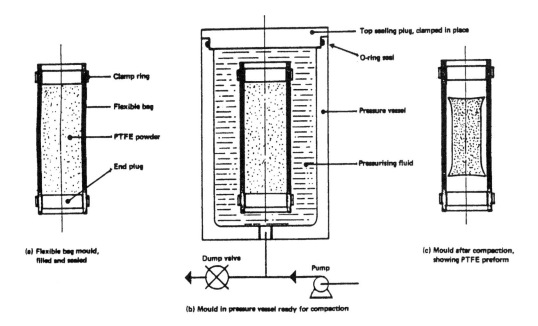

Figure 7.41 Basic isostatic compaction process.[8]

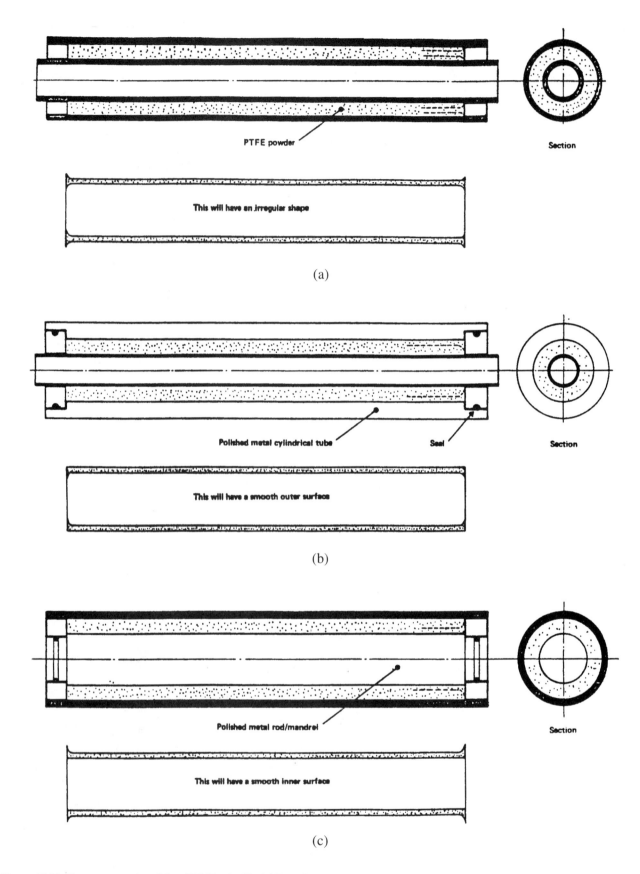

Figure 7.42 Three ways of molding PTFE tube.[8] *(a)* Two flexible bags. *(b)* Inner flexible bag, outer rigid cylinder. *(c)* Outer flexible bag, inner rigid rod.

7.5.5 Wet and Dry Bag Isostatic Molding

Wet- and dry-bag are two techniques for isostatic molding which are principally identical but operationally different. The *wet bag* process is similar to the basic molding procedure (Sec. 7.3.7.3) in which the mold is submerged in the pressurization fluid. In the dry bag technique, the mold and the bag are fixed in place and the functions of the mold and the pressurization vessel are combined. The pressurization fluid is introduced through a high-pressure liquid supply system behind the flexible bag (Fig. 7.43). The mold assembly is designed to withstand this pressure. The term *dry bag* contrasts the absence of submerged mold and wet mold assembly.

The dry bag process has advantages over the wet bag process. The operation of placing the mold in the pressure vessel has been eliminated. Sealing and unsealing of the mold and the pressure vessel have been reduced to just sealing and unsealing the mold. In addition to cycle time reduction, the risk of contamination of the preform with the pressurization fluid has been eliminated. The dry bag process can be automated and is an excellent method for the large-scale production of parts. The disadvantage of this process is the large cost differential between the dry over the wet bag molds. Dry bag molds must be able to withstand high pressure and in effect act as a pressure vessel. These molds must not be modified without reviewing the mold design because of safety considerations. It is important to take into account the need for mold modification in its initial design.

7.5.6 Isostatic Mold Design

A number of factors are fundamental to isostatic mold design disregarding the process type. Some of these factors include:

    Preform shape

    PTFE powder type

    Surface finish of the part

    Uniformity of compaction

    Ease of the process

    Pressure direction

It is important to satisfy all or as many of these considerations as possible to assure manufacture of good quality parts. The internal mold dimensions can be calculated from the finished size of the part, shrinkage and compaction ratio of the resin. The rest of the design depends on the strength required to withstand wet or dry bag pressure. Three factors have the most important effect on the configuration and surface finish of the mold:

1. Preform shape.
2. Surface finish of the part.
3. Pressurization direction.

7.5.6.1 Preform Shape

The shape of the preform is determined by the bag, the rigid part of the mold, if any, and the compaction pressure. The rigid part impacts the shape by blocking the progression of compaction and imparts smoothness to the adjacent surface. The flexible bag assumes a shape affected by a number of factors: the configuration of the mold, bag properties such as membrane thickness, residual stress and modulus of elasticity of the elastomer, surface finish of the metallic parts, PTFE characteristics, and the rate of pressurization and maximum pressure. Mold filling and variables such as interaction of the rigid and flexible parts and the stability of the pressurization also have some influence. Preform shapes can be fabricated reproducibly by controlling these variables. Robustness of the process must be assured which means that a small change in a variable such as powder density or pressure should be dampened and diminished.

Some of the variables are difficult to control, thus affecting the shape of the preform. A good example is the bag; it ages with use which changes its modulus of elasticity, therefore, compaction. Adjustments to the pressure can correct the elasticity changes of the bag up to a point. Resin flow, particle size, and mold filling are other variables which must be controlled for manufacture of reproducible preform shapes

It is not practical to attempt to produce preform shapes with a high degree of dimensional accuracy due to the flexible nature of the bag, but close approximations are feasible to manufacture, particularly in the case of simple parts. Excess PTFE can be charged to the mold to account for deviations from the desired shape. The excess material can later be removed from the part. Another problem is achieving sharp corners using a flexible bag. One possible, though difficult, way to obtain sharp edges and corners is the incorporation of ridges in the bag similar to Fig. 7.44.

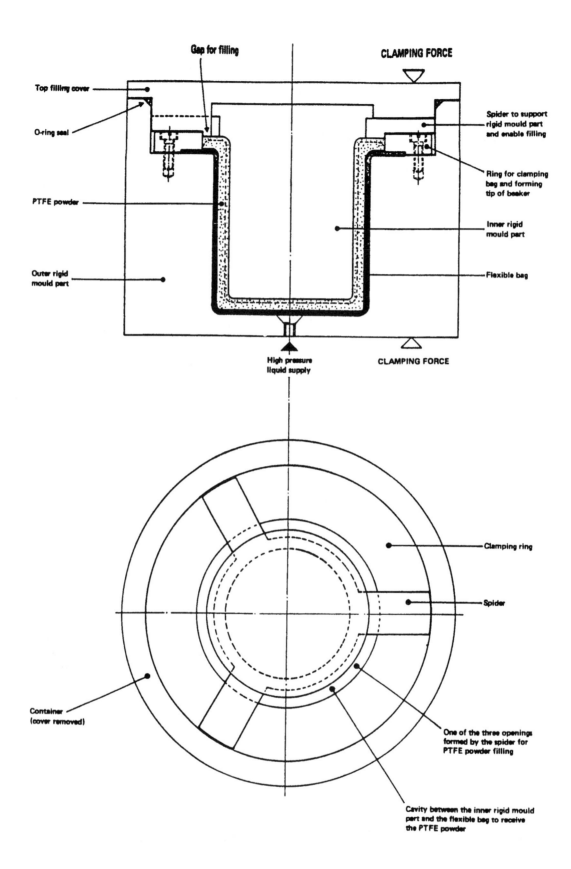

Figure 7.43 Mold for the production of a beaker using an outer flexible bag and inner rigid part. *(Courtesy ICI.)*[8]

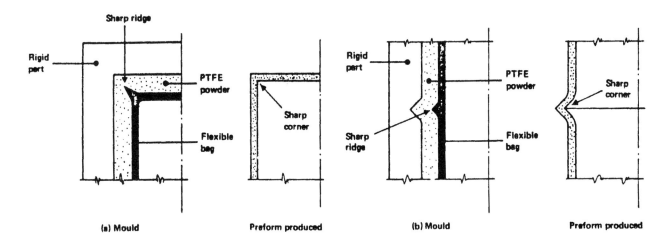

Figure 7.44 The use of bag ridges to produce sharp corners.[8]

Another important factor affecting the preform shape is stability of the compaction process. The powder exhibits two types of behavior as the pressure increases during the compaction process. At low pressures, the powder can move inside the flexible bag with little resistance. At higher pressures, after the polymer has undergone deformation, the resistance to pressure sharply increases. Both the flexible bag and the powder present resistance against pressure. The lower the resistance of the bag, the stronger the role the resin plays in the formation of the final shape. In general, when the pressure direction is from inside to outside (Fig. 7.42b), the compaction is stable. The powder in an outward compaction has little chance to bridge, that is to form agglomerates which do not break up and spread. Conversely, the inward compaction is unstable because pressurization brings the powder particles together and bridging is possible. After the bag has moved, the powder begins to develop resistance and stabilize the process. The approach to lessen the instability of the process involves reduction of bag movement which can be achieved by using powders with higher bulk density or reducing the thickness of the PTFE part.

Direction of pressure is particularly important to thick parts and little impact is experienced on thin parts. In terms of stability using two flexible bags, Fig. 7.42a represents the worst case for molding a hollow part such as a pipe liner. Little control can be expected because of the strong dependence of the process on the characteristics of the bags. The preferred configuration for an isostatic mold is a flexible bag and a rigid segment. Outward pressurization from the bag side to the rigid side is recommended.

Resin properties, mold filling, pressurization cycle, bag characteristics, ambient temperature, and fluid viscosity and temperature are the process key variables. Free flow granular PTFE is the resin of choice for this type of molding. Flow, particle size distribution, particle shape and particle surface finish can each affect the isostatic molding process to varying degrees. Any changes in these properties can affect the mold filling and powder particle packing inside the mold. One solution to overcome any variability is to vibrate the mold assembly during the loading stage, which will force the particles to move and fill. Vibration is especially important to molding complex shapes since it assures all the small cavities producing the intricate details of the part are filled.

A significant factor in reproducing preform shape is the shape of the bag at the time of powder loading. The shape of the bag can vary each time the preform is removed, especially if it is unsupported. This variability increases with increasing thickness of the flexible membrane. An empty bag forms buckles and ridges when pressurization is from outside to inside. It is highly recommended to support the bag without regard to the mold configuration. For example, a cage can be installed when the direction of the pressure is from outside to inside. Frequently vacuum is applied to bring the bag to its initial shape because of changes due to aging in the elastic properties of the elastomer membrane from which the bag is made. This procedure will assure uniform mold cavity shape and size and therefore uniform mold filling.

7.5.6.2 Surface Finish of Preform

The surface that PTFE forms during isostatic molding depends on the type of surface against which it is formed. The mandrels are usually made of polished metal and yield a smooth preform surface. The more polished the mandrel, the smoother the surface of the preform. A flexible bag forms a surface that is normally not smooth because the bag tends to conform to the individual particles of resin. A softer bag tends to form less smooth surfaces than a harder bag. Surface finish is important in a number of applications. For instance, it is desirable to have liners with smooth surfaces for chemical processing service to minimize buildup on the surface. These liners are usually made by inside mandrel isostatic molding where the interior surface is formed against a polished metal mandrel.

7.5.6.3 Pressurization Direction

There is little choice but to pressurize from outside to inside when solid parts are being molded. There are two alternatives for pressurization in molding hollow objects: inside-to-outside and outside-to-inside. From a stability standpoint, the former is the preferred route but other considerations may dictate the latter choice. For example, pipe liners are often long (6 meters) and have small diameters (2.5–30 cm) which makes machining the inside surface impractical. The outside surface can be machined easily if necessary. In this case, outside-to-inside pressurization is selected. In the case of a fitting such as a T piece, machining internally is more convenient and inside-to-outside pressure direction should be selected. If no machining is required, then the best molding technique can be selected.

Other processing factors such as preform removal must be considered in the design of the mold configuration. While the bag is flexible and can often be removed easily, the rigid parts must be designed in segments so that they can be disassembled for preform removal. The external rigid part can consist of two parts but an internal component may have to be broken into several sections. Another consideration is the ease of mold filling, particularly for complex shapes.

Another limiting factor in selecting the direction is the minimum internal diameter which can be obtained by pressurization from inside to outside. The parameters which determine this minimum value include: *(a)* maximum allowable elastic tensile strain of the bag, *(b)* compaction ratio of the resin, and *(c)* the desired thickness of the PTFE preform.

The minimum internal diameter can be calculated[8] from the following equation:

Eq. (5.5) $$d_m = D\left[\frac{(1+\varepsilon_m)^2(C-1)}{(1+\varepsilon_m)^2 C - 1}\right]^{0.5} = k_p D$$

d_m = Minimum internal diameter, cm
D = External diameter of tube, cm
ε_m = Maximum elastic strain of the bag, cm/cm
C = Compaction ratio of resin, normally provided by the PTFE supplier
k_p = Limiting coefficient

The ideal bag will have infinite elastic strain, which may intuitively suggest that limitations to minimum internal diameter can be eliminated. Equation 7.5 can be rewritten (7.6) if one assumes ε_m to be infinity:

Eq. (7.6) $$d_m = D\left[\frac{(C-1)}{C}\right]^{0.5} = k_i D$$

k_i = Limiting coefficient for an ideal bag with infinite elastic strain

Equation 7.6 indicates that a minimum internal diameter exists, even for an ideal bag. The following example illustrates this point.

Assume that a tube with an external diameter of 100 mm is to be molded in a mold with an ideal bag, using a PTFE powder with a compaction ratio of 2.2. Equation 7.6 can be used to calculate the minimum diameter:

$$d_m = 100\left[\frac{(2.2-1)}{2.2}\right]^{0.5} = 73.8 \text{ mm}$$

Wall Thickness = 100-73.8 = 26.2 mm

If we assume that the bag is made of a typical elastomer, say with a 500% maximum strain, then Eq. 7.5 can be used to calculate the minimum internal diameter

$$d_m = 100\left[\frac{(1+5)^2(2.2-1)}{(1+5)^2 2.2 - 1}\right]^{0.5} = 74.3 \text{ mm}$$

Wall Thickness = 100-74.3 = 25.7 mm

Surprisingly, little change in the minimum internal diameter is experienced by the change in the maximum strain of the bag from infinity to 500%. For a bag with relatively poor elastic strain, say 100%, the minimum diameter is 78.4 mm (wall thickness = 21.5 mm). This example illustrates that increasing the maximum allowable strain of the bag through material choice or bag enlargement has little impact on increasing the wall thickness. Deterioration of bag material, however, has a pronounced effect on reducing the wall thickness, as illustrated by the last case where maximum elastic strain was assumed to be 100%.

The maximum diameter which can be achieved when pressing from outside to inside can be calculated[8] from Eq. 7.7. The size of the mold and the wall thickness of the preform are the factors limiting the external diameter.

Eq. (7.7) $$D_m = \left\{\frac{1}{C}\left[D_i^2 + (C-1)d^2\right]\right\}^{0.5}$$

D_m = Maximum preform external diameter, cm
D_i = Maximum internal bag diameter, cm
d = Internal diameter of the preform, cm

7.5.6.4 Other Design Factors

A number of other factors impact the design of the mold including bag thickness, bag attachment to the rigid part, damage to the bag, protection of preform and compaction of PTFE.

The recommended bag thickness is 4% of its major dimension, diameter for a tube. A thicker bag will have increased longevity and may not require support. It will require more pressure to deform and will have poorer definition of part geometry. The likelihood of damage of the preform by the bag upon pressure removal and recovery increases with increasing thickness. One advantage of a thick bag is that it is less susceptible to buckling when pressurized from outside to inside.

The flexible bag has to be connected to the rigid part of the mold to seal the powder against the pressurization fluid, in both dry and wet bag processes. It is difficult and often impossible to achieve the desired shape at the point of attachment where transition from flexible to rigid matter takes place. The best choice is to strive to make this attachment at a point where the part is not critical or can be machined.

An example of the difference that the type of connection between the bag and the solid part of the mold make can be seen in Fig. 7.45. Thinning of the preform wall can be avoided by designing the bag to have a large radius at the corner as opposed to a tight radius.

The bag is exposed to repetitive motion in the mold leading to abrasion and tearing if care is not taken in mold design. The metal and other rigid surfaces of the mold should be finished as finely as possible, preferably better than 0.8 µm CLA.[8] All corners should be rounded to >1 mm to avoid snagging the bag. Deep and small holes and grooves lead to overstretching and possible failure of the bag. Diameter of holes should be more than twice the preform thickness while the depth of a hole should be less than twice the preform thickness.

The preform can crack because of excessive shearing of the powder during the compaction. For example, PTFE particles can be rubbed against sharp edges and corners or due to excessive movement against the metal surface under pressure. Mold configuration should be designed to avoid both outcomes. The snap-back of the bag during its recovery while depressurization takes place can damage the preform. Any deformation of inserts in the preform will reverse upon depressurization and cause cracking of the preform.

Presence of rigid parts alters the properties of an isostatically-molded part. The mold should be designed to minimize introduction of solid parts and avoid deep grooves and holes.

7.5.6.5 Mold Configuration and Dimensions

The shape of the mold, the resin space, the design of the flexible bag attachment to the rigid part, the shape of the bag, etc., all depend on the design factors listed in Sec. 7.5.6. End use part requirements and isostatic processing considerations are the final determinants of the mold configuration. Weighted factors can be assigned to the design factors after the end use requirements of the part and the processing constraints have been considered and a compromise is reached. For example, the end use may require a deep groove in the part outside the criterion discussed in Sec. 7.5.6.4. One example of a compromise solution would be to select the most durable elastomer for the bag material and reduce bag thickness, which would allow the bag to operate at the limit of Sec. 7.5.6.4 criterion. The rest of the depth of the groove would have to be achieved by machining.

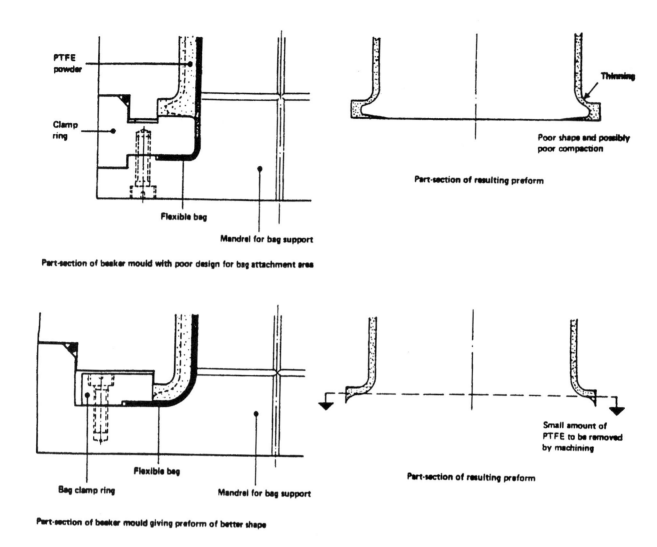

Figure 7.45 The choice of area of bag attachment to give good preform shape.[8]

Bag and rigid part dimensions can be classified into two types, those which have a direct bearing on the preform size, and a secondary group which has little or no impact. The first group includes the dimensions of the flexible bag and the rigid parts.

The size and dimensions of the bag depend on the individual dimensions of the part that is being fabricated. Calculation of the bag size is approximate and is only relevant to the initial shape of the bag. It is virtually impossible to predict all the changes that the shape of the bag may experience during the operation of the mold. This means that past experience should be considered in specifying the size of the bag, in addition to considerations discussed in previous sections, such as accuracy of the preform shape, damage to the bag, bag fastening to the rigid sections, and minimization of resin movement during pressurization.

For a thin walled part, it is possible to design the bag according to the geometry of the final part because one can assume little powder flow takes place during the compaction. It would be reasonable to assume that all the compaction takes place in the direction of pressure which is perpendicular to the hard surface of the mold. In a thick wall PTFE part, particularly one with complex geometry, trial and error is the best strategy because no other satisfactory procedure exists for predicting the shape of the bag necessary to produce an acceptable part. Compaction in the case of a thick wall part can take place in all directions if the movement has not been restrained.

Decisions about the size of the bag are more straightforward and can be estimated from the compaction ratio and shrinkage of the PTFE powder. Shrinkage and compaction ratio of the resin will have to be determined experimentally. Shrinkage depends on the type of powder, preforming pressure, and the sintering cycle. Preform pressure should be varied and the shrinkage and compaction measured. Uniform shrinkage may be assumed unless thick and thin parts are present together. Any restrictions in the compaction of a resin will lead to non-uniformity of shrinkage. The size of the resin space can thus be determined after the above data are obtained.

The shape and dimensions of the rigid parts are determined by the shape and dimensions of the desired PTFE component. Powder shrinkage during sintering and post-sintering machining must be taken into account in the design. In a perfectly isostatic operation, resin shrinkage is identical in all directions. This would mean the cubic root of the volumetric shrinkage could be assumed to be the "linear" shrinkage of the resin. Clearly this is not the case in the majority of actual cases, therefore requiring design corrections. The amount of correction is small if the mold has been designed well, that is, there are no restraints in the resin compaction.

Shrinkage is a strong function of preforming pressure in isostatic and other types of molding. Generally, shrinkage decreases when preforming pressure is increased. For example, in compression molding, shrinkage is much smaller in the pressurization direction than it is in the perpendicular direction. In isostatic molding, the mold design must include rigid parts which can prevent uniform compaction of the resin in all directions. For example, in the manufacture of a pipeliner, the powder is compacted radially and not axially. Shrinkage is independent of the pressure direction which could be from the inside-to-outside or outside-to-inside. Shrinkage also depends on other variables such as pressurization rate, dwell time and sintering cycle–particularly the cooling rate. The easiest approach to design is to measure shrinkage of simple shapes such as a solid rod under the same molding and sintering conditions. This approach will provide data which are closer to what the resin, and preform will actually experience in the desired (complex) shape. The benefit is to reduce modifications of the mold.

Modification of the mold should be anticipated in the design according to pressure direction. The rigid segment should be undersized for the inside-to-outside direction and oversized for the outside-to-inside pressure direction. Adjustments can be made by machining the rigid part of the mold after testing and making parts. In some cases, a new bag may be needed in the change of design if extensive machining would cause the bag to be overextended.

There are a number of dimensions of the bag and the mold, such as wall thickness, that do not affect the preform shape and size and have more to do with the properties of the construction materials. The designer should take into account the mechanical strength and durability required of the mold parts when specifying these dimensions. Other factors to consider include the weight and size of the mold.

In a wet bag operation, there is no net force on any of the parts of the mold because of submergence of the whole mold under the high-pressure fluid. In this case, the wall thickness of various parts needs to be only sufficient to seal the powder against the high-pressure fluid. This would not be valid if, for reasons such as loose fitness or motion during the pressurization, parts of the mold experience a net force. The mold design should take into account the forces exerted on the parts in determining dimensions such as wall thickness.

The dry bag mold is contained in a pressure vessel because it has to contain the high-pressure fluid. The issue of safety is especially important because of normal isostatic working pressures of up to 40 MPa. Dry bag mold requires stress analysis similar to any pressure vessel to assure that factors like pressure cycling, corrosion and poor welded joints do not contribute to its fatigue. The end plugs should be designed to accommodate the need for speedy closure of the mold and also withstand the force, which will be transmitted in the axial direction during the compaction. The use of C-clamps is one simple and quick means of fastening the end plugs to the mold. A new mold should be tested above the working pressure (up to 2x) prior to being placed in regular service. Any modification of the rigid part such as machining should be examined to insure structural integrity. Molds which have undergone large numbers of cycles (500–1,000 cycles) should be tested for early signs of failure.

7.5.6.6 Mold and Bag Sealing

In wet bag isostatic operation, the powder and pressurization fluid must be kept separate to avoid contamination of the powder or failure. The most effective location for the seal is around the outer edges of the bag. Additionally the plugs, covers and joints that allow the fluid to circulate inside the rigid part of the mold and behind the bag must be completely sealed to avoid fluid leakage. In this method, the bag and cover seals should not be combined because the stress to which the bag is exposed during repeated opening and closing of the mold can lead to its early failure. The other drawback is the change in shape of the bag prior to each filling. Two examples can be seen in the Fig. 7.46 and 7.47.

In dry bag molding, the same principles apply in addition to the need to seal the fluid from the ambient. The fluid-ambient and fluid-resin seals are often combined in the dry bag method. Typically, bags are fastened to the bottom part of the rigid section of the mold when pressurization direction is from inside to outside and the top part of the mold when pressurization is from outside to inside. These arrangements simplify the removal of the preform from the mold.

Pinching and dragging of the bag against sharp corners and edges should be avoided; otherwise the bag will deteriorate and have a short life. Restrictions can also lead to defective preforms because of the hindered movement of the bag and incomplete compaction of the powder.

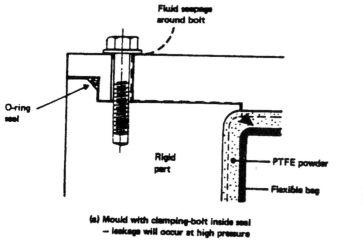
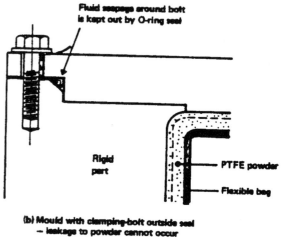

Figure 7.46 The effect of clamping-bolt positions on fluid sealing.[8]

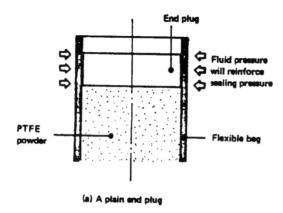

Figure 7.47 The use of fluid pressure to improve end-plug sealing.[8]

7.5.6.7 Tooling and Material: Rigid Part

Metals are the most common materials of construction for the rigid parts of the mold. Sometimes rigid plastics and cross-linked polymers such as epoxies can be used in wet bag molding, usually at low pressures for small parts. The important properties of the rigid part are strength to withstand process pressure, resilience to pressure and temperature cycling, and corrosion resistance. The material requirements of the dry bag process are much more stringent than the wet bag process. Metal selection should be made according to the guidelines used for pressure vessels.[9]

Design of the pressure vessel should be done while considering numerous factors including stress buildup in the body of the vessel and its closure, stress buildup at the vessel and closure interface, corrosion effect on metal strength, and pressure and temperature cycling effect on the life of the vessel at the maximum working pressure. Structural integrity of high-pressure vessels must be tested by techniques such as ultrasonic method, strain gauge test and over-pressurization prior to placing them in service. Design, fabrication, and inspection of pressure vessels have been documented by American Society of Mechanical Engineers in A.S.M.E. Boiler and Pressure Vessel Code. *Chemical Engineers' Handbook* (Perry & Green, sixth ed.) by McGraw Hill provides a summary coverage of pressure vessel design.

7.5.6.8 Tooling and Material: Flexible Bag

The bag is made from elastomers including natural rubber, polyurethane, RTV silicones, neoprene, nitrile rubber, butyl rubber and vinyl plastisols. Important properties of the bag are resilience, uniform elastic properties, physical strength, abrasion resistance, tear resistance and hardness above 45 Shore A. Fillers can be incorporated to increase the hardness of the elastomer. The elastomer chosen should be compatible with the pressurization fluid. Water containing a small amount of a corrosion inhibitor is the most common fluid selection, though mineral oils are occasionally used. Table 7.8 gives some of the properties of bag material.

Bag materials are chosen based on the complexity of the final part, physical properties of elastomer, cost, quality and the desired bag life. Well-designed thin bags made from a highly flexible elastomer can make satisfactory PTFE preforms because easy deformation of the bag under pressure tends to minimize the effect of the initial shape of the bag. The drawback is the short life of thin wall bags. A thick wall bag will be more durable but its structural strength can be so large that its deformation could be independent of the powder in the bag. A compromise has to be struck in deciding the bag thickness to achieve good quality preforms and durability, as illustrated in the following examples.[10]

Table 7.8. Typical Properties of Various Bag Materials[8]

Property	Natural rubber (molded)	Natural rubber (latex)	Nitrile	Neoprene	Butyl	Polyurethane	Silicone (RTV)	PVC
Tensile strength (kgf/cm^2)	210	210	105	140	140	280	70	140–210
(MN/m^2)	20.5	20.5	10.5	13.5	13.5	27.5	7	13.5–20.5
(lbf/in^2)	3000	3000	1500	2000	2000	4000	1000	2000–3000
Hardness range, Shore A	30–90	40	40–95	40–95	40–75	75–90	40–85	–
Tear resistance	Very good	Very good	Fair	Good	Good	Excellent	Poor	Fair to good
Abrasion resistance	Excellent	Good	Good	Good	Good	Excellent	Poor	Fair
Resilience	Excellent	Excellent	Fair	Good	Bad	Good	Excellent	Bad
Compression set	Good	Fair	Good	Good	Fair	Poor	Fair	Poor

For example, let us assume a long large diameter solid rod is to be made from PTFE by isostatic molding using a thick wall round polyurethane bag, which is fairly hard. The preform from this mold will probably have fairly round "dished-in" ends and oval cross sections throughout its length. The minor diameter of the oval will be smallest in the middle of the rod. This is the typical deformation observed when external force is applied to a closed end tube made of semi-rigid material, if the deformation is high enough to cause buckling in the bag. Low compression ratio materials such as metals or ceramic powders can be molded successfully with a heavy wall bag because compaction reaches its end state before deformation of the bag reaches buckling stage. Compression ratio of PTFE powders is too high for this to take place unless they are pre-compacted.

Suppose the same large diameter solid PTFE rod is to be made in an unstretched thin bag made from a very flexible elastomer. This time the preform will contain several ridges, which are created by localized folding and buckling of the bag. The high flexibility of the bag leads to multiple point buckling in contrast to the thick wall bag where localized buckling is prevented by its strength and only gross buckling of the whole bag takes place. In general, long thin wall bags produce more buckling ridges than short heavy wall bags.

It is important to prevent the bag from buckling in isostatic molding when compacting in the outside-in direction. The compressive stress of the bag should be kept below the critical level at which buckling occurs when the direction of pressure is from outside to inside. Bag strains are quite large when making large PTFE parts, which generates bag compressive stresses beyond the critical buckling value. A common way to get around this problem is to begin compressing while the bag is actually under tension. Selecting a slightly smaller bag and stretching it against a support shell accomplishes this.

7.5.6.9 Design Procedure

The design of an isostatic mold is complex and does not render itself to a purely logical method. The parameters in the design of an isostatic mold interact in ways which cannot be easily expressed in the form of an equation. Table 7.9 provides a list of considerations and corresponding factors for working through the design aspects of the mold.[8]

7.5.7 PTFE Resin Selection

Any type of powder can be used in isostatic molding, but processing difficulties can be expected with the high compaction ratio and low flow powders. Sometimes the finished part requirements override the powder selection criteria, for example surface finish and mechanical strength. Granular powders with excellent flow and relatively high apparent density (>500 g/l) work best in isostatic molding. Parts can be molded from both unfilled and filled powders.

7.5.8 Isostatic Processing of PTFE Resins

To produce isostatically-molded parts, several pieces of equipment are normally required. Mold, mandrels, elastic bags and mold accessories, vibrator, vacuum service and the sintering oven are needed. Pressurization fluid can be selected from a number of candidates but water along with a small quantity of soluble oil corrosion inhibitor is most common. Gases should never be used because of the hazards associated with high-pressure compressible fluids. Pneumatic or electric pumps are the usual means of pressurizing the fluid.

Isostatic processing is comprised of the following steps:

- Mold assembly
- Mold Filling and Closure
- PTFE Compaction
- Pressure Letdown and Mold Disassembly
- Preform Removal
- Preform Degassing

Sintering is the last step after the preform has been molded and prepared.

Before assembling the mold, all surfaces should be cleaned to prevent contamination of the resin and pressurization fluid in the case of wet bag molding. After assembly the powder is charged to the mold. Good complete filling of the mold cavity is critical to the bag movement and production of high quality preforms. The first requirement is a resin with consistent properties: apparent density, particle size distribution, particle hardness, flow and purity. The ideal procedure is to charge a fixed amount of resin to the mold to a predetermined level, accompanied by vibration. Vibration helps increase the effective bulk density of the resin; thus reducing its compaction ratio. Bag support and pulling vacuum on the liquid side of the bag during filling will help fix the initial

Table 7.9. Isostatic Mold Design Factors and Considerations[8]

	Considerations	Some of the Factors Involved
1	Details of PTFE article to be produced	Dimensions, tolerances, surface finish, physical/chemical properties, special conditions and quantity required.
2	Approximate cost evaluation—isostatic compaction in comparison with alternative methods of production	Comparison of isostatic compaction with other techniques and quantity required.
3	PTFE powder, compaction and sintering conditions to be used	Consideration (1), and suitability for processing
4	Compaction by wet- or dry-bag technique	Estimated equipment cost (may be partly offset against future work), mold cost, cycle time, quantity required and manufacturing costs.
5	Possible directions of pressing. If both pressing from the inside and pressing from the outside are possible, consider each separately.	Direction of pressing
6	Mold configuration, i.e., basic shape and layout of components.	Good uniform compaction of PTFE Compromise between accuracy of shape and complexity of mold. Ease of mold filling and preform removal (from bag and rigid part). Bag support (mandrel or container). Good sealing. Avoidance of PTFE and bag damage. Possible problems and need for special techniques.

7	Detail design of mold rigid parts (except bag support)—dimensions, material, location of parts, seals and clamping.	(a) Molding surface dimensions:
		    Dimensions and tolerances of required article (consideration 1).
		    Allowing for PTFE shrinkage (Consideration 3)
		    Possible need for machining (Consideration 10).
		(b) Material and overall dimensions.
		    General strength, subject to size limitations for wet-bag compaction; pressure vessel design for dry-bag compaction.
		(c) Seals and clamping

8	Detail design of flexible part (bag)—dimensions, material and thickness.	(a) Dimensions:
		    Assumed bag movement, compaction ratio and shrinkage for the PTFE (Consideration 3) and rigid molding surface dimensions (Consideration 7).
		(b) Material and thickness
		    Preform surface finish, ease of preform removal, need for special techniques (e.g., pre-tensioning and cost.

9	Detail design of bag support (mandrel/container)—dimensions, bag sealing and clamping.	(a) Dimensions:
		    Bag size and thickness (Consideration 8); pressurizing fluid flow rate for transmission channel dimensions.
		(b) Bag sealing and clamping:
		    Avoidance of bag damage.

10	Processing of PTFE after sintering, e.g., machining.	Dimensional tolerances and surface finish.
11	Detail estimate of costs.	Estimated equipment cost, mold cost, cycle time, quantity, scrap rate and manufacturing cost.
12	Estimate of time delay before full-scale production.	Time for prototype or development work, and to obtain equipment, mold set.
13	Choose most suitable powder and direction of pressing if choice still remains.	Cost, quality and time delay before full-scale production.
14	Prototype	
15	Modification	Cost of modification, in relation to cost savings.
16	Final design.	

shape of the bag and ensure reproducible volume. Taking these steps will ensure proper and reproducible mold operation and a good preform. Obviously, all of these actions may not be applicable to every isostatic molding system.

Compaction pressure of the powder influences the physical properties of the part. The value of the pressure is selected not only based on part properties, but dry bag mold cost, bag and mold life, and the cost of pressurization because of rapid wear at high pressures. The process should be operated at the lowest satisfactory pressure. The rate of pressure increase depends on the thickness of the part. The thicker the part, the slower the rate of pressure rise. The dwell time, i.e., the length of the compaction time, must be longer for thicker parts. The optimal values for maximum pressure, rate of pressure rise and dwell time must be determined experimentally.

Molding pressure is in the range of 10–35 MPa (1,500–5,000 psi) for unfilled free flow PTFE powders. The filled PTFE compounds often require a higher molding pressure (up to 70 MPa) which depends on the type and the amount of fillers. For example, a compound containing 60% bronze and 40% PTFE requires a higher molding pressure than a compound containing 15% glass fiber and 85% PTFE. After reaching pressure, a dwell time of 5 sec/mm is necessary, although very thin parts (<3 mm) do not require any dwell.[8]

At the completion of dwell, the mold must be depressurized at a rate which will not lead to cracking of the preform. This rate has to be reduced for thicker (2 MPa/sec for part thickness > 50 mm and 10 MPa/sec for part thickness <50 mm) and more geometrically complex parts. The rate should be sufficiently slow to avoid cracking of the preform because of the sudden expansion of entrapped air, preform or the bag. Some air is trapped in the preform in isostatic molding just like compression molding. Slow release of pressure allows diffusion of air and deformation of PTFE. Sudden release of fluid elastic pressure is more damaging to thick sections. A way to remove the trapped air is to apply vacuum to the resin after filling the mold.

After pressure release, demolding should begin by first wiping the fluid from the exterior surface of the wet bag mold. Parts should be carefully disassembled to avoid damaging the preform. Pulling vacuum on the liquid side can separate the bag. The preform is removed and placed on a rack which supports its shape. The preform should be permitted to degas, i.e., the trapped air diffuses out, before sintering (see Fig. 7.25). A well vacuumed (after resin filling and prior to molding) preform can be placed in the oven without degassing. Sintering of isostatically molded parts is done similarly to those made by compression molding. (see Sec. 7.3.3.2)

7.6 Ram Extrusion

Ram extrusion is the only continuous process for fabrication of parts from suspension polymerized (granular) PTFE powders. All the required steps of granular processing are performed in one machine called a *ram extruder*. The most common shapes are solid round rods and tubes. Rectangular rods, L-shaped cross sections and other ram extrudable profiles are occasionally fabricated.

7.6.1 Introduction to Ram Extrusion

The basic steps for processing granular resins are:

- *Compaction* of the powder to make a preform
- *Sintering* the preform, which consists of heating the preform above its melting point
- Air *quenching* or slow *cooling* the sintered part to allow controlled crystallization of PTFE.

These three steps are carried out inside the ram extruder continuously using a free flowing resin which is often a special presintered ram extrusion grade or a general purpose free flow powder (Fig. 7.48). These resins behave differently during extrusion. The commercial presintered resins have been specially designed for ram extrusion over a wide range of extrusion conditions and can be converted into a wide range of parts such as round rods with a diameter 2 mm to 400 mm. These parts have excellent physical properties and have high resistance to fracture at the interface of charges (or doses) called *poker chipping* in the industry. Presintered resins can undergo much higher pressures during extrusion than ordinary free flow granular powders, making them especially suitable for small diameter rods and thin wall tubes. General purpose free flow granular powders are more suitable for larger rods (>2 cm diameter) and thick wall tubes.

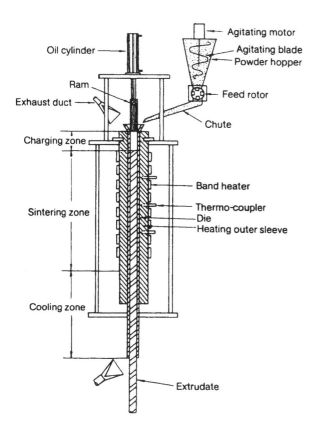

Figure 7.48 Ram extruder.[8]

7.6.2 Ram Extrusion: Basic Technology

Two types of common commercial equipment are vertical and horizontal extruders where the direction of the ram motion and the extrudate are, respectively, vertical and horizontal. The fundamental working principles of the two pieces of equipment are the same. The key difference between them is the method of extrudate support. In horizontal ram extrusion, a tray or other similar means can support the extrudate. In ram extrusion, especially a vertical machine, a mechanical *brake* may be needed to protect the extrudate and provide backpressure for coalescence. This brake is usually a chuck, collet or a gland that grips the extrudate and applies a controlled amount of pressure.

In both vertical and horizontal machines, a metered quantity of granular powder is charged to the feed section of the die. This end of the die is cooled to allow easy flow of the resin into the die. PTFE particles and agglomerates become sticky and powder flow suffers when heated above approximately 25°C. The next step is compaction of the powder and pushing it into the heated segment of the die by the forward action of the ram. Repetition of these steps advances the compacted resin through the heated length of the die where sintering of PTFE takes place. The adjacent charges of the resin are welded to each other under pressure while sintering is taking place.

The essential elements of ram extrusion processes are resin feed and compaction, sintering, and cooling. These elements are further explored in the following sections.

7.6.3 Ram Extrusion: Resin Feed

The main function of the feed section is to provide repeated individual charges of resin with uniform weight during every cycle of the process. The resin must be distributed evenly in the feed port at the cold end (beginning) of the die (Fig. 7.49).

It is important to water cool the feed section and the beginning section of the die where ram penetrates, preferably to 21°C. Temperature must be kept above the dew point, particularly if the extruder is operated in a humid area. Failure to maintain the temperature above dew point will lead to condensation of moisture in the feed cup and result in defects in the part. Increasing the temperature of the resin accelerates the decrease in resin flow.

Both uniform distribution and uniform charge weight of the feed resin are important. The uniformity of the distribution is especially critical for the extrusion of thin wall (< 2 mm thickness) tubing. The presence of the mandrel in the middle of the die forms an annulus, which is where resin must enter. The mandrel blocks the free movement of the resin. An increase in the size of the mandrel to produce larger diameter tubing intensifies the importance of uniform resin distribution. The other factor is the poor flow of resin under pressure, which means that little flow can be expected inside the die during compaction by the ram. To extrude a high quality tubing, the resin must be uniformly fed to the die which requires excellent resin flow in addition to uniform charge weight.

A solid rod die, where no mandrel is present to restrain flow, is less sensitive to the uniformity of resin distribution than a tubing die. The rod properties are sensitive to the uniformity of the weight of each resin charge. Variations in resin distribution across the cross section can also lead to the curving of the rod, which should be, normally straight. Variations in the charge weight cause variability in the extrusion rate and loss of productivity. Variation of apparent density will give nonuniform properties

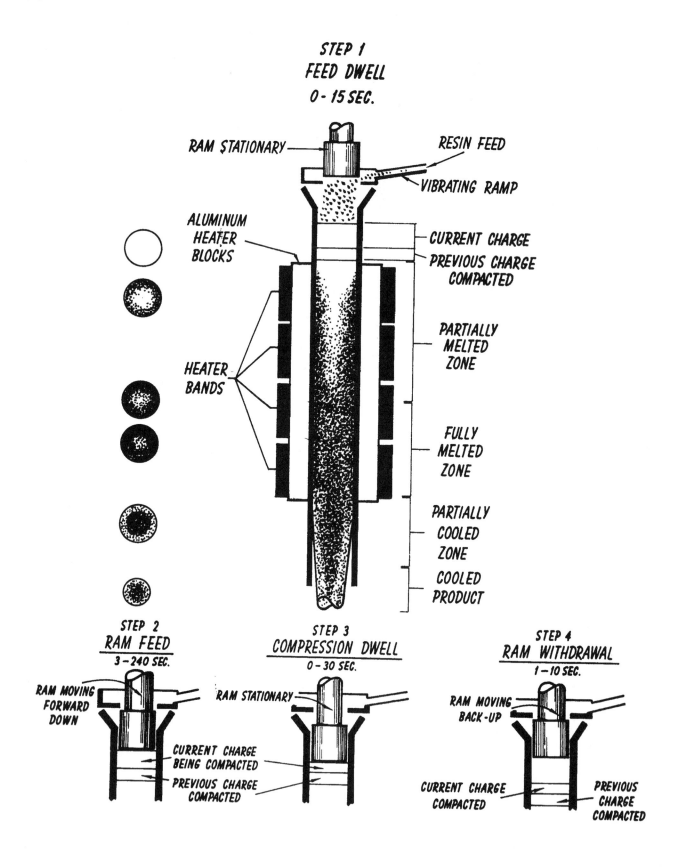

Figure 7.49 Ram extrusion cycles. *(Courtesy DuPont.)*

7.6.4 Ram Extrusion: Compaction

The second step in producing a ram-extruded part is compaction. The ram has two modes of motion—advancing and retracting. After the resin has been fed into the feed section the advancing motion of the ram compresses it. The single charge of resin is squeezed between the bottom surface of the ram and the top surface of the previous charge (Fig. 7.49). This is the mechanism of preform development from the resin inside the die.

The force of the ram is not the key factor in generating the compaction pressure as long as the ram has sufficient tonnage or force, which is determined by the hydraulic piston/cylinder capability. The main determinant of the preforming force is *backpressure*, which is generated by what occurs further down inside and outside the die. Backpressure can be defined as the resistance of the part inside and outside the die to advancing movement. This resistance is comprised of two forces: the friction force developed between the molten PTFE and the inside wall of the die, and the mechanical break which can be installed near the exit point of the part from the die (Fig. 7.50). At some point the resin does not compact and the ram force overcomes the backpressure because of the continued advancing movement to complete its stroke. At this point, the part is advanced by the length of one resin charge, thus the process name *ram extrusion*.

The compaction pressure of the ram should be in the right range. Estimation and experimentation best define this range. The pressure required to compact the powder into a dense preform with minimum void content defines the lower end of the range. The upper end of the range is determined by the pressure at which *poker chipping* (fracturing at the interface of adjacent charges) in the part begins. This means that increasing the pressure does not help the welding of the charges to each other beyond certain limit. The pressure limits vary from one resin to another. Acceptable operating compression pressures of presintered PTFE powders may be as low as 5 MPa to over 100 MPa. General purpose resins can exhibit poker chipping at pressures as low as 10 MPa.[11]

General purpose free flow resins tend to build up too much backpressure (compression pressure) which causes poker chipping. Presintered resins do not build high compression pressure under most conditions. Even at high pressures, parts made from these resins resist poker chipping. Parts look grainy or chalky when compression pressure is insufficient because of the presence of voids. The same problem will be encountered if the resin is not given sufficient time to sinter, that is, the extrusion rate is too high. Naturally, presintered resins cannot overcome an extrusion rate which is too rapid.

Another way to enhance the weld strength between the charges is to keep a small length of the die below the end of the ram stroke cool. This will keep the top surface of the previous charge, against which the resin is being compacted, cool. Other benefits include the prevention of certain types of polymer skin formation, that is, polymer buildup on the die wall, and the elimination of "springback" of the resin after the removal of the compaction force at the onset of ram retraction. *Springback* is simply the expansion of the compressed resin opposite the pressure direction, therefore, opposite the extrusion direction. Presintered resins have a greater springback than the general-purpose powders. It also increases with an increase in the ratio of cross-section to the perimeter of the part. This means that springback is more important with large diameter rods and tubes.

7.6.5 Ram Extrusion: Sintering and Cooling

After compaction, the preform is advanced into the first heated zone of the die for sintering (Fig. 7.48). Sufficient heat must be supplied to the preform to raise its temperature to above the melting point of the polymer. There are two considerations in the selection of the die temperature. First, melting temperature of PTFE increases as a function of the polymer pressure (Fig. 7.33). Second, the chosen temperature should be high enough for PTFE to melt and sinter completely during the residence time in the heated length of the die. Adequate pressure should also be applied to the part to eliminate the voids and obtain a strong welded bond to the adjacent charges. Finally, the part must be cooled at a rate that will lead to the desired crystallinity content. Each of these steps is discussed below in further detail.

7.6.5.1 Ram Extrusion: Heat Requirements of Sintering

Heating up in ram extrusion takes place by a conduction mechanism. Thermal conductivity and heat capacity of PTFE, die temperature, and preform size determine the time required reaching the sintering temperature. The preform has to be held at the sintering temperature for a period of time until densification and void elimination processes have been completed. The extrusion rate can be calculated by dividing the

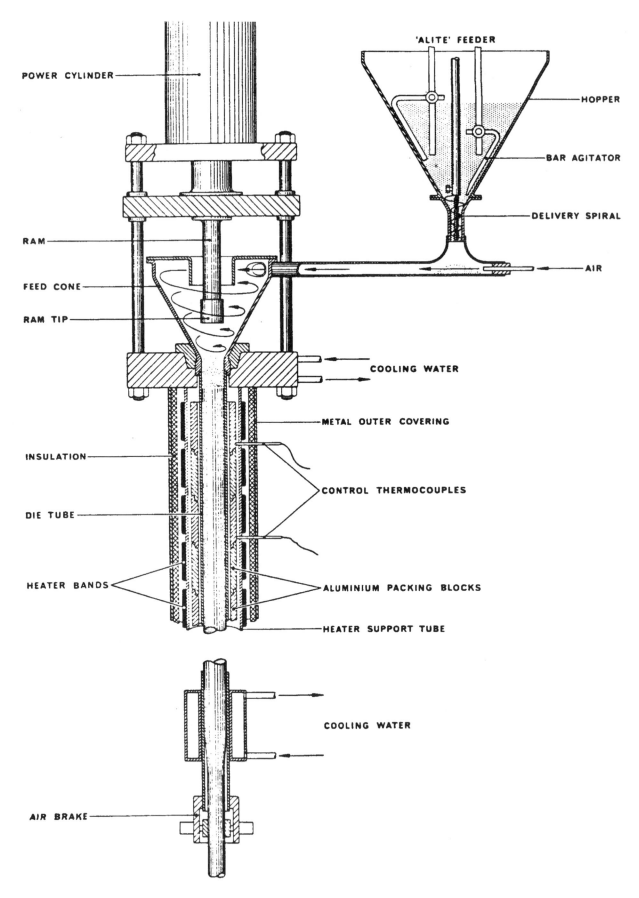

Figure 7.50 Ram extruder mechanical break.[12]

heated length of the die by the minimum time required for achieving acceptable sintering. For example, assume the minimum required sintering time for a solid rod is 2 minutes at 398°C and the heated length of the die is 600 mm. The maximum extrusion rate is 300 mm/minute (600 mm divided by 2 minutes) or 18 m/hour. Typically extrusion rates in the industry are 50-60% of the maximum rate (Table 7.10). The excess residence time helps reduce the temperature gradient in the part, thus, minimizing residual stress. It also allows void reduction and elimination of charge lines.

Table 7.10 gives a summary of minimum sintering times required for solid PTFE rods from 3–25 mm in diameter. The minimum required time increases with the diameter of the rod, as does the shrinkage. The rise in the minimum time is not linear with diameter but proportional to the square of the diameter. The selected temperature of the die is 395°C in this table. The recommended range is 370–425°C but the temperature should be kept below 400°C to minimize the degradation of the polymer. Overly long residence time in the die can also lead to degradation.

Sintering temperature and residence time have profound effects on the quality of parts made by ram extrusion. Figure 7.51 shows tensile strength of a rod, which is a measure of quality, as a function of extrusion rate decreases. Residence time has an inverse relationship with extrusion rate and clearly has a strong positive effect on tensile strength. Residence time is also a function of the shape of the part being extruded in addition to size and extrusion conditions. For example, extrusion of a rectangular part with the same size side as the diameter of a rod requires more heat or a higher residence time (20–30%) than the rod.

7.6.5.2 Ram Extrusion: Pressure Requirements

Pressure is required while the resin is being heated for the sintering to take place properly. The magnitude of pressure needed is much smaller than the preforming pressure in compression or isostatic molding techniques. Heat and pressure act interchangeably, that is, lower pressure is required when the resin is being compacted at an elevated temperature. A simple way of visualizing the interaction of pressure and temperature is to say that raising the temperature softens the polymer, which allows the application of a lower pressure.

Another aspect of ram extrusion is the much shorter sintering time. Here is how the sintering takes place. The resin expands axially and builds up internal pressure as it is heated without the ability to expand radially because of the die. Expansion of the resin is driven by a 15% increase in the specific volume of the resin from 0.434 cm^3/g for 100% crystalline phase to 0.500 cm^3/g for 100% amorphous phase. (Presintered resin has a crystallinity of approximately 50%.) The expansion occupies the voids and reduces the void content of the preform. This internal pressure becomes the driving force for rapid coalescence of the molten particles and consolidation of the preform into a mass of gel. The internal pressure of the resin helps in the welding of the charge-to-charge interfaces. Insufficient pressure leads to voids being left in the part. For example, the center of a solid rod will look chalky which is an evidence of voids; away from the axis of the rod there will be less voids. A void free rod appears somewhat translucent.

Table 7.10. Average Shrinkage, Required Heating Times, and Representative Die Lengths for Ram Extruders Used with Teflon® PTFE Fluoropolymer Resins for Ram Extrusion[11]

Rod Diameter, d mm (in)	Average Shrinkage, S %	Minimum Required Heating Time, min, 395°C (743°F)	Representative Die Length, mm (in)*	For Representative Die Length	
				Maximum Extrusion Rate, m/h (ft/h), 395°C (743°F)	Typical Extrusion Rate, m/h (ft/h), 395°C (743°F)
3 (0.12)	6	1	250 (10)	15 (50)	10 (33)
6 (0.25)	8	1.25	500 (20)	24 (80)	12 (40)
13 (0.50)	11	5	1,000 (40)	12 (40)	8 (25)
19 (0.75)	12.5	11	1,500 (60)	8 (25)	4.5 (15)
25 (1.0)	13.5	20	2,000 (80)	6 (20)	3 (10)

* Die lengths considerably greater or smaller than the values given can be used with ram extrusion resins. Generally, shorter dies must be used when ram extruding general-purpose resins.

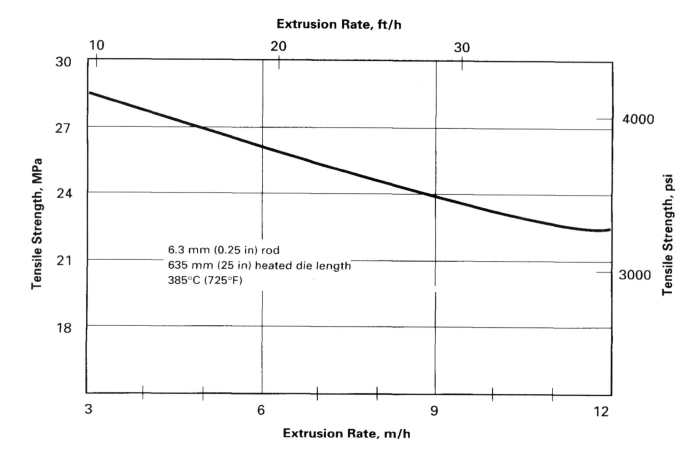

Figure 7.51 Tensile strength vs extrusion rates.[11]

It is important to compact the resin sufficiently to reduce its void content in the feed end of the die. It is virtually impossible to sinter a preform which has a large void content.

The molten polymer restrained from expansion wets the surface of the die and exerts a significant friction force on the surface resisting motion. Adequate ram force (or pressure) to overcome the frictional force developed between the molten PTFE and the die wall will allow the extrusion to take place and not stall. This force is the source of *backpressure*, which can be calculated by dividing the force by the area of exertion.

The resistance force of a mechanical brake is added to the backpressure if one is present in the system. Ram force must equal the total resistance presented by backpressure and the brake.

Backpressure depends on the friction force which itself depends on several factors:

1. Temperature profile of the hot length of the die, surface smoothness of the die, and the type of metal from which the die is constructed all impact the coefficient of friction which, along with the internal pressure developed by the resin, determine the quantity of the friction force.

2. The pressure buildup of the polymer upon melting which depends on temperature and void content of the resin.

3. An increase in the area of the contact between the PTFE part and the die, excluding the cooling zone, elevates backpressure. In the cooling zone, the part shrinks and separates from the die wall, therefore developing no backpressure. Higher backpressure is obtained by an increase in the length to inside die diameter (L/D) of any of the three zones of ram penetration, cooled length below the penetration zone, and the heated length of the die.

4. Increasing the extrusion rate results in a higher backpressure.

5. The longer the charge length, the higher the backpressure, particularly if the ram penetration zone is water-cooled.

Temperature of the die has a fairly complex effect on backpressure development. As temperature increases, backpressure decreases, as discussed previously about the entry end of the die. The force necessary to extrude the powder at room temperature is very high because of the large backpressure. It will begin to rise at some point above the melting temperature of the polymer, on the order of 40°C. This value varies depending on the type of the resin and other extrusion parameters.

In vertical ram extrusion of large cross section solid parts, supporting the weight of the extrudate is important because it is suspended and its weight counteracts the backpressure. For example, a 50 mm diameter extrudate is heavy and must be either cut in small sections or it must be supported by a higher backpressure. External means such as a mechanical brake would be required if longer lengths are desired. In the absence of weight support, the weight of the extrudate strains the molten polymer, which can lead to melt fracture or rupture of the melt.

Extrusion of hollow parts such as pipe liners requires the use of a mandrel, as seen in Fig. 7.52. There are two kinds of mandrels: the ram and mandrel are connected and reciprocate together (driven) or the ram reciprocates and the mandrel is fixed (Fig. 7.53). The latter is generally used to extrude presintered PTFE. Backpressure is a strong function of the ratio of the area of contact between the resin and the die-to-ram cross section area, in the extrusion of tubing. Most of the same factors affecting the backpressure of the rod also impact the tubing backpressure.

There are unique mandrel effects absent from the extrusion of solid rods and other parts. The length of the mandrel can affect the backpressure profoundly. It often has a different temperature profile from the die. For instance, consider the case where the length of the mandrel exceeds the length of the die. Near the end of the die, the tube begins to cool and shrink away from the die and towards the mandrel. At some point, backpressure begins to build up because of the friction force between the mandrel surface and the tube. Choosing the appropriate length of the mandrel can control this type of backpressure. No backpressure from this source will be experienced if the mandrel ends at the beginning of the die-cooling zone.

Another method for controlling the backpressure of thin wall tubing is to retain the cooling zone temperature above 200°C, which limits the extent of shrinkage. The tube is sufficiently cool to shrink away from the die but not cool enough to shrink tightly onto the mandrel. The rest of the cooling can be accomplished outside the die, assuming the mandrel does not extend more than 5–10 cm beyond the die exit. The best setup for extruding a thin wall tube is to employ a non-tapered, non-reciprocating mandrel which extends a few centimeters outside the die. The shrinkage of the tube onto the mandrel over a short length does not increase the backpressure significantly but improves the appearance of the inside surface of the tube.

If the length of the mandrel is so short that the extrudate passes over the end of the mandrel while the polymer is molten, poor dimensional uniformity is obtained. The cross section may not be round and the diameter of the tube may alternate between larger and smaller than the desired tube diameter, known as "hourglassing." A similar phenomenon is observed when the tube exits the die, supported by the mandrel, in gel state. The surface defects of hourglassing may resemble poker chipping. The severity of the defects grows with a decrease in the tube wall thickness.

Tube stock shapes should be ram-extruded from presintered polymer. If tubing is to be made from general-purpose powder, the following machine designs will help reduce backpressure and prevent poker chipping:

1. Stationary mandrels can be tapered, that is, have a smaller diameter at the exit end than at the entry point. It is also possible to construct the mandrel for two cylinders joined by a taper. The taper is designed such that the molten resin would be able to expand inward without touching the mandrel.

2. The mandrels can reciprocate with the ram or independently. A mandrel moving with the ram should have a taper (slight) to assure that it does not pull back the tube during reverse motion. An independently moving mandrel can have constant diameter. The mandrel simply advances by the resin during the advancement of the ram. It retracts by one length of charge, which it had advanced, during the dwell while the ram is stationary at maximum penetration. The cycle is repeated after the retraction of the ram.

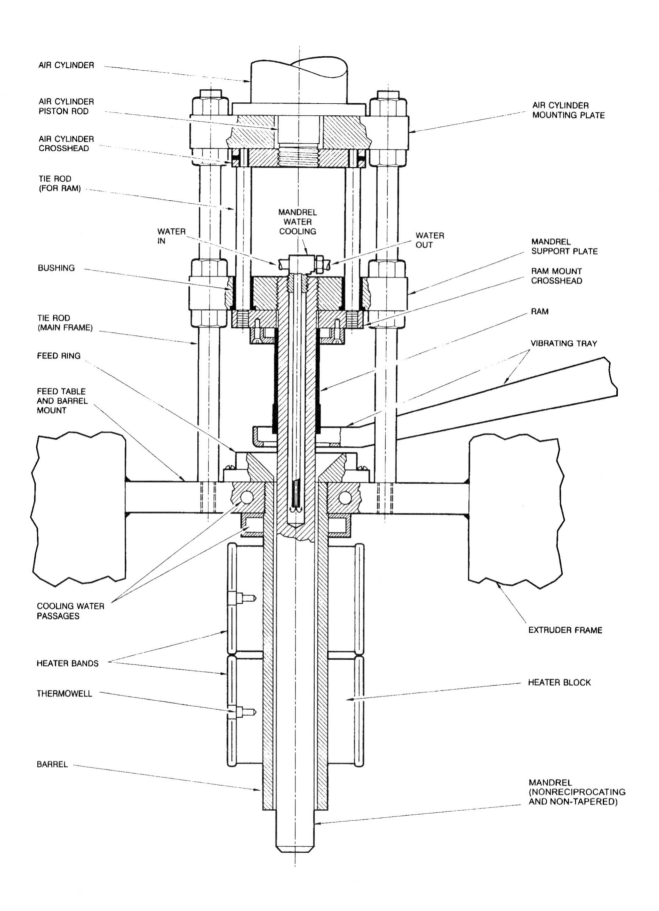

Figure 7.52 Vertical ram extruder for making tubing of PTFE fluoropolymer resin.[11]

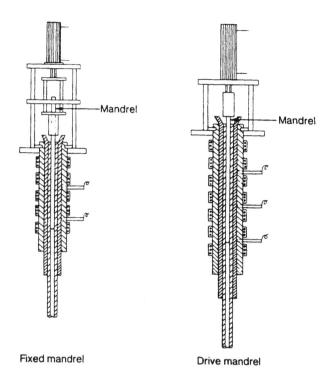

Figure 7.53 Pipe extruder.[3]

These designs lower backpressure by: *(1)* a reduction in the friction force exerted on the die and mandrel surface by the polymer pressure generated during the melting process, and *(2)* a reduction in the contact area between the stationary die surfaces and the resin.

Producing large diameter solid rods in short dies (low die length to rod diameter ratio) presents a special problem because of the difficulty of obtaining sufficient backpressure. All the strategies previously discussed may have to be employed to increase the backpressure adequately for the compaction of the powder. They include cooling a length of the die below the entry point, tapering the die to have a smaller diameter at the exit than at the die entry, and application of a mechanical brake to the extrudate outside the die.

7.6.5.3 Ram Extrusion: Cooling

Cooling is the last step in ram extrusion as in other fabrication techniques. It determines the crystallinity and shrinkage of the part. Crystallinity determines the properties of the rod, tube and other cross sections. Shrinkage determines the final dimensions of the parts. Often air quenching is the means of cooling in which the gel leaves the die and is exposed to forced air. This method yields minimum crystallinity and shrinkage. Typically, crystallinity is in the range of 55–60%.[11]

Shrinkage of the solid rods ranges from 10 to 14% of the die cavity diameter. It can be held constant by adjusting the die temperature profile, extrusion pressure, cooling rate and sintering time. Presintered resins shrink less than the general-purpose free flow powders. Hollow parts such as tubing shrink less (4–6%) than solid parts such as rods.

Internal stresses build up in ram extruded parts depending on the cooling rate, part geometry, die temperature profile and other process variables. Length and temperature profiles of the cooling section of the die can be designed to achieve controlled cooling rates. Large rods (>50 mm) are susceptible to shock and even internal fracture if cooled too rapidly. Control can be achieved by extending the cooling section of the die by adding an insulated tube. This tube may have to be heated to assure a controlled cooling rate.

It is possible to design the cooling step to obtain precise sizes required for parts which cannot be machined or ground to size.

In some applications, particularly when precise machining is to take place, the products of ram extrusion can be annealed in a second process step. The residual stresses in the part are reduced or eliminated by annealing which may also cause some size alteration of the part.

7.6.5.4 Ram Extrusion: Die Skin Formation

An inevitable event during the ram extrusion of polytetrafluoroethylene is the buildup of a thin polymer skin on contact surfaces of the die and the mandrel. The polymer, which is fairly soft even in the solid phase, passes through the die under high pressure to overcome the backpressure generated by friction. This force gives rise to a shear stress at the interface of the die wall and the resin surface. PTFE has fairly low shear strength and cannot stand up to the wall shear stress, particularly, at the minute pores arising from the imperfect smoothness of the die wall. Consequently, polymer is sheared off from the gel body and is deposited on the die wall. The same argument can be applied to the mandrel surface, where shear stress exceeds the shear strength of the molten PTFE.

The polymer coats the wall of the die, mandrel, and other contact surfaces. Once the surface pores have been coated, no more coating will take place until some of the polymer is degraded and is removed. At this point, rough pores are exposed which are refilled by shearing off new polymer. This skin formation is normal and affects the die because it resides in the surface pores. PTFE degradation generates corrosive products such as hydrofluoric acid, which can corrode the die to different extents depending on the material of construction and the temperature of the die. The skin immediately next to the metal wall turns black over time because of carbonization of PTFE. The layers closer to the extrudate are renewed continuously, thus stay degradation free.

The formation of normal skin during the normal operation of the ram extruder does not interfere with process performance or product quality. After shutdown and cooling, the skin shrinks and pulls away from the die wall, taking away parts of the immediate black buildup. If the extrusion is resumed without cleaning the die, the skin will be carried off by the extrudate and appears as black spots on the PTFE part. After a new skin is established, the black spots will disappear and the extrudate appearance will return to a white color.

In commercial operations, the small dies (less than 2–3 cm diameter) are cleaned, at the beginning of the extrusion, by manually inserting a small cylinder of a soft metal wool in the die ahead of the resin. The extruder is then started and the metal wool is allowed to be driven by the resin through the die. One or two doses of metal wool usually abrade most of the die wall build-up and, after a short transition, the extruded product will be back to normal. Larger dies would have to be cleaned off-line by mechanical techniques or by placing them in a hot oven and burning the skin off at a temperature in excess of 450°C.

Normal operations take place when the die and ram clearing is roughly in the range of 100–200 µm.[11] If the resin contains particles that are smaller than the ram clearing, they may enter the clearing space and get smeared, that is, highly sheared by the motion of the ram. The repetition of this process will lead to the buildup of these smeared particles at the entry area of the die. The sheared particles will enter the resin being compacted and form defect sites because of difficulty of coalescing these damaged particles. The flaws form voids and weaken the extrudate mechanically. The best way to prevent this problem and avoid excessive ram clearing is to cool the die entry section with water and control the particle size distribution of the resin to avoid particles smaller than the clearing.

Another type of skin formation is observed in the form of periodic bands of loose particles around the extrudate. The particles are highly sheared and accumulate somewhere in the sintering section of the die. Typical symptoms observed sometimes before any bands are seen include a reduction in the diameter of the extrudate and a roughening of its surface. The cause of the formation of these bands is primarily cooling the gel too rapidly. Other factors include certain die metals such as unplated steel, too much ram clearing and presence of fines in the powder.

7.6.6 Ram Extrusion Equipment Design

A key decision in the design of a ram extruder is to select the type(s) of resin that may be extruded using the equipment. It is clear from the discussions so far that general-purpose free flow resins require short dies because of the large backpressure associated with these resins. High backpressure leads to poker chipping in the extrudate, which renders it commercially useless. Short dies can only be operated at low extrusion rates because of sintering time limitations. For example, a 13 mm diameter solid rod requires a minimum sintering time of 5 minutes and cannot be extruded in a die longer than 500 mm. This means that a maximum rate of 100 mm/min can be reached which is about double (see Table 7.10) the practical rate that allows production of rod with acceptable quality. In contrast, the maximum die length for a presintered resin to make the same rod is 1,000 mm. This indicates that for a minimum sintering time of 5 minutes the maximum rate is 200 mm/min, which is double that attainable with the general purpose powder. More interestingly, presintered resins can withstand much higher backpressure; 100 MPa backpressure has been reported in a die with a die cavity-surface to cross-section-surface ratio of 400:1 without encountering poker chipping.[11]

Table 7.10 can be used as a reference for selecting the die length for various solid rod sizes for extruding presintered PTFE powders. Ordinary free flow resins must be extruded in shorter dies at slower rates. Conversely, presintered resins can be extruded through short dies, as long as sufficient backpressure can be generated to compact the resin. Cooling the die below the entry section can be used to heighten the backpressure for short dies. The recommendations of Table 7.10 can be altered by varying the process conditions (e.g., cooling techniques) to produce parts with good properties from presintered powder.

Two types of equipment are commercially common: vertical extruder (Fig. 7.52) and horizontal extruder (Fig. 7.54). In principle, both types are capable of producing rods. Each has certain advantages and disadvantages. For example, it is easier to center the

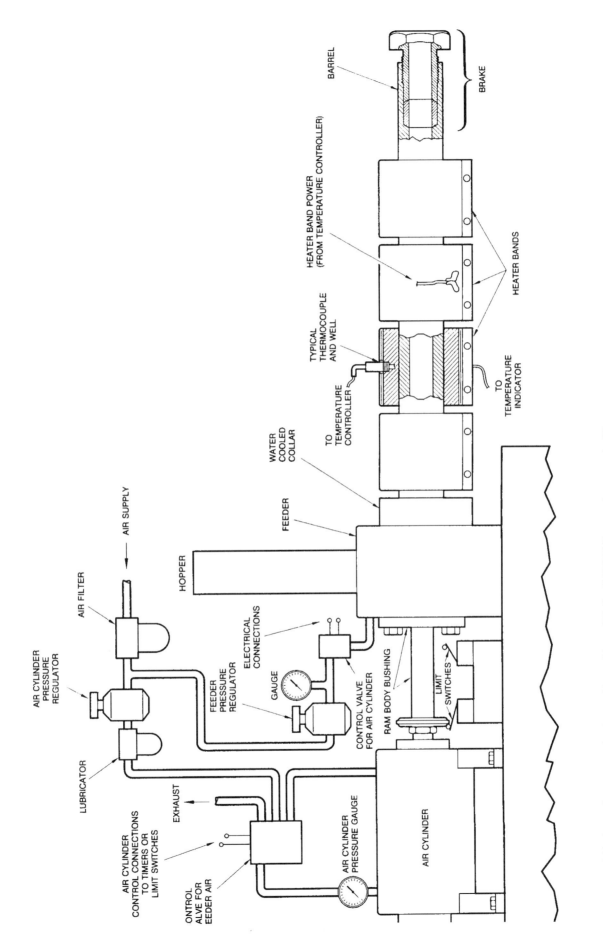

Figure 7.54 Horizontal ram extruder for making tubing of Teflon® PTFE fluoropolymer resin.[11]

mandrel in a vertical machine, by gravity, than in a horizontal piece of equipment. Horizontal extruders allow production of very long parts more easily than using a vertical extruder. Table 7.11 summarizes a comparison of the capabilities of the two types of extruders.

The choice of the extruder type depends on all these factors and one cannot be automatically selected over the other.

Disregarding type, a ram extruder consists of four basic components:

- Drive system (hydraulic or pneumatic)
- Ram
- Die and mandrel (if one is needed) and heaters
- Resin feed system

Water cooling systems should be designed as part of the feed system, ram drive and the die. All four components must be equipped with the appropriate instruments to control the resin feed rate, ram timing and die temperatures. Other components such as brakes and cooling tube may be required

7.6.6.1 Drive System

The drive system, whether hydraulic or pneumatic, must have enough force to overcome the backpressure generated in the die. The pressure should be sufficient to compact the resin to minimize voids. A pneumatic system with an inlet air pressure of 0.6–0.7 MPa and a cylinder of 100 mm in diameter can generate sufficient force to extrude all cross sections up to 25 mm diameter.[11] The pressure of the air cylinder can be monitored as an estimation of the backpressure.

Timing of the drive is an important factor in the movement of the ram. There are two ways to monitor the timing of the ram. A simple method is to install limit switches (Fig. 7.54) which activate when the ram reaches extreme advancing or retracting positions. The preferable way is to equip the drive with a timer to allow ample time for resin feed and compaction of the resin in the die. The timing method produces a more uniform product than the limit switch technique. Compaction pressure and dwell behave inversely. Generally, the longer the powder is held during the compaction, the lower the necessary pressure will be. A larger charge requires a longer dwell.

Air is interspersed with the powder. For air to escape from the resin, it is helpful to delay the ram in the withdrawn position. The length of this delay depends on the size of the resin dose fed but it is normally a few seconds long. Another advantage of the delay is that it allows the top surface of the last charge to relax. This is more important in the extrusion of general-purpose resins than presintered powders. Care must be taken to prevent springback of the previous charge, which can reduce the extrusion rate. Cooling the die below the feed zone will increase the friction between the extrudate and the die wall preventing springback. A delay in the forward position of the ram, during compaction, will eliminate springback.

7.6.6.2 Extrusion Ram

The ram is attached to the air cylinder's piston and must be aligned as close to perfect as possible. Misalignment causes poor quality product and shortens the life of the ram and the die because of wear, rubbing and scoring. Another key issue is the proper clearance between the ram and die wall. The ram typically consists of a standard size round stock attached to a tip with the appropriate size and shape. The tip is the front most segment of the ram and should be only as long as the ram diameter.

The ram and tip penetrate the die and must be smaller than the die bore. A clearance of 30 µm is recommended for extrusion of solid rods. In tubing extrusion, the clearance between the ram (hollow) and the non-reciprocating mandrel should be around 50 µm. Satisfactory tubing can be made by using non-reciprocating, non-tapered mandrels by ram extruding presintered PTFE powders.

Table 7.11. Comparison of Capabilities of Ram Extruders

Extruder	Floor Space Required	Producing Long Parts	Centering Mandrel	Feeding Resin	Maintenance	Cooling the Die
Vertical	Less Space	Less Easily	More Easily	More Easily	Less Easily	Less Easily
Horizontal	More Space	More Easily	Less Easily	Less Easily	More Easily	More Easily

Mandrels can be solid or hollow and should extend the full length of the die. The die and the mandrel should be water-cooled at the resin entry end. An effective technique to reduce the heating and residence time of the resin in tube manufacturing is to equip the mandrel with internal heaters.

7.6.6.3 Extrusion Die

Ram extrusion dies are metallic tubes with a uniform internal cross section throughout their lengths. The die must be constructed from a corrosion resistant metal because of the generation of a small amount of degradation products in the temperature range to which the die is heated. The selection of the metal depends on expected useful life and consistency of the part quality throughout the life of the die. Cost is an important consideration because more durable metals are more expensive. The die can be simply made from chrome-plated steel or stainless steel. Examples of the latter include Monel® 400, Hastelloy® C, Xaloy® 306 and Inconel® 625. The use of unplated stainless steel may cause polymer skin formation.

The recommended inner diameter of the die for rod extrusion is to make it 14% larger than the desired finished diameter of the rod, to take into account the shrinkage of PTFE. The final size of the rod diameter can be manipulated by altering the process variables such as resin type, pressure, die temperature profile and extrusion rate. For tubing extrusion, the inside diameter of the die should be 5% larger than the outside diameter of the finished tube. Die wall thickness in the range of 1–1.5 cm is recommended. Wall thickness should be increased for larger diameter dies to account for higher internal pressures.

Heating of the die is normally done with electrical band heaters. It is best to subdivide the heater into several zones and install independently controlled heaters on the die, spaced apart. This scheme allows the various parts of the die to be controlled independently. Spacing the heaters apart will for the most part prevent communication between adjacent heaters. To ensure excellent heat transfer, heater bands should be installed on top of aluminum bands, which tightly embrace the die. Temperature non-uniformity of the heated section of the die is minimized because of the higher thermal conductivity of the aluminum.

7.6.6.4 Resin Feed System

Different techniques must be used for polymer feed to vertical and horizontal ram extruders. Volumetric or weight-based methods may be used for either technique. The volumetric approach involves filling the feed cavity after the retraction of the ram. A vertical extruder can be fed by gravity conveyance where the resin flows into a weighing device or a shuttle, which will charge the specified weight or volume of powder, respectively. A variety of feeding methods based on these two principles have been developed including screw feed systems, mechanical loaders, force feed systems and shuttle boxes. These techniques aim to uniformly distribute a constant powder weight to the extruder. Contamination and mechanical damage to the polymer powder must be avoided no matter what feeding technique is use. Horizontal ram extruders can be fed gravitationally similar to vertical machines.

7.6.7 Other Ram Extrusion

Extruding multiple ends using the same extruder, especially for small diameter (<1.5 cm in diameter solid rods) can increase productivity of ram extrusion. Multiple identical dies are installed in the machine, which is equipped with multiple rams. All dies should be under the same process conditions such as pressure, temperature, residence time, temperature profile, etc. On the other hand, no individual die cavity can be independently controlled. There are no changes in the requirements that have been discussed for single end ram extrusion.

Filled compounds (see Ch. 11) of PTFE are produced by blending a variety of fillers with fine cut polytetrafluoroethylene powder. Free flow and presintered resins can be made from the filled compounds. Ram extrusion of these resins is possible with some modifications of the preforming and sintering processes and the equipment. Continuous extrusion of filled compounds should be done in a hardened steel die because of the higher wear rate of these compounds. These compounds generate more backpressure, in general, but increased backpressure may be necessary to assure production of low void content parts. Presintered filled PTFE resins are capable of tolerating more backpressure than molding grade compounds. Cooling of the parts under pressure is also helpful to reduce porosity. Little pressure is applied to the extrudate during cooling because it loses contact with the die due to shrinkage. To increase pressure during cooling, a mechanical brake can be used.

7.6.8 Typical Resin, Process and Property Data

Selection of resin for ram extrusion depends on the size of the part. Small diameter tubes and rods (< 1 cm in diameter) have to be extruded using presintered grades of polytetrafluoroethylene. Poker chipping is avoided by proper fabrication of smaller size parts from these resins. Larger rods and thick wall tubes can be ram-extruded from any free flowing PTFE powder. A number of commercial grades of presintered and free flowing powders have been described in Ch. 6.

Tables 7.12 through 7.14 show temperature profile and properties of a number of different parts extruded using presintered resins.

Table 7.12. Molding Conditions and Tensile Properties of Cylindrical Rods[3]

Rod size	Extrusion pressure	Cahrge length	Length of heating zone	Sintering temperature (°C/°F)			Extrasion rate	Tensile properties	
				Upper	Middle	Lower		Tensile strength	Elongation
(mm)	(kgf/cm²)	(mm)	(mm)				(m/h)	(kgf/cm²)	(%)
22	60	60	750	400/752	—	390/734	2.0 ~ 2.6	200	210
30	40	60	900	390/734	390/734	350/662	1.6 ~ 2.0	200	220
40	33	60	1100	390/734	390/734	350/662	1.2 ~ 1.5	190	230
50	25	70	1100	385/725	385/725	350/662	0.7 ~ 1.0	190	230

Note: Ram speed: 30 mm/sec

Rod diameter	Extrusion pressure	Charge length	Length of heating zone	Sintering temperature (°C/°F)			Extrusion rate	Tensile properties	
				Upper	Middle	Lower		Tensile strength	Elongation
(mm)	(kgf/cm²)	(mm)	(mm)				(m/h)	(kgf/cm²)	(%)
18 × 4	70	40	610	390/734	390/734	390/734	2.4 ~ 2.8	200	220

Pipe size (O.D.) × (I.D.) (mm)	Extrusion pressure (kgf/cm²)	Charge length (mm)	Length of heating zone (mm)	Sintering temperature (°C/°F)			Extrusion rate (m/h)
				Upper	Middle	Lower	
30 × 20	60	60	900	390/734	390/734	350/662	2.0 ~ 2.6
50 × 35	35	60	1100	385/725	385/725	350/662	1.2 ~ 1.5

Pipe size (O.D.) × (I.D.)	Extrusion pressure	Charge length	Length of heating zone	Sintering temperature (°C/°F)			Extrusion rate	Tensile properties	
				Upper	Middle	Lower		Tensile strength	Elongation
(mm)	(kgf/cm²)	(mm)	(mm)				(m/h)	(kgf/cm²)	(%)
30 × 20	90	30	420	390/734	—	390/734	2.0 ~ 2.6		
136 × 130	55	30	220	395/743	—	395*/743*	1.6 ~ 2.0	//250 ⊥270	//250 ⊥270

Notes: //: Parallel to direction of pressure
⊥: Perpendicular to direction of pressure
* Mandrel temperature

Table 7.13. Conditions and Properties of M-24 Molded Cylindrical Rods[3]

Rod diameter (mm)		9	13
Die dimensions (length of sintering zone) (mm)		810	1150
Extrusion conditions	Sintering temp., upper part (°C/°F)	400/752	400/752
	Sintering temp., middle part (°C/°F)	—	380/716
	Sintering temp., lower part (°C/°F)	390/734	350/662
	Charge length (mm)	390/734	90/194
	Cycle (s)	10	17
	Extrusion rate (m/h)	7~8	6~7
	Extrusion pressure (kgf/cm^2)	100~150	70~100
Properties of molded articles	Specific gravity	Approx. 2.15	Approx. 2.15
	Tensile strength (kgf/cm^2)	Approx. 190	Approx. 200
	Elongation (%)	Approx. 240	Approx. 250

Table 7.14. Molding Conditions for M-24 Multiple Ram Extrusion Molding (3-Rod Extrusion)[3]

Rod diameter (mm)		5	9
Die dimensions	Total length of die (mm)	670	1110
	Length of sintering zone (mm)	420	810
	Length of cooling zone (mm)	200	250
Extrusion conditions:	Sintering temp., upper part(°C/°F)	400/752	400/752
	Sintering temp., middle part(°C/°F)	—	390/734
	Sintering temp., lower part(°C/°F)	390/734	380/716
	Length of charge (mm)	70	70
	Cycle (s)	11	11
	Extrusion rate (m/h)	6~7	6~7
	Extrusion pressure (kgf/cm^2)	100~200	100~200
Properties of molded articles	Specific gravity	Approx. 2.14	Approx. 2.14
	Tensile strength (kgf/cm^2)	Approx. 190	Approx. 190
	Elongation (%)	100~200	Approx. 210

References

1. Mascia, L., "Thermoplastics," *Materials Eng.,* 2nd ed., Elsevier Applied Science, New York, 1989.
2. Lontz, J. F., *Fundamental Phenomena in the Material Sciences,* (L. J. Borris and H. H. Hansner, eds.), Plenum Pres, New York, 1:37, 1964.
3. "Fluorocarbon Polymers of Daikin Industries," Daikin-Polyflon® TFE Molding Powders, Daikin Industries, Ltd., Osaka, Jpn, Mar. 1992.
4. Frenkel, J., *J. Physics (USSR),* 9:385–391, 1945.
5. US Patent 5,420,191, E. G. Howard, Jr. & A. Z. Moss, assigned to DuPont, May 30, 1995.
6. US Patent 5,512,624, E. G. Howard, Jr. & A. Z. Moss, assigned to DuPont, Apr. 30, 1996.
7. "Isomatic and Isostatic Presses," Bulletin Loomis Products Company, Levittown, Pa., 19057.
8. "Fluon® Polytetrafluoroethylene, Isostatic Compaction of PTFE Powders," Technical Service Note F 14, Molding Powders Group, ICI Plastics Division, Welwyn Garden City, U.K., Aug., 1973.
9. American Society of Mechanical Engineers, Boiler and Pressure Vessel Code, *Chemical Engineers' Handbook,* (Perry and Green, eds.), 6th edition, McGraw Hill, New York, 1984.
10. "Teflon® PTFE Fluoropolymer Resin, Isostatic Processing Guide," DuPont, Geneva, Switzerland, Apr., 1994.
11. "Teflon® PTFE Fluoropolymer Resin, Ram Extrusion Processing Guide," DuPont, Wilmington, Del., Sept., 1995.
12. "Fluon® Polytetrafluoroethylene, The Extrusion of Granular Polymers," Technical Service Note F2, 2nd ed., Vinyl Group ICI Plastics Div., 1966.

8 Fabrication and Processing of Fine Powder Polytetrafluoroethylene

8.1 Introduction

This chapter discusses the fabrication of coagulated dispersion polymerized tetrafluoroethylene known as *fine powder* or *coagulated dispersion powder* into shapes and articles. The most common fabricated commercial forms include rods, tapes, wire insulation, tubing, sheeting, and other profiles. Tube diameter ranges from a fraction of a millimeter to almost a meter with wall thickness of 100 µm to a few millimeters. Rods up to 5 cm can be produced and calendared, prior to sintering, to produce tapes. Unsintered tapes are broadly applied as thread sealant tape in pipe fittings. Unsintered PTFE can be fabricated into cable, sheeting, and pipeliner by wrapping the wire and mandrels followed by sintering.

This form of PTFE is unique, highly crystalline (96–98%) with a high molecular weight. The crystalline form of PTFE changes from a triclinic to a hexagonal lattice at 19°C. Above this temperature, fine powder PTFE becomes softer and more deformable which is important to its processing. Modified resins, that is, those which contain a small amount of comonomer, have a lower transition temperature. At 30°C, transition to phase I takes place which consists of the same 15 carbon helical conformation at 19°C but molecules have increased rotational orientation about their long axis.[1] This change further softens the PTFE, therefore, processing of fine powder PTFE is carried out above 30°C.

Because it does not melt and flow, fine powder PTFE is fabricated by a technology adopted from ceramic processing called *paste extrusion*, where PTFE powder is first blended with a hydrocarbon lubricant (hence the term *paste*) which acts as an extrusion aid. It is then formed into a cylindrical preform at a fairly low pressure (1–8 MPa) and placed inside the barrel of a ram extruder where it is forced through a die at a constant ram rate. The extrudate is passed through multiple ovens and a cooling device where it is first dried, then sintered, and finally cooled. The lubricant can also be removed by extraction in a hot solvent bath.[2]

A major requirement of paste extrusion is that, up to the point of sintering and coalescence, the extrudate must possess sufficient strength to withstand the extensive handling that takes place during the process. The tendency of fine powder to *fibrillate* (form a web of strong filaments between particles) when extruded provides the needed strength and the unique characteristics of fine powder articles. The extrudate is dried in an oven to remove the hydrocarbon lubricant prior to sintering. Cooling is the final step of the process. Cooling rate, similar to parts made from granular PTFE, controls crystallinity of the part.

Reduction ratio, the ratio of the cross sectional surface areas of the preform and the extrudate, is an important variable in paste extrusion. For a given extruder barrel, the smaller the cross section of the final product is, the higher the reduction ratio will be. The size of the preform can be varied in a fairly limited range, therefore, resins must be able to undergo the reduction during the extrusion. Different fine powder grades have been developed by the suppliers to accommodate the wide range of reduction ratios of commercial operations.

8.2 Resin Handling and Storage

Fine powder PTFE is susceptible to shear damage, particularly above its transition point (19°C). Handling and transportation of the containers could easily subject the powder to sufficient shear rate to spoil it if the resin temperature is above transition point. A phenomenon called *fibrillation* occurs when particles rub against each other, in which fibrils are pulled out of the surface of PTFE particles. Uncontrolled fibrillation must be prevented to insure good quality production from the powder. Premature fibrillation leads to the formation of lumps which cannot be broken up easily. Fibrillation and its role in fine powder processing are discussed in Sec. 8.3.

To ensure that the resin does not fibrillate, it should be cooled below its transition temperature prior to handling and transportation. A typical commercial container (20–30 kg) should be cooled 24–48 hours to <15°C[3] to assure temperature uniformity throughout the container. In practice, drums of resin are stored and transported at <5°C. Specially designed shallow cylindrical drums are used to minimize lump formation, compaction, and shearing of the resin.

Individual particles of PTFE form agglomerates which are roundish and average several hundred microns in size. Figure 8.1 shows agglomerates of a typical fine powder. Closer examination at higher magnifications reveals that each agglomerate comprises of many small (<0.25 µm) primary round particles which should have the same shape as when they were polymerized. This means that the post-polymerization isolation and drying processes should not affect the appearance of the particle. Any deformation of the resin particles or its fibrillation should be taken as an indication of potential defects in the fabricated part.

Fine powder can be compacted to a small extent during transportation and storage, even when refrigerated and handled gently, thereby creating lumps. Sifting the resin through coarse wire mesh will help break up the majority of the lumps. The size of the sieve should not be smaller than 10-mesh; 4-mesh is preferable. The resin should never be scooped out of the container but poured over the sieve to avoid its shearing. The wire mesh should be vibrated gently up and down to avoid shearing as opposed to side way movement. The remaining lumps of the powder should not be removed from the sieve but poured into a wide mouth plastic jar. After the bottle is one third full, it should be shaken gently to break apart the lumps.[4] It would be wise to process this part of the powder separately by making a different preform to minimize the risk of adding damaged powder to the rest of the resin.

Treatment of fine powder PTFE with utmost care is paramount to fabricating high quality parts at high production yield. It is best to minimize handling of the resin, such as avoiding screening when not required.

8.3 Paste Extrusion Fundamentals

The structure of the individual particles shown in Fig. 8.1 is critical to paste extrusion of fine powder PTFE. The structure of a single particle is depicted in Fig. 8.2 where almost all the polymer chains are packed in a crystalline lattice. The orderly packing of completely linear polymer chains[5] takes place during the polymerization, monomer by monomer or brick by brick. This is precisely the reason that nearly perfect crystallinity is achieved despite very high molecular weights. The chains fold after reaching certain length which creates areas with less order at the point where the molecule bends.[6]

One of the characteristics of PTFE crystals is that they are loosely packed because of the relatively low van der Waals attraction forces between the chains in

Figure 8.1 Agglomerates of a typical fine powder. *(Courtesy Dupont.)*

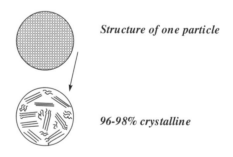

Figure 8.2 Structure of one PTFE fine powder particle.

contrast to other polymers such as polyethylene. The transition from a triclinic to a hexagonal cell unit corresponds to 1.3% increase in volume.[7] This means that chains can be removed from the surface of particles by the application of a fairly small force at above the transition temperature. Decreasing the temperature increases the shear force required for chain pullout. The chains abstracted from the particle are called *fibrils*. The ease with which chains can be pulled out at higher temperatures is the main reason that fine powder PTFE is handled, stored and transported below its transition point.

Figure 8.3 exhibits the fibrils in the structure of a paste extruded fine powder PTFE using scanning electron microscopy at 20,000 times magnification. The thickness of these fibrils is less than 50 nm and they connect multiple particles. The particles have preserved their round shape and do not appear deformed. Figure 8.4 contains four SEM micrographs taken from a cross section of a paste extruded and calendared unsintered tape. The micrographs taken at 1,000 and 30,000 times magnification exhibit extensive fibrillation in the machine direction, i.e., the direction of extrusion, extensively connecting the particles and conferring strength to the tape in this direction. Some deformation of the round particles can be observed, due to the calendaring that the tape has undergone.

A resin which has been subjected to shear stress prematurely contains particles which have fibrillated in arbitrary direction(s). These particles are referred to as *abused* in the industry parlance. They will not be able to adequately fibrillate into other particles during the processing of the resin and will appear as defects in the final product. Hose leakage and wire spark-out are typical examples of failure which may occur at the point where abused particles are located and will prevent the full realization of the part's properties.

The following principals for fine powder processing can be summarized:

- Fine powder PTFE is sensitive to mechanical shear, especially above its 19°C transition point.
- Shear stress causes fibrillation of fine powder particles, which is the removal of a few chains of PTFE from the crystalline phase upon application of shear force, generated by the extruder pressure, to a particle. Resin fibrillation increases when extrusion pressure is raised.
- All transportation, storage and handling of the powder must take place below its 19°C transition temperature.
- Paste extrusion should take place at above the 30°C transition temperature of the polymer.
- A hydrocarbon lubricant is added to PTFE to aid in processing. It is removed prior to sintering the article.

Figure 8.3 Close-up view of individual fibrils extending among multiple particles. *(Courtesy DuPont.)*

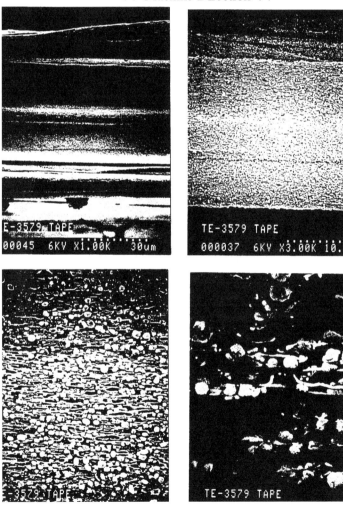

Figure 8.4 Fibrilation of fine powder PTFE in an extruded and calendared tape (freeze fractured). *(Courtesy DuPont)*

- The extrudate develops strength in the direction of extrusion as a result of fibrillation, permitting its handling in the process.
- Extrusion pressure is a function of the molecular weight of the polymer and lubricant content of the preform for the same processing conditions. The higher the molecular weight, the higher the extrusion pressure will be. The higher the lubricant content, the lower the extrusion pressure will be.
- Extrusion pressure is a function of several variables which are listed below:
  - Resin Type
  - Reduction Ratio
  - Lubricant Content
  - Lubricant Type
  - Die Cone Angle Length
  - Die Land Length
  - Extrusion Speed
  - Temperature

The impact of these variables is discussed as processes for extrusion of different shapes are described.

8.4 Extrusion Aid or Lubricant

An extrusion aid is added to fine powder PTFE as a *lubricant* to enable smooth uniform extrusion. The extrusion aid must easily coat the resin yet be readily removable from the extrudate. It should also not leave a residue which could alter the color of the product. The volatilization temperature of the lubricant should be lower than the sintering temperature of the polymer. The other requirements of lubricants include high purity, low odor, low polar components, high auto-ignition temperature, low surface tension and low skin irritation. Common lubricants are synthetic isoparaffinic hydrocarbons available in a wide boiling range. Some of the commercial lubricants include Isopar® solvents (available from Exxon Corp.), mineral spirits, and VM&P Naphtha (available from Shell Corp.).

Table 8.1 and 8.2 list the characteristics of a number of solvents. They have a range of boiling points which inversely correlates with vapor pressures. The higher the boiling point the more slowly the solvent will leave the extrudate. Ideally, the lubricant should have a lower surface tension than the critical surface tension of PTFE, which is a very low 18 dynes/cm. While this is not practical for PTFE systems, most solvents have fairly low surface tension, which helps their spreadability on polytetrafluoroethylene. Surface tension of Isopar® series rises with increasing boiling point, which adversely affects their spreadability. For example, Isopar® G has a surface tension of 23.5 dynes/cm and spreads more easily than Isopar® V with a surface tension of 30.8 dynes/cm.

The amount of the lubricant in the compound depends on the type of the product, equipment design, and the desired extrusion pressure. Its content should be as low as possible but not so low that the extrusion pressure would be excessively high. A less volatile extrusion aid is often recommended for the manufacture of an unsintered tape.[10] The range of lubricant content is 15–25% of the total weight of the compound.

Petroleum solvents are volatile and their vapors can be ignited causing flash fires or explosion. They must be kept away from heat, sparks, and open flame. Table 8.3 presents flammability data for Isopar® solvents. Fire can be avoided by controlling vaporization, controlling air-vapor concentrations, eliminating sources of ignition and eliminating spills. To reduce the risk of fire, the solvent with the highest flash point should be selected for paste extrusion. Another safety measure is thorough ventilation of the vapors from the workspace, which is also helpful in minimizing human exposure. Static electricity may act as a source of spark and can be important in the blending of powder and solvent. The measures to reduce chances of spark from static charge buildup include grounding of the equipment, avoiding splashing and vapor build up in the workplace and transferring solvent through conductive hose and nozzle.

8.5 Wire Coating

One of the important applications of fine powder polytetrafluoroethylene is wire insulation primarily for automotive, aerospace and industrial applications where high temperature rating (>250°C) and resistance to chemicals are required. The main use of PTFE insulated wire is for hookup in electronic equipment in the aerospace and military industries. Coaxial cable made by paste extrusion or tape wrapping is the other large volume consumer of fine powder PTFE. PTFE insulated wire is also found in airframe and computer applications. In this section, processing of this resin into a wire coating is described.

Table 8.1. Properties of Isopar® Solvents[8]

Grade	Isopar C	Isopar E	Isopar G	Isopar H	Isopar K	Isopar L	Isopar M	Isopar V	Test Method
Solvency									
Kauri-Butanol Value	27	29	28	27	27	27	27	25	ASTM D 1133
Aniline Point, °C (°F)	78 (172)	75(167)	83(181)	84(184)	84(184)	85(185)	91(196)	93(199)	ASTM D 611
Solubility Parameter	7.2	7.3	7.3	7.3	7.3	7.3	7.2	7.2	Calculated
Volatility									
Flash Point, °C (°F)	-7(19)	7(45)	41(106)	54(129)	57(135)	64(147)	91(196)*	129(264)*	ASTM D 56
Distillation, °C (°F)									ASTM D 86
IBP	98(208)	118(244)	160(320)	178(352)	177(350)	191(376)	223(433)	273(523)	
50%	99(211)	121(250)	166(331)	182(360)	185(365)	195(383)	238(460)	288(550)	
Dry Point	104(219)	137(279)	174(345)	188(370)	197(386)	207(405)	252(487)	311(592)	
Vapor Pressure, mm Hg @ 38°C (100°F)	98	52	14	6.2	5.7	5.2	3.1	0.3	ASTM D 2879
General									
Specific Gravity @ (60°/60°F)	0.699	0.722	0.747	0.758	0.760	0.767	0.788	0.817	ASTM D 1250
Density, lb/gal	5.82	6.01	6.22	6.31	6.33	6.39	6.56	6.80	Calculated
Color, Saybolt	+30	+30	+30	+30	+30	+30	+30	+30	ASTM D 156
Viscosity, cP @ 25°C (77°F)	0.48	0.62	1.00	1.29	1.39	1.61	2.70	7.50	ASTM D 445
Auto-Ignition Temp., °C (°F)	399(750)	382(720)	293(560)	349(660)	349(660)	338(640)	338(640)	210(410)	ASTM D 2155
Bromine Index	<5	<5	<10	<10	10	<10	5	500	ASTM D 2710
Composition, mass%									
Saturates	100	100	100	100	100	100	99.9	99.5	Mass Spectrometer
Aromatics	0.01	0.01	0.01	<0.01	0.01	<0.01	<0.05	<0.5	UV Absorbance
Purity, ppm									
Acids	None	None	None	None	None	None	None	None	Exxon Method
Chlorides	<3	<2	<1	<3	2	<1	—	7	Exxon Method
Nitrogen	—	<2	<1	<1	<1	<1	—	—	Exxon Method
Peroxides	0	0	Trace	<1	<1	<1	<1	0	Exxon Method
Sulfur	<2	<2	1	<2	<2	<2	<2	1	Exxon Method
Surface Properties									
Surface Tension, dynes/cm @25°C (77°F)	21.2	22.5	23.5	24.9	25.9	25.9	26.6	30.8	duNuoy
Interfacial Tension with water, dynes/cm @25°C (77°F)	48.9	48.9	51.6	51.4	50.1	49.8	52.2	44.9	ASTM D 971
Demulsibility	Excellent	Excellent	Excellent	Excellent	Excellent	Excellent	Excellent	Excellent	Exxon Method

The values shown here are representative of current production. Some are controlled by manufacturing specifications, while others are not. All may vary within modest ranges.
* ASTM D 93

Table 8.2. Properties of Lubricants Supplied by Shell Corporation[9]

Property	VM&P Naphtha HT	Shell Sol 340 HT	Shell Sol 142 HT	Mineral Spirit 200 HT	Odorless Mineral Spirits
Solvency					
Kauri-Butanol Value	34.6	32.0	30.0	32.1	26
Aniline Point, °C	61.1	67.7	70.6	67.8	84.4
Volatility					
Flash Point, °C Distillation, °C	12.8	39.4	61.7	43.9	51.7
Initial BP	119.4	159	187	162	179
Dry Point	138.9	176	206	206	204
Vapor Pressure, mmHg @ 20°C	9.8	1.4	0.4	1.1	0.5
General					
Specific Gravity, (15°C/15°C)	0.753	0.773	0.788	0.777	0.759
Density, kg/l @ 15°C	0.75	0.77	0.79	0.77	0.76
Color, Saybolt	30	30	30	30	30
Composition					
Paraffins, % vol.					96.5
Naphthenes, % vol.					0
Olefins, % vol.	<0.01	<0.1	<0.1	<0.1	3.5
Aromatics, % vol.	0.01	<0.1	0.25	<0.1	<0.1
Benzene, ppm	<30	<1	<1	<1	<1

Table 8.3. Solvents Flammability Data[8]

	Flash pt., TCC ASTM D 56 °C(°F)	Auto-ignition temperature ASTM D 2155 °C(°F)	Flammable limit vol. % in air @ 25°C(77°F) lower	upper	Vapor vol. per unit vol. liquid 25°C (77°F)	100°C (212°F)	Vapor density, air=1	Recommended TLV■	Density kg/m³	lb/gal	Approx. avg. mol. wt.
Isopar C	-7(19)	399(750)	0.9	7.0	150	188	3.9	400▲	700	5.83	114
Isopar E	<7(<45)	382(720)	0.9	7.0	143	179	4.2	400▲	722	6.01	123
Isopar G	41(106)	293(560)	0.8	7.0	123	153	5.1	300▲	747	6.22	149
Isopar H	54(129)	349(660)	0.7	7.0	115	145	5.5	300▲	758	6.32	160
Isopar K	57(135)	338(640)	0.7	7.0	113	142	5.7	300▲	761	6.33	164
Isopar L	64(147)	338(640)	0.6	7.0	110	137	5.9	300▲	768	6.39	171
Isopar M	91(196)●	338(640)	0.6	7.0	100	125	6.6	300▲	788	6.56	191
Isopar V	129(264)	210(410)◆	0.9	7.0	101	126	6.8	200▲	818	6.81	197

- ■ TLV is a registered trademark of the American Conference of Governmental Industrial Hygienists. It is the threshold limit value or occupational exposure limit -- the time-weighted average concentration for a normal 8-hour work day, 40-hour work week, to which nearly all workers may be exposed repeatedly without adverse effect. Refer to the most recent Material Safety Data Sheet for latest recommended maximum exposure limit for each solvent.
- ▲ A TLV has not been established for this product. The value shown has been recommended by Exxon Corporation medical research based on consideration of available toxicological data.
- ● Pensky-Martens Method, ASTM D 93.
- ◆ ASTM Method E 659. According to ASTM, auto-ignition, by its very nature, is dependent on the chemical and physical properties of the material and the method

The important attributes of PTFE insulated wire include:

- Lowest dielectric constant (2.1) and dissipation factor (3×10^{-4}) of any insulation material
- Flame resistance and low smoke generation
- Continuous service temperature range of -260°C to 260°C
- Resistance to all common chemicals, solvents and moisture
- Good electrical properties over a frequency range of $10^2 - 2 \times 10^{10}$ Hz
- High volume (10^{18} Ω cm) and surface resistivity (10^{16} Ω per square)
- High dielectric breakdown strength (20-160 kV/mm)
- High surface arc resistance >300 sec
- Ability to color by inorganic pigments
- Laser-markability of PTFE filled with titanium dioxide

8.5.1 Blending the Resin with Lubricant

Blending the lubricant, PTFE, and pigment must be conducted in a clean enclosed area where the temperature is below resin's transition temperature (19°C). This area should be controlled at a relative humidity of 50%. Safety features should be designed into the blending room. They include antistatic flooring and clothing, explosion-proof lighting and grounding for all equipment. All clothing and other fabric should be lint-free to avoid contamination of the extrusion compound.

Blending is performed most frequently by two methods: bottle or jar blending and motorized blenders. Neither technique has a clear advantage over the other. The bottle process is suitable for modest scale manufacturing. Large scale blending is usually done in a V-cone blender such as the units offered by Patterson-Kelley Corporation, East Stroudsburg, Pennsylvania, USA.

The bottle or jar method (see Fig. 8.5) requires a wide mouth polyethylene or polypropylene bottle for easy (low shear) powder loading. The jars must be sealed tightly to prevent the loss of the lubricant by evaporation. The following steps should be taken to prepare the paste extrusion compound:

1. Weigh the powder after screening and carefully load into the bottle.
2. Create a cavity by giving the bottle a rapid twist.
3. Pour the lubricant into the cavity in the middle of the powder.
4. Close the lid and place the bottle on rollers (15 rpm) for 20–30 minutes.
5. Allow the blend to age for at least 12 hours at 35°C to allow complete diffusion of the lubricant into the polymer particles.
6. Any small lumps should be broken by sieving and the lubricated powder re-rolled for 3–5 minutes.

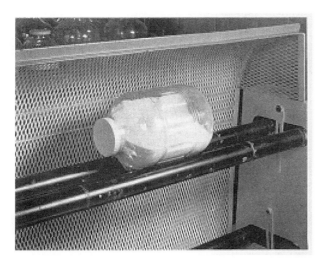

Figure 8.5 Blending fine powder PTFE by bottle rolling. *(Courtesy DuPont.)*

For large quantities of resin (25–70 kg), a twin-shell V-blender (see Fig. 8.6) may be used to incorporate the lubricant. The following steps are taken:

1. Load the powder carefully into the V-blender to avoid shearing the resin.
2. Add the lubricant evenly to the resin.
3. Set the blender to tumble at 24 rpm for 13 minutes for a 25 kg batch of resin. Longer rotation times may be required for larger batches of resin.
4. Screen the compound to break up loose lumps and separate those that do not break easily.
5. Empty the blender and store the lubricated resin in a jar or the original drum and make sure the lid is sealed tightly. Allow the blend to age similarly to step 5 above for jar blending.

Other blenders such as Turbula® Shaker Mixer (Turbula® is a registered trademark of Bachofen Maschinenfabrik, Utengasse, Switzerland) can be used for large charges of polymer.[11] This blender consists of cylindrical container held in mechanical arms with a complex motion pattern. This mixer is reported to minimize shearing of PTFE by avoiding the slamming of the powder against its walls. It does consume significantly more energy than V-blenders.

Figure 8.6 Patterson-Kelley twin-shell liquid solids blender. *(Courtesy Patterson-Kelley Co., Div. of Harsco Corp.)*

8.5.2 Pigment Addition

Pigments in dry form or as dispersion can be added to PTFE to color the insulation, primarily for the ease of differentiation of wires. Pigment dispersions (*liquid pigments*) are preferable for critical applications such as thin-wall wire insulation and spaghetti tubing. Dispersions reduce the formation of flaws due to undispersed pigment. These flaws weaken the extrudate and frequently lead to the electrical breakdown of the insulation. Pigments can be dispersed in hydrocarbons using dispersants and most are commercially available in dispersion form. Pigment loading of any type in the final insulation should not exceed 1% because of its detrimental effect on the dielectric properties.

Inorganic pigments should be selected for coloring PTFE because almost all organic pigments will completely degrade at the sintering temperatures of the polymer. Table 8.4 contains examples of inorganic pigments that are compatible with PTFE. Dry pigment should be first screened before being added to the dry PTFE. After adding the pigment to the resin, the container should be rolled or tumbled for 5–10 minutes to mix the pigment throughout the dry resin. The lubricant should be added afterwards by following the procedures described in Sec. 8.5.1. It may be necessary to dry the pigment by heating it in a vacuum oven to remove moisture and other volatile substances prior to use.

Liquid pigments should be well dispersed prior to addition to the resin for even dispersion in the polymer. This can be achieved by shaking or rolling the dispersion for several hours before use. The lubricant and pigment dispersion should be mixed and quickly added to the PTFE powder because the settling of pigment particles occurs rapidly after the addition of the lubricant. The amount of additional lubricant should be adjusted for the hydrocarbon content of the pigment dispersion.

8.5.3 Preforming

After lubricant and pigment have been added and the compound aged, preforming takes place. This step, which usually takes place at the room temperature, shapes the compound into a billet with the same shape as the barrel of the ram extruder. The rule of thumb is to compact the resin to one third of its initial height.[3][11] Preforming removes the air from the PTFE powder and, by compaction, maximizes the quantity of material available for extrusion. The objective is to extrude the maximum possible length of wire.

Preforms are made in a cylinder equipped with mandrel and a pusher similar to the schematic in Fig. 8.7. The mandrel is positioned in the center of the cylinder and the resin is charged in the annular space. The diameter of the preform and the center hole are designed so that:

1. The outer diameter is 0.2–1.3 mm less than the inside diameter of the barrel.
2. The core diameter is 0.25 mm larger than the extruder wire guide (mandrel).

The surfaces of the barrel wall and mandrel should be smooth and free of scratches and nicks. A surface finish of finer than 63 RMS is recommended.[3]

The aged lubricated resin without lumps is loaded into the preform cylinder and is evenly distributed around the core mandrel to ensure uniform compaction throughout the preform. The pusher is put on top of the cylinder and compaction commences. Resin compression can begin at a fairly rapid rate but has to be reduced at the latter stages of compaction. This rate reduction is size dependent (Table 8.5) and is aimed at the prevention of air entrapment, otherwise the preform may crack.

Low pressure should be used to compact the powder in the cylinder. At the initial stage, 0.5–1 MPa pressure is applied which should be increased to 2 MPa by the end of the compression cycle. The criteria for pressure selection is to compress the resin at a sufficiently high pressure to push the air out and prevent preform cracking. Yet pressure should not be so high that a large portion of the lubricant would be squeezed out of the resin without increasing compaction. A small amount of lubricant tends to be pushed out of the preform even when pressure is adequate and should not raise concern.

The preform is quite weak and can easily break or deform, therefore it requires care during removal from the cylinder. The preform can be loaded in the extruder immediately after removal. It can also be stored at ambient temperature in a plastic tube prior to the extrusion to avoid contamination and damage, and to avoid lubricant loss.

Table 8.4. Examples of Pigments Compatible with PTFE

Color	Pigment	Supplier
White	TiPure R-101	DuPont
Black	Regal 400-R	Cabot
Red	Dark Red V-8540	Ferro
Green	F-5687	Ferro
Yellow	Krolar Yellow Light KY-781-D	Dominion Color
Orange	Orange V-8810	Ferro
Blue	Monastral Blue G BT-465-D	Ciba-Geigy
Purple	Purple V-8291	
Brown	Yellow Brown F-6109	

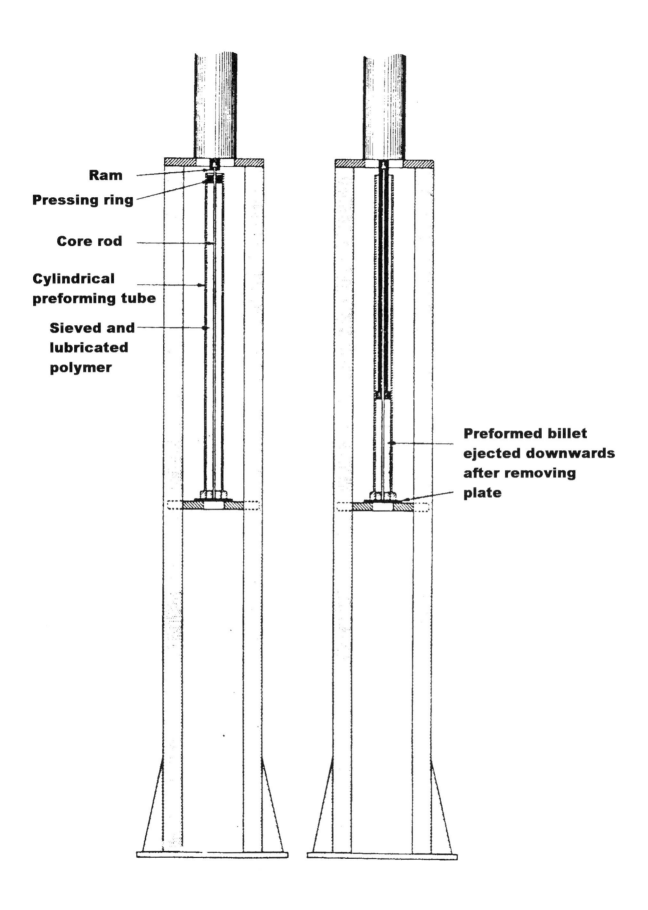

Figure 8.7 Preforming equipment.[11]

Table 8.5. Preforming Compaction Rate (cm/min)[3]

Compression Stage	Preform Diameter, cm	
	< 8 cm	> 8 cm
Initial	25	25
Middle	50	50
Final	50	1

8.5.4 Extrusion Equipment and Process

Figure 8.8 shows a schematic of a paste extrusion line for wire insulation. The wire is passed through the paste extruder where it is coated with PTFE while moving through the die. It then enters a vaporizing oven where it is stripped of the lubricant by evaporation. Vaporization temperatures depend on the type of the lubricant, heavier hydrocarbons requiring higher drying temperature. Next the dried and coated wire goes into the sintering zone, which usually consist of several individual ovens placed in series. Temperatures of the ovens are set to heat the polymer above its melting point quickly. After leaving the ovens, the wire is cooled and passed through a *spark* tester where the insulation is subjected to a voltage for detection of the flaws. The last step is winding the wire on a spool, which is done by a motorized take-up system.

The coating that forms the insulation consists of a thin-wall tube, which is paste-extruded onto moving conductors, proceeded by lubricant removal and sintering. The extrusion of the small insulation tube around the wire requires the preform to be *reduced* by forcing the paste through a small die. This reduction gives rise to an important parameter called *reduction ratio*, which is a characteristic of fine powder PTFE; it will be defined in Sec. 8.5.5. Reduction ratio refers to the ratio of the paste cross section area in the extruder barrel to the cross section area of the extrudate. Wire coating requires resins capable of moderate to high reduction ratio, depending on the thickness of the wall of the insulation. The extrusion of the preform is conducted above room temperature (30–100°C) to take advantage of the deformability of PTFE at higher temperatures. Electric heating bands equipped with independent controls heat the extruder barrel and the die.

8.5.4.1 Extruder

The ram extruder for this process is a special unit and can be either horizontally or vertically oriented, which refers to the direction of the ram movement. The extruder consists of a heated barrel where the preform is loaded, and a hydraulic or screw-driven ram. The conductor is drawn by a power system through a hollow mandrel located at the center of the barrel. The mandrel terminates in a wire guide tube. It can be adjusted to alter the position of the tip of guide tube relative to the die. Figure 8.9 and 8.10 exhibit examples of two commercial paste extrusion units.

There is an option of prepressing the preform prior to the onset of the extrusion. The advisability of prepressing depends on the assembly condition of extrusion equipment. Prepressing could be helpful in driving out the air if the entrapped air cannot escape through the back plate seal and the die seal during the extrusion. This means that the seals are fairly airtight. Otherwise, it is likely that the number of faults could increase because of prepressing. The escape of high-pressure air through the insulation could leave holes and voids which are too large to be closed during the sintering.

The wire payoff system is usually motorized and is equipped with an adjustable tensioning device to keep the wire from slackening or being too tight. The speed of the wire and the ram must be coordinated to produce insulated wire. In commercial extruders, a control system synchronizes the changes in the wire and ram speeds.

The preform is forced through the die by the force of the ram. It is important to be able to control the speed of the ram on a continuous scale and keep it literally constant at a set speed. The uniformity of the thickness of the coating is critically dependent on the constancy of the ram speed. The hydraulic or mechanical drive system must be capable of supplying the force necessary to extrude the preform. Ram pressure capability up to 150 MPa may be required for extrusion at high reduction ratios.

8.5.4.2 Die

The preform fibrillates in the die under ram pressure and forms an extrudate, which should have the right thickness and smoothness. The design of the die (Fig. 8.11) plays a key role not only in the property and quality of the coating but the extrusion pressure. The preform is pressed through the extruder cylinder (barrel) with little pressure development until it reaches the die where the cross section area for the passage of the preform decreases by the angular design of the wall. At this point the polymer particles are forced to compete for flow through this increasingly smaller area, thereby rubbing past each other. This phenomenon gives rise to fibrillation as described in Sec. 8.3.

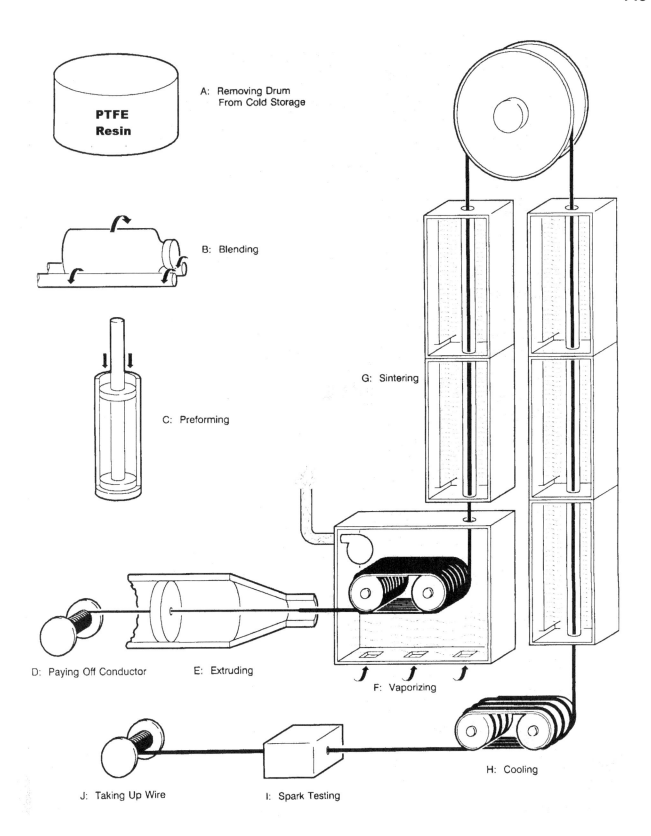

Figure 8.8 Schematic of PTFE wire coating paste extrusion line.[3]

Photo courtesy of Davis Electric Wallingford Corp.

Figure 8.9 A Davis Electric Vertical Extruder.

Photo courtesy of Jennings International Corp.

Figure 8.10 A Jennings International Horizontal Extruder.

The angle of the die wall (cone) affects the surface smoothness of the extrudate. The range of the angle is 15–60° but an angle of 20° for thin coatings on fine conductors and an angle of 30° for thicker coatings have been recommended.[3] Surfaces of the cone and the die land areas should be polished to a mirror finish of 0.1 microns R_a.[11] Importance of surface finishes increases at higher extrusion speeds and pressures.

The conductor (wire) emerges from the guide tube at a tip (Fig. 8.11) which guides it through the segment called the *die land*. The location of the tip is

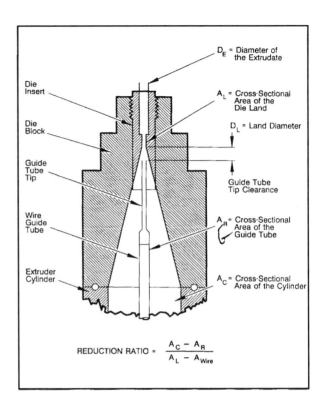

Figure 8.11 Details of master die.[3]

critical to the insulated wire quality because it affects how the wire and PTFE are brought together. Adjusting the position of the tip can control coating thickness, stripability of the PTFE (tightness) and the number of flaws. If the clearance is too small, the tip constrains the movement of the paste and extrusion pressure rises. Small clearances also reduce the tightness of the insulation around the conductor and lead to low strip force. If the clearance is too large the wire tension increases because of the drag of the paste on the conductor. It also elevates the tightness of the insulation and may lead to cracking. Larger clearance will allow the paste to extrude backwards into the tube. The guide tube should be snug around the wire with a maximum clearance of 25–50 μm.[11] It is important to experiment with the tip clearance to obtain the best extrusion condition for a given configuration.

The diameter, length, and temperature of the die land area influence the coated wire properties. Allowance should be made for swelling ("blow up") of the coating when it leaves the die. Relaxation of built-in strains leads to die swelling of the polymer, which is a transient effect. The diameter of the die should accommodate 3–10% die swell. Die land length of 6–13 mm and die temperature >35°C have been found to produce smooth extrudates.[3]

Die design is a fairly complex process and should be tried after an in-depth understanding of the effect of its parameters on the extrusion process and product quality. An iterative process by trial and error can be costly. Benbow and Bridgwater[12] offer an excellent source for the design of paste extrusion dies.

8.5.4.3 Drying

The polymer coating containing the lubricant is fairly fragile and susceptible to mechanical damage. The lubricant content of the insulation is nearly 40% by volume, which must be removed prior to sintering. The coating may crack, if it contains a large amount of the lubricant and reaches the sintering zone. Any remaining hydrocarbons will degrade at the sintering temperatures and leave a colored residue.

There are two configurations of drying ovens: internally heated with tubular design and horizontal heated console design; both are vented to remove the vapors. A typical tubular oven is about 3 m long with a diameter of 150–200 mm. One or two 3 m ovens are required for complete drying. Temperature at the oven entrance is 150°C and at the exit 300°C. In the heated console type oven, several 10 meters of the wire are wound around multiple sheaves which allows longer residence time in this oven than the tubular kind. Temperature is lower in console ovens and ranges between 90°C and 200°C. The exact temperature depends on the speed of the wire and lubricant type. A capstan drive pulls the wire forward synchronously with the ram speed.

The drying process removes large volumes of flammable hydrocarbons and must be capable of reducing the concentration of these vapors below the flammability range. Most Isopar® solvents have a room temperature flammability range of <1% to 7% (Table 8.3). This requires a large volume of air to be blown by the exhaust system to reduce the vapor below the lower limit of the flammability range. A reduction of vapor concentration to 0.2–0.4% will reduce the risk of ignition of hydrocarbons.

8.5.4.4 Sintering and Cooling

The wire enters the sintering zone immediately after it leaves the drying oven. It still lacks strength because of its porous unsintered structure. PTFE undergoes coalescence and void elimination during sintering. Detailed description of the coalescence of polytetrafluoroethylene can be found in Ch. 7. After sintering and cooling, the insulation assumes its permanent dimension and ends at 50–60% crystallinity.

The polymer must be heated to, at least, its melting point of 342°C for a brief period of time for melting to actually occur. In practice, higher temperatures well above the PTFE melting point are used to reduce melt creep viscosity of the polymer for rapid void closure. The oven temperatures are typically set at 400–600°C, based on a number of variables including the speed of the wire and thickness of the coating. Heat transfer to the polymer accelerates at these temperatures, but care must be taken to prevent exposure of PTFE to temperatures above 380°C where degradation begins to speed up. The sintering ovens should be equipped with an exhaust system to remove the toxic by-products of PTFE degradation.

Multiple ovens are used for sintering, sometimes as many as eight. The most common ovens are tubular with a typical diameter of 250 mm and about 1 m long. Older ovens have radiant electrical heating elements. The newer ovens are electrically heated with quartz lining, which emits infrared radiation.

Cooling of the coated wire is relatively easy because it is thin compared to granular PTFE parts discussed in Ch. 7. The exiting wire insulation is in the molten state, solidifying upon contact with the ambient air. The wire is usually allowed to cool by natural convention in the ambient air. Blowers can be installed to move the warm air away from the area. The crystallinity of the PTFE is about 50%. It is possible to quench the coating by blowing cold air or passing it through a cold water bath. In such cases crystallinity can be driven below 50% which can have a measurable impact on the properties of thicker coatings.

The wire enters a spark tester, which is a dielectric breakdown tester, and conceptually performs in a manner similar to ASTM Method D149. It subjects the wire insulation to a known voltage continuously. The objective is to measure the number of spots in a length of wire which are too weak to stand up to the test voltage. The failure makes a dielectric arc accompanied with a buzzing sound, thus called *sparks*. The number of sparks can be automatically measured and recorded. Sparks are a measure of the quality of the wire, directly measuring its functional value. The voltage applied depends on the thickness of the coating. For example, assume a coating has a thickness 0.25 mm. Short term dielectric breakdown resistance of PTFE is 24 kV/mm, therefore calculating to a voltage of 6 kV/mm for a 0.25 mm thickness. In practice, half this voltage (3 kV/mm) may be used for testing since 6 kV/mm would represent a typical maximum value for the polymer.

Tables 8.6 and 8.7 show examples of manufacturing process variables for coating of wires, which comply with four US Military (MIL) Standards, using two commercial resins.

8.5.5 Reduction Ratio

Earlier a qualitative description of reduction ratio was given. In this section we will expand the discussion of this important characteristic of the polymer and paste extrusion process.

Figure 8.11 shows the basic design of a die for wire extrusion. Reduction ratio ($R.R.$) is defined as the ratio of the cross section of the polymer before extrusion to that after extrusion. This ratio can be written as Eq. (8.1).

$$\text{Eq. (8.1)} \quad R.R. = \frac{A_C - A_G}{A_L - A_W} = \frac{\pi D_C^2 - \pi D_G^2}{\pi D_L^2 - \pi D_W^2}$$

A_C = Cross sectional area of the extruder cylinder (barrel), cm^2

A_G = Cross sectional area of the guide tube (mandrel), cm^2

A_L = Cross sectional area of die land, cm^2

A_W = Cross sectional area of the wire (conductor), cm^2

Equation (8.1) can be simplified, shown in Eq. (8.2).

$$\text{Eq. (8.2)} \quad R.R. = \frac{D_C^2 - D_G^2}{D_L^2 - D_W^2}$$

D_C = Diameter of the extruder cylinder (barrel), cm

D_G = Diameter of the guide tube (mandrel), cm

D_L = Diameter of die land, cm

D_W = Diameter of the wire (conductor), cm

Table 8.6. Recommended Tooling and Processing Conditions for Extruding E223 Constructions of Teflon® CFP 6000 Fluoropolymer Resin[3]

Variable	Unit	Barrel Size, in (mm)		
		1.75 (44.5)	2.0 (50.8)	2.5 (63.5)
Mandrel, O.D.	in (mm)	0.625 (16.9)	0.625 (16.9)	0.75 (19)
Conductor, O.D.	in (mm)	0.0296 (0.75)	0.0296 (0.75)	0.0296 (0.75)
ISOPAR G Concentration	wt%	16.5	18.0	18.5
Die Size	in (mm)	0.056 (1.42)	0.056 (1.42)	0.056 (1.42)
Tip, I.D. × O.D.	in (mm)	0.032 × 0.042 (0.81 × 1.06)	0.032 × 0.042 (0.81 × 1.06)	0.032 × 0.042 (0.81 × 1.06)
Tip Clearance	in (mm)	0.08–0.10 (2.03–2.53)	0.08–0.10 (2.03–2.53)	0.08–0.10 (2.03–2.53)
Blow-Up, O.D.	in (mm)	0.057–0.058 (1.45–1.47)	0.058–0.059 (1.47–1.49)	0.058–0.059 (1.47–1.49)
Finished O.D. (Hot)	in (mm)	0.050 (1.26)	0.050 (1.26)	0.050 (1.26)
Wire Speed	ft/min (m/min)	260 (79)	260 (79)	260 (79)
Pressure	psig (MPa)	8,000 (57.6)	9,000 (64.8)	11,000 (79.3)
Reduction Ratio	—	1250:1	1700:1	2500:1
Vaporizer	°C (°F)	177 (350)	177 (350)	177 (350)
Sintering Oven Zone	°C (°F)			
1		482 (900)	482 (900)	482 (900)
2		538 (1,000)	538 (1,000)	538 (1,000)
3		566 (1,050)	566 (1,050)	566 (1,050)
4		593 (1,100)	593 (1,100)	593 (1,100)
5		482 (900)	482 (900)	482 (900)

Table 8.7. Examples of Electric Wire Insulation Molded from Daikin PTFE[4]

Item	F-201 I	F-201 II	F-201 III
Core wire structure (no. of strands/diameter mm)	7/0.320	19/0.127	7/0.127
Wire plating	Silver	Silver	Silver
Wire outside diameter (mm)	0.96	0.64	0.38
Insulation thickness (mm)	0.25	0.25	0.15
MIL standard	E-20	E-24	ET-28
Extruder Cylinder diameter (mm)	38	38	38
Mandrel diameter (mm)	16	16	16
Die angle (degree)	20	20	20
Die tip diameter (mm)	1.60	1.321	0.762
Guide tube diameter (mm)	1.067	0.686	0.406
Reduction ratio (R.R.)	732	899	2751
Amount of extrusion aid blended (Weight part)	19.0	21	22
VM8P Naphtha (% by wt)	15.9	17.3	18.0
Preforming pressure and time (kgf/cm^2 × min)	25 × 5	25 × 5	25 × 5
Guide tube/guide tip clearance	0.8	0.6	0.3
Calculated value (mm)	(0.78)	(0.76)	(0.62)
Die temperature (°C)	50	50	50
Extruder ram speed (mm/min)	18.3	19.0	11.0
Haul-off speed (m/min)	8.2	14.0	18.2
Extrusion pressure (kgf/cm^2)	615	500	1015
Oven temperature #1 (drying) (°C)	95	95	95
#2 (drying) (°C)	205	205	205
#3 (sintering) (°C)	400	400	400
Molded product dimensions Outside diameter (mm)	1.52	1.10	0.68
Insulation thickness (mm)	0.28	0.23	0.15
Number of sparks	None	None	None
(Test voltage) (KV)	(3.4)	(3.4)	(1.5)

Smaller cylinder, larger mandrel, larger die land and smaller wire diameter can each reduce the reduction ratio. Practically, the size of the wire is fixed, which leaves the preform as the only practical variable. To make a long length of wire, a reasonable size of preform is required which is why high reduction ratios similar to those in Tables 8.6 and 8.7 are encountered. To be sure, smaller barrels are used where possible. Resin manufacturers have developed polymers for very low to very high reduction ratios, some of which are capable of undergoing reduction ratios as high 4000:1. The most important polymer property affecting the operating range of the reduction ratio is molecular weight, which can be easily manipulated during polymerization.

Higher reduction ratio increases the extrusion pressure, which can be reduced by the type of lubricant used and increasing its content. Figure 8.12 illustrates the effect of reduction ratio on extrusion pressure for two commercial resins. The relationship is close to linear, e.g., a doubling of reduction ration nearly doubles the extrusion pressure.

8.5.6 Conductor

A high quality wire has a strong impact on the coated product that is produced. A poor wire can lead to a large number of flaws in the PTFE insulation. It often results in a poorly performing process. Stranded wires should be free of loose strands or high wire; the latter means extra strands in the stranded bundle. Both lead to snagging and process disruption. One likely solution may be a larger diameter guide tube. This will only add to the sensitivity of the already critical guide tube position. Cleanliness of the wire is another important factor in insuring good quality insulation. The wire should be free of all contaminants such as oil, grease, or dust.

8.6 Extrusion of Tubing

The majority of tubes made from PTFE by paste extrusion have fairly thin walls (<8 mm) and are produced in a wide size range from a fraction of millimeter to several centimeters in diameter for applications ranging from fluid transfer in healthcare to fuel and hydraulic transfer in jet engines. Tubing is divided into three categories based on the size and wall thickness, for various applications. Table 8.8 summarizes the size and applications of each type. Pressure hoses are composite devices of one or two layer PTFE lining reinforced with overbraiding, usually metal wire, to increase its pressure rating. Each of these tubes requires a somewhat different processing method from the others because of significant disparity of their sizes.

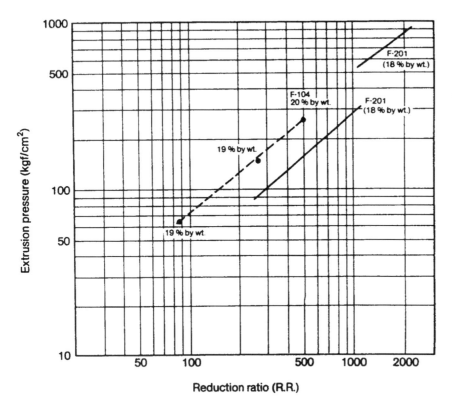

Note: Extrusion aid used: Super VM & P Naphtha (Shell)
Extruder used: Jennings extruder

Figure 8.12 Relation of reduction ratio and extrusion pressure.[4]

8.6.1 Blending Lubricant and Pigment and Preforming

Resin lubrication, pigmenting, and preforming should be done similarly to preparations described for wire coating in Sec. 8.5.1, 8.5.2, and 8.5.3. The amount and type of the lubricant and pigment may change, but the basic methods for blending, mixing, and handling remain the same. Tubing preforming is also done the same way but sizes may be different from those used for wire insulation. Any differences in preparations have been pointed out wherever necessary.

Table 8.8. Types of Tubing and Applications Made from Fine Powder PTFE

Type of Tubing	Diameter, mm	Wall Thickness, mm	Applications
Spaghetti Tubing	0.2–8	0.1–0.5	Electrical Insulation, Fluid Handling in Medical Equipment and Chemical Applications
Pressure Hose	6–50	1–2	Fuel and Hydraulic Transfer in Aerospace, Chemical and Gas Transfer in Chemical Processing
Pipe-liner	12–500	2–8	Lining Metal Pipes and Fitting for Chemical Processing

8.6.2 Extrusion of Spaghetti Tubing

A small vertical paste extruder can be used to manufacture spaghetti tubing. The small size of this kind of tubing eliminates most of the handling problems associated with larger ones. A vertical machine can extrude upwards or downwards. Frequently the extruder is placed at a height of 10–15 meters above the shop floor, which allows the drying, sintering and cooling to be accomplished during the downward fall of the tubing. A motorized windup unit on the shop floor collects the tubing.

A die design for tubing extrusion can be seen in the Fig. 8.13. It resembles the design for wire coating (Fig. 8.11) except that instead of a guide tube there is a core pin which extends into the die land. Wire coating paste extruders could be set up in a modified arrangement to make tubing. In this case the guide tube should extend beyond the die land exit to prevent its blockage by the polymer. The core pin or the guide tube diameter should be sized to give the correct internal tubing diameter. The ovens should be long enough to allow straight tube production. A horizontal or an upward machine will require motorized take-up to move the tubing through the ovens.

Extrusion conditions are quite similar to those used in wire coating and Sec. 8.5.4 should be consulted. Extrusion pressure can reach very large values (100–150 MPa) because high reduction ratios are necessary to obtain small diameter and wall thickness. This means that the extruder barrel and the die must be structurally sound to withstand the pressure. The lubricant has to be removed prior to sintering as in the wire coating process. Tables 8.9 and 8.10 provide examples of extrusion conditions for three commercial resins.

A frequently encountered problem in the fabrication of spaghetti tubing is start-up threading problems. The tubing is small and mechanically weak in the gel state. It tends to break under the fairly moderate tension required to pull the tube through the process. The trick that is used with larger tubing is to connect a leader of wire or chain to the tubing and guide it through

Table 8.9. Typical Extrusion Conditions for "Spaghetti" Tubing[11]

Polymer	Fluon® CD509	Fluon® CD509
Type of lubricant	ISOPAR H	VM&P Naphtha
Lubricant content (% of total mix)	19.0	19.0
Extrusion cylinder diameter (in)	1.75	2.75
Mandrel diameter (in)	0.625	0.650
Core pin tip (in)	0.040	0.694
Die land diameter (in)	0.055	0.759
Die included angle (°)	20	20
Reduction ratio	1900:1	75:1
Die temperature (°F)	150	100
Extrudate speed (ft/min)	30	1
Sintered outside diameter (in)	0.054	0.683
Sintered inside diameter (in)	0.040	0.625
Sintered wall thickness (in)	0.007	0.029

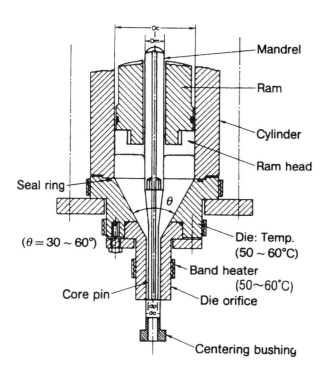

Figure 8.13 Schematic of tube extrusion die.[4]

Table 8.10. Examples of "Spaghetti" Tubes Extruded from Daikin-Polyflon TFE[4]

Item		F-201 I	F-201 II	F-201 III
Die diameter (outside, mm)		1.32	1.60	1.90
Core pin diameter (inside, mm)		1.08	1.27	1.50
Clearance (thickness, mm)		0.12	0.165	0.20
Reduction ratio (R.R.)		2112	1266	870
Extruder	Cylinder diameter (mm)	38	38	38
	Mandrel diameter (mm)	16	16	16
	Die angle (degree)	20	20	20
	Die temperature (°C)	50	50	50
Amount of extrusion aid blended *1				
(Weight part)		23	22	21
(wt.%)		18.7	18.0	17.4
Preforming pressure (kgf/cm^2)		25	25	25
Extrusion pressure (kgf/cm^2)		880	600	540
Extrusion speed (m/min.)		22.0	18.0	16.0
Drying zone temperature *2(°C)				
#1		100	100	100
#2		250	250	250
Sintering zone temperature *2(°C)				
#3		400	400	400
Product dimensions				
Outside diameter (mm)		1.08	1.27	1.54
Inside diameter (mm)		0.94	1.07	1.24
Thickness (mm)		0.07	0.10	0.15
Tensile strength				
Longitudinal direction (kgf/cm^2)		720	720	
Elongation				
Longitudinal direction (%)		350	350	

Note: *1 Extrusion aid used; Super VM & P Naphtha (Shell)
*2 Corresponding to Jennings extruder (30-ton)

the line. The weight of the leader can be too much for fine tubing and break the tube. One approach that is usually successful is to thread the tubing before sintering ovens have reached the melting temperature of PTFE. The preform, before melting, has sufficient strength to withstand the tension of threading. The oven temperatures can be raised afterwards with minimal material loss because the small size of the tubing will allow it to heat up rapidly to the sintering temperature.

The die angle is about 20° in the majority of spaghetti tubing dies. The core pin is made of a hard plastic material such a polyacetal (such as Delrin® by DuPont). The flexibility of the plastic (as opposed to a metal) allows the tube to have better concentricity than if a metal core pin is used.

8.6.3 Pressure Hoses

This class of tubing serves critical purposes in a number of industries. In this section, their applications and requirements have been covered followed by a discussion of their manufacturing process. The normal size range is from 6 to 50 mm in diameter. The tube sizes in the United States have a special designation. They are called by – (dash) followed by a dash number such as – 4 (dash four) or –12 (dash twelve), in the industry parlance. To obtain the diameter of the tube in inches, the number must be multiplied by $1/16$, thus, – 4 is equal to $1/4$ inch and –12 is equal to $3/4$ inch.

PTFE tubes form the inside liner component of these hoses and come in contact with the fluid. Chemical resistance and durability at extreme temperatures are supplied by polytetrafluoroethylene but mechanical integrity has to be fortified. The reinforcement of the tubing results in significantly higher operating pressure than can be achieved with the tube alone. Figure 8.14 represents one example of braiding by which the tube is fortified. Stainless steel wire (filament) is braided in the pattern seen in the Fig. 8.14(a). Braiding can be done in double or triple layers using other fibers such as glass, polyester, polypropylene or high performance polyaramid yarns such as Kevlar®. Figure 8.14(b) illustrates an example of a double braided tube with stainless steel filament. The tubes are fitted with metal inserts and fittings for connection.

Some applications require flexibility to bend and curl the hoses. To accomplish this, the extruded and sintered tubes can be convoluted in a separate step. The tube is passed through a heated die, which melts the PTFE and creates a spiral peak and valley pattern into the tube. A key requirement of convolution process is to assure that the wall thickness remains uniform, in other words, the tube is not stretched. Any thinning of the wall will weaken and reduce the burst pressure of the hose. Figure 8.14(c) shows an example of convoluted tube, which is partly braided with stainless steel.

Another issue in high velocity transport of hydrocarbons such as jet fuel is the build up of static charge on the interior layer of the PTFE tube. The discharge of static charge in the absence of oxygen can lead to a failure in the form of a pinhole in the wall of the tube and subsequent leakage. In the presence of oxygen, static discharge could act as an ignition source. To overcome this problem, the inner layer of the tube is made from a 1–2% carbon-filled PTFE, which would then allow surface drainage of static charge through the metal fittings. The inside part of the preform is made of carbon filled PTFE by essentially partitioning the mold.

Pressure hoses find applications where corrosive liquids and gases are transported. Examples include:

— Hydraulic fluid and fuel transport in aerospace industry

— High pressure air, fuel and hydraulic transfer in automotive industry

— Chlorine, steam, acids and organic compounds in chemical processing industry

— High purity transfer lines in pharmaceutical industry

The hose must meet numerous requirements in the various high pressure applications:

— Low permeability
— High flex life
— Good mechanical properties
— Chemical resistance
— Service at extreme temperatures
— High purity

Special grades of polytetrafluoroethylene have been developed to meet all these stringent requirements. Lower permeability and higher flex life require a polymer with minimal void content after sintering. The latter can be accomplished by lowering the molecular weight to reduce melt viscosity, thus improving coalescence. Lowering the molecular weight normally

(a)

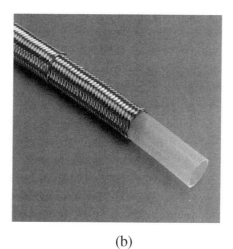
(b)

(c)

Figure 8.14 Reinforcement of PTFE tubing. (a) Braiding of PTFE tubing with stainless steel. (b) Hose in TEFLON 62 with double braiding for high pressure applications. (c) Overbraided convoluted tubing—natural and antistatic.[13]

results in higher crystallinity, therefore, reduced flex life. To overcome this dilemma, the resin suppliers have modified[13] PTFE by copolymerizing a small amount of another perfluorinated monomer such as perfluorpropylvinylether (PPVE) with a pendent group and a lower molecular weight. This small amount of the comonomer is sufficient to disturb the crystallinity of the chains, which leads to a more amorphous phase.

Mechanical strength increases because of the larger content of the amorphous phase. Flex life increases while flex modulus, which is a measure of the ease by which the tube can be bent decreases (see Table 8.11). Lower melt viscosity allows improved coalescence of the polymer particles and nearly complete void closure and elimination. The leading commercial PTFE grades consumed in the fabrication of tubes for pressure hose have been listed in Table 8.12.

Table 8.11. Stiffness as a Function of Crystallinity[13]

% Crystallinity	Specific gravity	Flexural modulus, psi	Flex life (cycles to failure)
50	2,14	40 000	4×10^6
53	2,15	54 000	$1-4 \times 10^6$
75	2,22	150 000	6×10^4
85	2,25	170 000	1×10^3

Table 8.12. Major Commercial Grades of PTFE for High Pressure Hose

Manufacturer	Resin Grade
DuPont	Teflon® 62
DYNEON	Hostaflon® TFM-2001
Imperial Chemical Industries	Fluon® CD-086
Daikin America	Polyflon® F-303
Ausimont	Algoflon® DF-381

8.6.3.1 Blending Lubricant and Pigment and Preforming

Resin lubrication, pigmenting, and preforming should be done similarly to the preparations described for wire coating in Sec. 8.5.1, 8.5.2, and 8.5.3. The amount and type of the lubricant and pigment may change but the basic methods for blending, mixing and handling remain the same. Preforming is also done the same way but preform sizes are larger than those used for wire insulation. Larger preforms take significantly longer time to mold than smaller ones.

8.6.3.2 Extrusion, Sintering and Cooling

The basic die, extruder, and oven design are similar to those used for spaghetti tubing except for the scale. Much larger components are needed because of the size of the tubing. Process conditions, insofar as the ram extruder and the die are concerned, are quite similar.

Pressure hoses are extruded by continuous and batch processes, vertically and horizontally. The advantage of the continuous process, vertical for > 2 cm diameter tubes, is that the fragile preform is not handled before sintering. It has the disadvantage of requiring a tall building. The horizontal extrusion can be operated without much height. It involves handling of the preform in order to sinter it in a separate oven. Both processes have to utilize large extruders to accommodate large size tubes and relatively thin walls. The vertical process begins with the downward extrusion of a preform from a large extruder located several stories above the shop floor. The preform is dried and sintered in several in-line ovens. The batch process has the advantage of isolating the cooling step from the extrusion step while in the vertical process the two are combined.

Independently controlled ovens are used for drying and sintering in vertical extrusion. Temperature is most closely controlled in quartz infrared ovens each around 1 m long. Half the zones are devoted to dry the preform. The total number of drying and sintering zones depends on the wall thickness and the rate of extrusion. Oven temperatures are normally set to only partially melt the polymer in the initial zones in order to avoid subjecting the melt to the weight of a long length of the tube. Most of the coalescence takes place in the last two or three zones. Hot air is blown countercurrently into the ovens to remove the lubricant vapor and off-gases.

The tube leaving the last oven is in melt state and must be cooled rapidly to minimize crystallinity. This is necessary due to the importance of flex life and flexibility (low modulus) in pressure hose applications. Crystallinity of modified PTFE is less sensitive to the cooling rate than the unmodified polymer. Quenching in cold water will keep crystallinity close to 50%,[13] which is about the practical lower limit that can be achieved in manufacturing. At this level of crystallinity, the hose should have excellent flex life, assuming that no other flaws are present.

Horizontal extrusion is almost always a batch process because of the difficulty of handling the fragile preform. For very small tubes, it is possible to do in-line (continuous) drying and sintering. Larger tubes can be extruded inside a metal pipe in small sections. The preform can be extruded over a metal mandrel for very large sizes, again in small lengths. The pieces can then be dried, sintered and cooled in batches.

The process of choice for larger diameters (2–5 cm) has been continuous vertical extrusion where the preform is extruded in a downward direction without subjecting the tube to any bending prior to the completion of the cooling step. The problem with bending a preform or even a partially sintered tube is that mechanical damage (e.g., cracking) to the part done by bending can not be corrected by the sintering process.

Selection of the number and length of the ovens must be made according to the wall thickness and the extrusion rate. Quartz infrared ovens provide more uniform heating to the tube and easier temperature control. Temperature settings of the ovens have a powerful impact on the process. Temperature of the first 1–3 zones, depending on the tube wall thickness, should be set such that the extrudate is completely dried before PTFE begins to melt. This is to avoid trapping and degrading the lubricant in the polymer. Lubricant vapors inside the tube must be removed by forced air or by another mechanism.

The weight of the vertical column below the drying zones is supported by the wet and dry extrudate. It is important that the unsintered part has sufficient mechanical strength to handle the suspended tube weight. This requires a high level of fibrillation, particularly for large tubes. Adequate fibrillation of modified PTFE can be achieved by running the extruder at a sufficiently high pressure.

It is important to remember that, after PTFE melts, its ability to bear load significantly decreases. The fibrillated structure that provides mechanical strength disappears upon reaching gel state, indistinguishable from non-fibrillated structures. Completion of the melting PTFE and its coalescence must take place as close to the cooling zone as possible. This arrangement minimizes the length of the tube that is supported by the PTFE in gel state.

The ideal cooling technique is to quench the tube instantly in very cold water to minimize crystallinity. Practically speaking, modified PTFE can be quenched in cool water without a sizable increase in its crystallinity. The adequacy of sintering and cooling can be determined by measuring the heat of fusion by differential scanning calorimetry. Unsintered polymer will manifest itself by a peak above the first melting point of PTFE (342°C). Oversintered polymer can be detected by the large value of the second heat of fusion (*ca.* >28–30 J/g). The intensity of quenching is observed by the value of first heat of fusion (ca. 20–23 J/g).

It is important to recognize that there is a large body of proprietary technology and knowhow that processors of fine powder polytetrafluoroethylene hold. Successful fabrication of tubing requires both the correct practice of the fundamental principles and access to the information that processors have developed over the last few decades. This book and other books can familiarize the reader with fundamental principles and the information that is available in the public domain. It is unlikely that literature alone could enable one to practice the art without the benefit of commercial knowhow.

8.6.3.3 Quality Control of Pressure Hoses

Stress cracking is the main reason that high pressure overbraided hoses fail. The mechanical stress at fittings, chemical environment, process variables and the polymer type are the parameters on which this type of cracking depends. A number of industrial and military aerospace specifications have addressed themselves to the questions of testing and requirements of pressure hoses. An example is MIL-H-25579, which is used by the United States Air Force for military aircraft hose specification. The types of measurements required by such specifications are lengthy and time consuming. There are a number of rapid tests, which simplify the task of assessment of hose quality.

These tests do not replace any tests mandated by military or civilian aerospace specifications. A tube which meets requirements of the simplified tests is highly likely to meet these specifications:

a) Stretch Void Index (SVI): This test is an indication of the number of voids present in the PTFE tube.

b) Weep Test: This method is used to determine the minimum pressure (WP) at the onset of leak of a military fuel through the tube.

c) Orientation Index (OI): This index is a measure of the degree of orientation in the machine direction (longitudinal) versus that of the cross direction (transverse).

A high quality aerospace hose should meet the following criteria:

Total SVI < 2%

WP > 0.70 (Burst Pressure)

OI < 0.1

The tests to measure these indices are described in the following paragraphs.

a) Stretch Void Index (SVI). SVI provides an indication of the number voids in the part. It indicates how well the sintering and coalescence have eliminated small voids, which can be present because of the processing technique or the properties of the resin. Voids directly affect the performance of a tube in the end-use. For example, a void free or low void content part will have a longer flex life and resist flex fatigue than a part containing more voids.

This index relies on the change in the specific gravity of a specimen, which has been elongated in an *extensometer*. The application of tensile stress stretches the polymer and enlarges any voids that may be present in the sample. An abundance of small voids alters the appearance of the sample to a dull or blushed look. The procedure for measuring SVI can be found in ASTM Method D4895. It basically consists of preparing a microtensile bar and stretching it to 200% of its initial length in 1–1.5 minutes. Next, specific gravity of both stretched and unstretched samples are measured and the SVI is calculated from Eq. 8.3.

Eq. (8.3)

$$SVI = \frac{SG(\text{unstretched}) - SG(\text{stretched})}{SG(\text{unstretched})} \times 100$$

SVI should be measured in both machine and transverse directions to determine the extent of isotropy in the tube. If the SVI values are very low (<1%) in both directions or equal, the sample is isotropic. This means that the sample has been adequately sintered and stresses have been relieved. If the values are both high, sintering has been poor. If the transverse value is high and the machine direction value is low, then orientation is primarily in the machine direction and it is unbalanced.

b) Weep Test. In this test, unbraided samples of the tube are pressurized by filling them with a military fuel (MIL-S-3136, Type III and MIL-H-25579) consisting of 70% isooctane and 30% toluene. A 15 cm length of the tube is filled with the fuel with a small amount of red dye added to it to improve the visibility of the leak. The tube is pressurized to an initial pressure calculated from Eq. 8.4.[13]

Eq. (8.4) Initial Pressure = $\dfrac{16.9t}{d}$

t = Wall thickness, mm

d = Outside diameter, mm

The factor 16.9 is the standard burst strength in the peripheral direction in MPa unit. At this point the pressure is increased in 0.035 MPa increments until seepage is observed. The results are recorded as the Weep Pressure and calculated as a percentage of the peripheral burst pressure.

c) Orientation Index (OI). This factor is concerned with orientation of the polymer in the machine and transverse directions which builds strength in the tube. In the majority of paste extruded tubes, tensile strength is higher in the machine direction than in the transverse direction. OI provides a numerical value to monitor the disparity of the tube strength in the two directions.

If a thin-walled cylinder is subjected to a given pressure P, Eqs. (8.5) and (8.6) can be used to calculate the stress in the longitudinal and peripheral direction.[14]

Eq. (8.5) $P_l = P \dfrac{d-t}{4t}$

Eq. (8.6) $P_t = P \dfrac{d-t}{2t}$

P_l = Stress in the longitudinal direction, MPa.

P_t = Stress in the transverse direction, MPa.

d = Outer diameter of the tube, mm

t = Wall thickness of the tube, mm

A comparison of the formulas shows that the stress experienced by the tube in the transverse direction is twice as large as that in the longitudinal direction. It is desirable, ideally, for the tube to have twice as much tensile strength in the transverse direction than in the machine or longitudinal direction. The orientation during the paste extrusion, however, takes place predominantly in the machine direction, that is, the direction of extrusion, of the unsintered tube. Sintering the tube properly closes the voids (35–40% of the volume occupied by the lubricant) and eliminates the pre-sintering orientation. After sintering, molecular orientation is much more random in the molten phase. To freeze the orientation, the molten tube must be quenched rapidly in cold water. Slow cooling will allow the crystalline phase to orient itself back in the direction of extrusion, because of the memory of the polymer molecules.

Orientation of the tube can be measured by x-ray diffraction. Practically, a comparison of the tensile yield strength of the tube in the two directions provides a measure of the orientation. The Orientation Index is, thus, defined as:

$$\text{Eq. (8.7)} \qquad OI = 1 - \frac{\gamma_t}{\gamma_l}$$

γ_t = Yield strength in the transverse direction, MPa

γ_l = Yield strength in the longitudinal direction, MPa

It can be seen that an orientation of zero means that tube is randomly oriented, which is ideal. A value of one indicates that all orientation is in the longitudinal direction, which is the worst case.

8.7 Unsintered Tape

Major applications of unsintered polytetrafluoroethylene are as tape in thread sealing and wrapping electrical cables, and as rod and tape in packings. Important properties of PTFE like chemical resistance, broad service temperature, low friction, flexibility, high machine direction strength, and deformability in the cross direction make unsintered fine powder PTFE ideal for these applications.

Thread sealant tape is used in pipes and fittings in a variety of industries including water pipe, chemical, pharmaceutical, semiconductor manufacturing, food processing, and others. Electrical grade tape is wrapped around the cable or wire, and then the construction is sintered to obtain good insulation and electrical properties. Some tape is sintered, treated to impart adherability and then coated with pressure sensitive adhesive for wrapping objects to reduce friction or provide quick release properties.

Another important area is oriented tapes and webs which are used in the fabrication of porous fiber, fabric, tube, and sheet. These porous articles find applications for protective clothing, waterproof and weatherproof fabrics, gaskets, filter bags, and many other items.

The unsintered tape is not directly produced in the final thickness and width. Rather, it is made by the extrusion of either a round or rectangular bead (thick ribbon) followed by calendaring which converts it to a thin tape. Calendaring is necessary to obtain tapes that are thin enough to be conformable to the substrate. Normally, the lower limit of thickness is 50–75 µm, but 25 µm thick tape can be made by calendaring. The important properties of electrical wrapping tape include adequate physical properties for handling, appropriate thickness, and excellent layer to layer adhesion (which is a property of the resin). The plies of the tape, which have been wrapped around the cable, melt during the sintering cycle and must bond together for good property development in the insulation. The choice of a lower molecular weight PTFE, such as a modified grade, which has a lower melt creep viscosity will improve interlayer adhesion.

Thread sealant tape, on the other hand, must perform after it has been applied to the pipe thread. Lower density enhances the drapability (deformation) of the tape around the threads. A balance between the tensile strength in the machine and transverse directions in the range of 15:1[15] is required to achieve the desirable deformability. Elongation of 100–200 in the transverse direction and better than 800% in the machine direction are desirable. The amount of fibrillation determines the tensile properties and deformability of the tape. Too little fibrillation will yield a tape with insufficient tensile strength while too much fibrillation will create a hard tape lacking enough deformation.

8.7.1 Blending Lubricant and Pigment and Preforming

Resin mixing with lubricant and preforming should be done according to the procedures described previously in the Sec. 8.5.1, 8.5.2 and 8.5.3. The choice of the lubricant should be made based on the succeeding process that the extrudate will undergo. For example, if the rod or the ribbon is intended for the end application without further change in its characteristics, such as packing, the lubricant should be fairly volatile to be easily removable. Isopar® C or E or other isoparaffin with equivalent volatility (for distillation temperature range, see Table 8.1) are good lubricant selections. Both can be removed in a convection oven at fairly low temperatures.

When the rod or the thick ribbon is intended for calendaring to produce thin unsintered tape a different lubricant choice must be made. A less volatile lubricant is needed to assure that it remains in the extrudate and on its surface without significant evaporation prior and during the calendaring operation. Presence of 18–20% of a heavier lubricant like Isopar® M in the extrudate is critical for proper calendaring.

8.7.2 Extrusion of Round and Rectangular Bead

The general principles of extruding a bead from a PTFE paste are similar to those in wire and tubing fabrication. Figure 8.15 illustrates a typical die and extrusion cylinder for manufacturing a round bead (rod). The ram forces the polymer through the die, which is a simple orifice. There are no guide tubes or mandrels because the bead is solid. The die land length to diameter ratio influences the quality of the extrudate. Its value is not critical and is typically in the range of five to ten. Die cone angles in the range of 30–60°C are common. Rectangular beads follow the same principles except for the shape of the die land, which is rectangular.

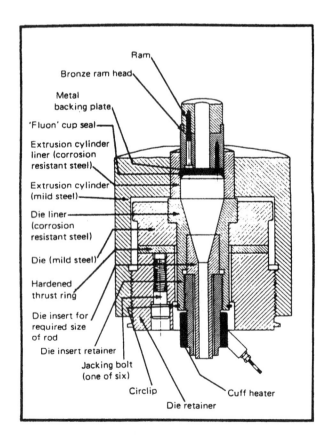

Figure 8.15 Extrusion cylinder and die for rod.[13]

Extrusion conditions affect the quality of the rod produced. It is important to operate both the barrel and the die above the transition temperature of the resin (19°C) to obtain a smooth and coherent extrudate. Temperature of both units should be in the range of 25–35°C. The amount of work (extrusion pressure) put into the extrudate determines the hardness or deformability of the bead. The more the polymer is worked, the harder it will turn. A bead intended for calendaring into a tape should be significantly softer than a rod aimed at packing application. A bead made at an extrusion pressure which is too high will be stiff and will not calendar well. A bead that is too soft, made at extrusion pressure too low, will yield narrow tape weak in the transverse direction. Typical conditions for the extrusion of a bead are given in Table 8.13.

Reduction ratio for extruding rods is usually fairly low. For example, the reduction ratio of 100 is required to make a 1 cm diameter bead in an extruder with a barrel diameter of 10 cm. This means that extrusion pressure is fairly low (10–20 MPa). A lower lubricant content may be needed to increase the pressure, if the bead is too weak for handling. Another useful change is to increase the die cone angle to 60°, which will increase the extrusion pressure. An added benefit of increasing this angle is reducing the length of the die cone, which decreases the wasted material.

8.7.3 Calendaring

Calendaring is the process by which a lower thickness is obtained from a thicker bead. Manufacturing of unsintered tape from the bead, round or rectangular, involves calendaring followed by removal of the extrusion aid, slitting, and winding into rolls (Fig. 8.16). Calendaring process reduces the specific gravity of the tape by about one third to 1.4–1.5 g/cm³ because of the lubricant and voids. Removal of the lubricant leaves the volume that it was occupying in the form of voids. In a sintered part, melting and coalescence of the polymer eliminate the voids. Stretching after calendaring introduces additional voids in the tape; the porous structure, thus, having a lower density. Lower cost is the main purpose of density reduction.

Calendaring a large bead into a much wider tape followed by slitting allows higher productivity than calendaring into the finished width. It also reduces the amount of scrap produced in the process. Single or multiple stage calendaring is possible, though the latter has no advantage over the former. Two rolls with a defined gap are used to "squeeze" the thicker bead into a thin tape. Heating the rolls assists the thinning process. The normal lower limit of the thickness of the final tape is around 50 µm.

Table 8.13. Typical Extrusion Conditions for Extrusion of a Rod*[11]

Type of Lubricant	Mineral Oil (Boiling Range 260–385°C) Petroleum Ether (Boiling Range 95–116°C)	Low Odor Paraffin (Boiling Range: 191–246°C)	VM&P Naphtha (Boiling Range: 118-139°C)
Rod Application	Calendaring into Tape	Calendaring into Tape or Packing	Packing
Lubricant Content, weight %	20±2	20	16
Extruder Barrel Diameter, cm	6.2	11.3	3.8
Die Land Diameter, cm	1.27	1.27	0.127
Reduction Ratio	25:1	80:1	900:1
Die Cone Angle, °	30	30	20
Die Temperature, °C	35	35	29
Extrudate Speed, cm/min	91.5	183	1,982
Extrusion Pressure, MPa	6.2	6.9	75.9

* Imperial Chemical Industry Resin Grade Fluon® CD-1 was extruded in these trials.

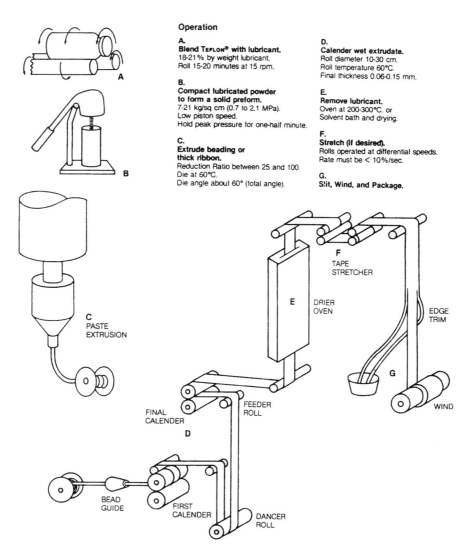

Operation

A. Blend TEFLON® with lubricant.
18-21% by weight lubricant.
Roll 15-20 minutes at 15 rpm.

B. Compact lubricated powder to form a solid preform.
7-21 kg/sq cm (0.7 to 2.1 MPa).
Low piston speed.
Hold peak pressure for one-half minute.

C. Extrude beading or thick ribbon.
Reduction Ratio between 25 and 100.
Die at 60°C.
Die angle about 60° (total angle).

D. Calender wet extrudate.
Roll diameter 10-30 cm.
Roll temperature 60°C.
Final thickness 0.06-0.15 mm.

E. Remove lubricant.
Oven at 200-300°C, or
Solvent bath and drying.

F. Stretch (if desired).
Rolls operated at differential speeds.
Rate must be < 10%/sec.

G. Slit, Wind, and Package.

Figure 8.16 Outline of process for making thread seal tape.[15]

8.7.3.1 Calendaring Equipment

Calendaring can be done by any twin rolls with adjustable gap in the thickness range. The width of the rolls is determined by the width of the tape to be made. More than one set of calendaring rolls can be used to reach the desired thickness. In the case of multiple calendars, only the gap in the last calendar would have to be precisely controlled. The rolls are made from a hard rigid material such as chilled cast iron to prevent deflection under the substantial pressures that can develop during the calendaring.[11] Roll finish of the final calendar should be smooth (0.25 µm) while the initial rolls can be rough. A finish which is too fine will prevent sufficient grip between the bead and the rolls, reducing the effectiveness of calendaring.

The diameter of the rolls depends on the desired width of the tape. Larger diameter rolls should be selected for wider tapes. For example, to calendar a bead into a 100–200 mm wide tape, the calendar diameter should be 150 mm and its width about 300 mm. The rolls should be equipped with internal water or oil heating capable of providing a temperature of 50–80°C to facilitate the calendaring of the bead. A continuous drive should provide roll speed in the range of 1.5–40 m/min.

Guides must be installed on the calendar so the bead can be safely fed into the nip point of the rolls. The bead containing lubricant is limp and slippery with a tendency to move sideways on the rolls. The guide should constrain erratic movement of the bead and the tape to maintain a smooth process. One type of guide is a tubing with a slightly larger diameter than the bead and 8–15 cm long.[15] Another type of guide is called "fishtail" and is quite effective. Figure 8.17 shows the schematic of a fishtail guide where a hole bored through the guide accommodates the bead. The guides can be made from a metal like aluminum or a hard plastic such as polyacetal (Delrin® by DuPont). The guide should be machined to match the curvature of the rolls at the feed point. The guide should be narrower than the width of the tape produced without restriction to control the movement of the tape. A dancer roll (Fig. 8.16) should be installed to supply positive tension to keep the tape moving. Excess tension will lead to stretching of the tape and a reduction in its width.

8.7.3.2 Calendaring Operation

It is important to keep the lubricant in the bead prior to calendaring if the extrudate is not immediately calendared. The bead can be kept in a sealed container or stored in a container of the lubricant until it is going to be used. Preheating the bead helps produce a wider and more uniform tape. The recommended method is to warm up the extrudate in an airtight plastic bag and immerse the bag in 45–60°C water.[15]

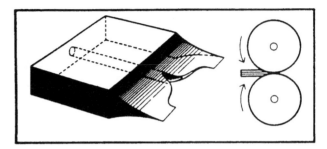

Figure 8.17 Diagram of fish-tail guide.[11]

Another issue is starting up calendaring which is difficult because of the slippery nature of the bead, particularly with a round bead or cross section. The first 20–30 cm of the bead can be flattened to make it easier to feed it into the gap between the rolls. Another method is to wrap the end of the bead in a rough paper so friction develops and the extrudate is carried through the nip. For a rectangular bead, the gap can be increased at the start up and afterwards reduced.

The variables of calendaring are roll speed and surface temperature of the rolls. In general, it is difficult to predict the impact of these variables on the tape properties because of their strong dependence on the characteristics of the bead. Sometimes an increase in the roll speed leads to a wider tape. Temperature increase, up to a point, is beneficial to soften the bead and reduce the calendaring pressure. Tape splitting and rapid lubricant loss are the outcomes of high roll temperature.

8.7.4 Drying and Stretching the Tape

The lubricant must be removed from the tape before stretching is done and before the completion of final product. The lubricant can be removed by direct drying in an oven or by the extraction of the lubricant by a more volatile solvent and a subsequent lower temperature oven drying. The extraction technique[2] is less common because of the need to use toxic solvents such as trichloroethylene to remove the lubricant. The evaporated solvent is usually removed from the oven effluents and recycled. It is sometimes processed in a pollution abatement device where the heat could be recovered and utilized in the process.

A variety of oven designs have been satisfactorily used in the removal of lubricant. They include internally heated flat plates, heated drum, resistance heaters and infrared lamps. The ovens must be thoroughly exhausted to assure that hydrocarbon concentration is well below its lower explosive limit. Figure 8.18 depicts a typical convection drying oven in conjunction with a stretching step. The temperature settings depend on the volatility of the solvent (150–300°C air temperature). It should be noted that the tape itself should remain well below (20–30°C) the melting point of PTFE (342°C). An even partially sintered tape, that is, when only some of the polymer has reached its melting point, loses its usefulness as a tape for wrapping wire.

Tapes are basically stretched uniaxially, which means that they are drawn in one direction. Two rolls with controlled variable speed and heating are the basic requirements of stretching. The usual arrangement is illustrated in the Fig. 8.18 where the oven has separated roll 1 and roll 2. In addition to increased production yield, stretching is reported to improve the tape performance. Typical uniaxial stretch is less than 150% and is done at less than 10% per second.[15] Biaxial and more extensive drawing will be discussed later in this chapter.

The difference in linear speed of the two rolls determines the amount of stretch which has been achieved. Equation 8.8 and 8.9 offers simple formulas to calculate the total stretch and stretch rate.

Eq. (8.8) $\text{Total Stretch} = \left(\dfrac{\text{Roll 2 Speed}}{\text{Roll 1 Speed}} - 1\right) \times 100\%$

Eq. (8.9)

$$\text{Stretch Rate} = \dfrac{\text{Total Stretch}}{\left(\dfrac{\text{Distance Roll 1 to Roll 2}}{\text{Roll 2 Speed} - \text{Roll 1 Speed}}\right)}$$

No stretch takes place if the rolls move at the same speed. An example where the tape is drawn is illustrated below, where the two rolls are 14 meters apart.

Roll 1 Speed m/min	Roll 2 Speed m/min	Total Stretch %	Stretch Rate % per minute
50	120	140 (from Eq. 8.4)	700 (from Eq. 8.5)

Quality control of the tapes is accomplished by measuring the following properties of the final product:

Width and Thickness

Density (ASTM D4895-97)

Tensile Strength and Elongation at Break (ASTM D882)

A record of these properties can often provide clues for troubleshooting the process. Typical properties of calendered tape made from a commercial resin are given in Table 8.14.

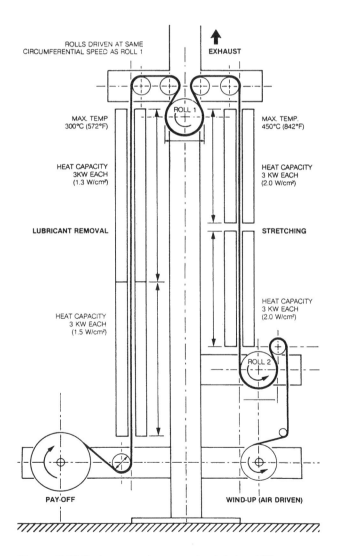

Figure 8.18 Drying oven for removing lubricant.[15]

Table 8.14. Typical Properties of Tape Produced by Calendaring[11]

Tape Properties	Calendar Roll Temperature, °C	
	50	80
Width, mm	95-100	115-120
Thickness, μm	75	75
Tensile Strength, MPa		
Machine Direction	11	16
Transverse Direction	0.8	0.7
Break Elongation, %		
Machine Direction	185	260
Transverse Direction	255	700

8.7.5 Final Tape Product

Whether stretched or not, the final tape product is slit to remove irregular edges and obtain the desired width. Slitting is sometimes done after drying and stretching, if any, has been completed. A separate machine is often used to slit the tape into the variety of the width required by the end use. A slitting machine is equipped with a payoff and a windup station and two parallel and adjustable knives. Speed and the tension of the tape are also adjusted to obtain the best conditions for slitting.

8.8 Expanded PTFE Manufacturing

One of the most unique and remarkable developments in the processing of homofluoropolymers is the expansion of the polymeric matrix without the use of soluble fillers, foaming agents or chemical additives. This invention constitutes the physical inducement of billions of small pores in the structure of an article of polytetrafluoroethylene resulting in new properties and significant savings in material consumption. W. L. Gore and Associates invented this technology in the early 1970's. The trademark Gore-Tex® is well known in lightweight waterproof and breathable fabrics, microfiltration membranes, medical implants, microwave carriers, industrial sealants, and high tensile fabrics.

The expansion process has enabled manufacture of flexible, high tensile strength cords, tubes and fibers, and thin and tough membranes with unique dielectric properties. A significant feature of the expanded PTFE is the drastic reduction in the creep, which is a weakness of this polymer. The characteristics of the Gore-Tex® material are both dependent on the chemistry of the polymer and the conditions of the expansion process.

The expansion process begins with paste extrusion of fine powder polytetrafluoroethylene using typical lubricants, as has been discussed in this chapter. The lubricant is completely removed by heat, similar to unsintered tape. The lubricant-free extrudate, which can be in the shape of a rod, tube, or tape, is the feed for the expansion process. This process consists of heating the unsintered PTFE anywhere from 35–320°C[16][17] restrained in a device capable of stretching it at high rates. The stretched part is heated to a temperature above 330°C while being held in a restraining device to prevent its shrinkage. After this heat treatment, called *amorphous locking,* for a period of time the expanded part is cooled and removed.

Expansion can be done by uniaxial or biaxial stretching of PTFE. The effect of the expansion can be seen in Figs. 8.19 and 8.20, which show a simulation and an actual scanning electron micrograph of the expanded structure. An examination of these figures reveals that the matrix is a complex interconnection of a large number of nodes by fibrils. The long axis of the nodes is perpendicular to the direction of stretch. The fibrils are reported to be wide and thin in cross section with a maximum width of 0.1 μm and a minimum width of one or two molecular diameter in the range of 0.005–0.01 μm.[16] The nodes vary in size from less than a micron to 400 μm, depending on the conditions of the expansion.

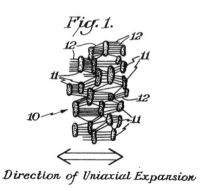

Figure 8.19 A depiction of the structure of a stretched PTFE membrane.[17]

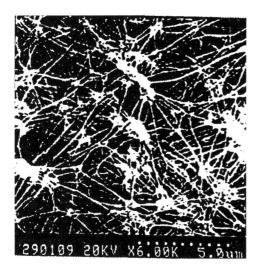

Figure 8.20 A scanning electron micrograph of a stretched PTFE membrane at 6000 times magnification.[17]

The process conditions are extremely wide where stretch temperature ranges from 35–320°C and stretch rate varies from 10% per second to 40,000% per second. The optimum amorphous locking (heat treatment) temperature range is reported to be from 350–370°C for anywhere from a few seconds to one hour.[17] Total stretch ranged from 50 to about 2,000 times the original length. Higher stretch rates and temperatures produce more uniform matrix. Polymers that are most suitable for the expansion process are 100% tetrafluoroethylene resins with high degree of crystallinity, 98% or more. Copolymers of TFE, even at very low comonomer level do not work well because of the defects that the comonomer creates in the crystalline structure of PTFE molecules.

An expanded web or rod can be generally described as a porous structure with significantly lower density. A sintered PTFE part has a density of about 2.15 g/cm^3 and an unsintered unexpanded part has a density of 1.5 g/cm^3. The density of an expanded part can be as low as <0.1 g/cm^3, with a porosity of 96%. The relationship of porosity with pore size and density can be seen in Fig. 8.21. Density and porosity have a linear relationship. Pore size is quite small, less than 1 µm, up to 90% porosity; larger size pores (1–6 µm) contribute to driving the porosity above 95%.

The heat-treated expanded web has a higher permeability than normally sintered PTFE to gases and liquids due to its porous structure. It can also act as semipermeable membrane by allowing the wetting liquids through while being impermeable to the non-wetting fluids. For example, a gas-saturated membrane in contact with the gas and water will allow gas through and keep water out as long as the pressure of water does not exceed the water entry pressure. Figure 8.22 indicates that above 90% web porosity, air permeability increases drastically while water entry pressure of the web decreases to a fairly small value. A comparison of Figs. 22(a) and (b) indicate that there is a range where porosity can be selected to balance the air permeability and water impermeability, which is useful to applications such as clothing.

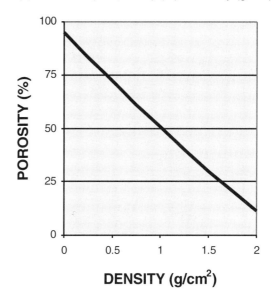

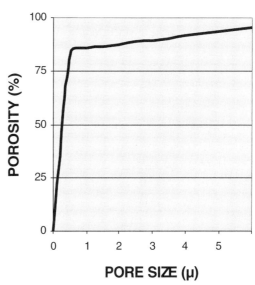

Figure 8.21 Typical porosity/pore-size relationships for expanded PTFE film.[19]

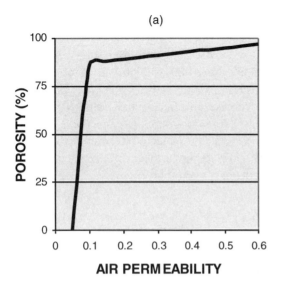
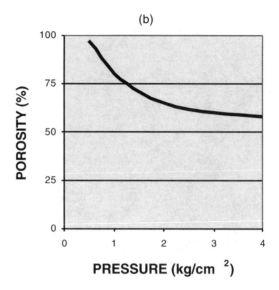

$$\text{Air Permeability} = \frac{V * H}{t * A * P}$$

V = Volume of air (cm³)
H = Thickness of film (cm)
t = Flow time (sec)
A = Area (cm²)
P = Pressure (cm Hg)

Figure 8.22 Typical air-permeability/water-entry pressure relationships.[19]

Table 8.15 presents a comparison of the properties of expanded and full density (unexpanded) polytetrafluoroethylene. Crystallinity of the *amorphously locked* PTFE is about 95%, which is significantly above the highest commercially attainable value with unexpanded parts. The most striking improvement is in the tensile strength, which is orders of magnitude over the full density material and has opened new applications for PTFE. Tensile strength of expanded PTFE is calculated for the matrix by multiplying the measured value of tensile strength by the ratio of the densities of the full to expanded PTFE. Flex life and maximum service temperature of expanded PTFE are both higher than those of the full density material.

Table 8.15. Comparison of the Typical Properties of Expanded vs. Full Density PTFE[19]

Property	Units	Full Density	Expanded
Specific gravity	G/cm³	2.1	0.1–1.0
Crystallinity	%	50–70	95
Porosity	%	<1.0	25–96
Poor size range	Microns	N/A	0.02–15
Matrix tensile strength	Mpa	20–30	50–800
Flexural fatigue resistance	Cycles to failure	1×10^6	3×10^7
Service temperature (max.)	°C	260	280
Thermal conductivity	KCal/m·hr°C	0.2	<0.1
Thermal expansion coefficient	Per °C	3×10^{-4}	1×10^{-4}
Resistance to cold flow	As creep	Poor	Excellent
Abrasion resistance	-	Moderate	Excellent
Chemical resistance	-	Excellent	Excellent

How does expansion build strength in the structure of polytetrafluoroethylene? To understand the mechanism for this increase we have to consider two parameters, *Poisson's ratio* (v_{xy}) and *Young's Modulus of Elasticity* (E_x). Poisson's ratio (Eq. 8.10) is defined as the ratio of the lateral strain (e_x) to the longitudinal extension caused by uniaxial tensile extension. In Fig. 8.23(a), assuming extension occurs in the x direction, Poisson's ratio is defined as the ratio of dimension change in the y direction to the extension in the x direction. Most material would experience *necking* or a reduction in the y direction, thus preserving their volume. For example, Poisson's ratio of rubber is in the range of 0.5–1 and for engineering polymers $v_{xy} > 1$. An increase in the y direction is rare and would be indicated by a negative sign.[18]

Young's Modulus of Elasticity is the relationship between stress and strain (Eq. 8.11) in a uniaxial extension.

Eq. (8.10) $\quad v_{xy} = \dfrac{\varepsilon_y}{\varepsilon_x}$

Eq. (8.11) $\quad E_x = \dfrac{\sigma_x}{\varepsilon_x}$

Expanded PTFE has a negative Poisson's ratio (Fig. 8.24) until significant extension or strain has been reached, that is, about the strain of 2000. Evans[18] has proposed a model to explain this phenomena. Figure 8.23(a) shows the structure of unexpanded paste extruded PTFE where the nodes lay flat and are connected by fibrils. As the stretching starts, the nodes begin to move. At some point they are pulled which tilt them up, thus increasing the width or bulk of the material. The tilting movement leads to the perpendicular orientation of the long axis of the nodes to the direction of draw. Scanning electron micrographs have corroborated this model (Fig. 8.20).

Zone I of Fig. 8.24 shows a linear Young's Modulus indicating an elastic behavior by the expanded material, possibly the fibrils acting as elastic bands, as proposed by Evans,[18] storing energy. At the end of Zone I, Poisson's ratio reaches zero and finally in Zone III levels at 1–2. The material behaves as a plastic in Zone III where the original anisotropic structure breaks up. In the expanded structure, the tensile load is borne by highly oriented fibrils with a virtually 100% crystalline structure which explains their much higher tensile strength.

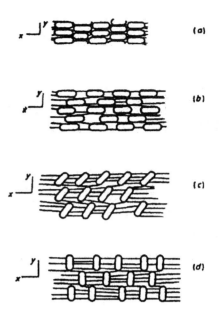

Figure 8.23 Schematic diagram of structural changes obseved in microporous PTFE undergoing tensile loading in the x direction.[19]

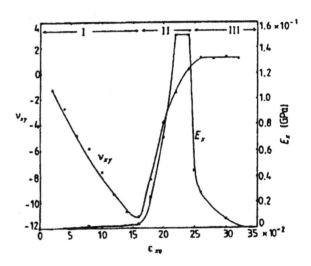

Figure 8.24 Plot of Poisson's ratio, v_{xy} and youngs Modulus E_x plotted against engineering strain ε_x showing the three regions of behavior: I, II, III.[19]

8.9 Fine Powder Resin Selection

A number of fluoropolymers are available for paste extrusion from manufacturers. How would one choose one or more candidates for evaluation? Selection of the resin type is heavily dependent on the part that is going to be made, the fabrication process, and use conditions. Severe chemical environment, particularly at elevated temperatures, is an example of an application where fluoropolymers would be required. The most important feature of the part is the size of its cross section. Paste extrusion, basically, consists of reducing a hollow cylinder of the paste into a preform of the final product by means of pressure. This has led to the definition of reduction ratio for a given part.

Reduction ratio is the ratio of the cross section of the paste in the extruder barrel to that of the extrudate. The resin selected to make a particular part must be reducible to the extrudate. This means that it should generate sufficient pressure for fibrillation of the resin, yet the pressure must not exceed the normal range of the available equipment. Reduction ratio is primarily related to the molecular weight of the polymer; the lower the molecular weight, the higher the reduction ratio. Most manufacturers specify the recommended range of reduction ratios for their fine powder grades.

The first factor to be determined is the reduction ratio of the part. Most extruders allow the use of different size barrels in a fairly limited range, therefore, allowing some adjustment of the reduction ratio. After the range of reduction ratio has been selected, one has to search the product literature of commercial fine powder resins and find those with matching ranges.

There are a number of overlapping reduction ratios where one could make more than one resin choice. In these cases, the type of application should be used as a guide to select resins. After the appropriate polymer options have been considered, the properties desired in the intended part, processing characteristics, end-use conditions, and cost are the final determining factors for the selection of candidate resins. The reader can refer to Ch. 6 to find the properties of some of the common fine powder resins. It is quite likely that at the end of this selection sequence, one would arrive at more than one resin supplied by different manufacturers. In such a case, it is best to experiment with each resin to determine which one is more suitable for the specific paste extrusion process and best fits the final application of the part. Other factors such as quality, reliability of supply, availability of support, and technical information from the resin supplier should be taken into consideration in making polymer selections.

References

1. Quinn, F. A., Roberts, D. E, and Work, R. N., "Volume-Temperature for the Room Temperature Transition in Teflon®," *J. Appl. Phys.*, 22:1085, 1951.
2. Luntz, J. F., Jaffe, J. A., and Robb, L. E., "Extrusion Properties of Lubricated Resin from Coagulated Dispersion, *Ind. Eng. Chem.*, 44(8)1805–1810, Aug., 1952.
3. "Teflon® PTFE Fluoropolymer Resin, Processing Guide for Fine Powder Resins," *DuPont Materials for Wire and Cable*, DuPont, Wilmington, Delaware, Dec., 1994.
4. "Fluorocarbon Polymers of Daikin Industries," *Daikin-Polyflon®TFE Fine Powder*, Daikin Industries, Ltd., Osaka, Jpn, July, 1986.
5. Gangal, S. V., "Polytetrafluoroethylene, Homopolymers of Tetrafluoroethylene" in E*ncyclopedia of Polymer Science and Engineering*, 2nd ed., 16:577–600, John Wiley & Sons, New York, 1989.
6. Makinson, K. R. and Tabor, D. "The Fiction and Wear of Polytetrafluoroethylene", *Proc. Royal Soc.*, 281:49, 1964.
7. McCrum, N. G., "An Internal Friction Study of PTFE," *J. Polymer Sci.*, 34:355, 1959.
8. "Isopar® Solvents," publication DG-1P from Exxon Corp., 1994.
9. SOLVENTS, *Shell Mineral Spirits*, Shell Corp. Pub. SC:1237–96, 1996 and Shell Chemical Co. Tech. Bulletin SC:1219–98R, *Shell Hydrated Hydrocarbon Solvents Typical Prop.*, 1998.
10. "The Processing of PTFE Coagulated Dispersion Powder," Technical Service Note F3/4/5, Molding Powders Group, ICI, 1975.
11. "The Processing of PTFE Coagulated Dispersion Powder, Fluon® PTFE Resins," Imperial Chemical Industries, Ltd., 1986.

12. Benbow and Bridgwater, *Paste Flow and Extrusion*, Clarendon Press, Oxford, England, 1993.
13. "Teflon® 62, Hose and Tubing," DuPont, Geneva, Switzerland, Feb., 1992.
14. ISO/TC 138 N 1081 and ISO 1167 Standards, International Standards Organization.
15. "Teflon® PTFE Fluoropolymer Resin, Thread Sealant Tape Processing Guide," DuPont, Wilmington, Delaware, Nov., 1992.
16. US Patent 3,953,566, Gore, R. W., assigned to W. L. Gore & Associates, Inc., Apr. 27, 1976.
17. US Patent 3,962,153, Gore, R. W., assigned to W. L. Gore & Associates, Inc., June 8, 1976.
18. Craddock, B. D. and Evans, K. E., Univ. of Liverpool, *J. Phys. D*. Applied 22:1877–1882, 1989.
19. Fluoropolymers Conference 1992, UMIST, "Expanded PTFE-Properties & Applications," by Norman, E. G. and W. L. Gore & Associates, Paper 9, published: RAPRA Technology, U. K.
20. Moynihan, R. E., "The Structure of Perfluorocarbon Polymers. Infrared Studies on Polytetrafluoroethyelne," *J. Am. Chem. Soc.*, 81:1045–1050, 1959.

9 Fabrication and Processing of PTFE Dispersions

9.1 Introduction

Tetrafluoroethylene is polymerized in an aqueous dispersion medium to produce *dispersion* PTFE products. In general, the key characteristics of this polymerization regime include ample dispersing agent and mild agitation at elevated temperature and pressure. The dispersion recovered from the reactor is finished by two different series of processes depending on whether a dispersion or a dry powder (fine powder) is the desired final product. Polymerization and finishing techniques are reviewed in Ch. 5.

This chapter discusses the coating of surfaces and fabrication techniques using *dispersions* of polytetrafluoroethylene (PTFE). It does not cover the topic of *PTFE Finishes*. These finishes are usually highly formulated and are applied as multicoatings which include special primers and intermediate layers. They may include pigments, additives, other resins and other fluoropolymers besides PTFE. The main applications of PTFE finishes are in well-known cookware, and industrial anticorrosion and high temperature uses. Coverage of this topic may well occupy a separate volume.

One can think of the PTFE coating process as impregnation of fibrous or porous materials such as glass fiber, woven glass cloth, and polyaramide fibers and fabrics. The composite product combines the properties of PTFE and substrate. The polytetrafluoroethylene coated or impregnated products are characterized by a number of common attributes (Table 9.1).

Polytetrafluoroethylene dispersions are aqueous milky dispersions consisting of very small particles (<0.25 µm) of resin suspended in water. This form of polytetrafluoroethylene is highly crystalline (96–98%) and is produced in a wide range of molecular weights. The monomer is polymerized by the dispersion (emulsion) method in which a surfactant is added to the aqueous medium prior to the start of polymerization. Chapter 5 provides a review of dispersion polymerization techniques.

The most common articles coated using PTFE dispersions are glass fabrics for architectural and industrial applications such as stadium roofs and conveyer belts. Formulating its dispersion into a form that can be spun through spinnerets can produce polytetrafluoroethylene fibers. Coatings and fibers of PTFE are usually sintered to improve mechanical properties. Other examples of applications of dispersions include cast films, packings, gaskets, bearings and polymer additives. PTFE dispersions have high utility due to their fluid nature. This is especially important because polytetrafluoroethylene does not flow after melting and does not dissolve in conventional solvents, therefore cannot be processed by melt or solution techniques. See Chs. 7 and 8 for discussions about the sintering process and mechanism of PTFE.

This chapter discusses basic aspects of coating technology insofar as they are applicable to the use of polytetrafluoroethylene dispersions. Readers who are unfamiliar with this topic are encouraged to consult the references cited for coating technology for an in-depth understanding of the subject.

9.2 Applications

The end uses of polytetrafluoroethylene dispersions are numerous due to the convenience of coating techniques. They can be classified in different ways based on the point of view of product attributes or processing techniques. Table 9.2 is a product type summary of dispersion applications. The focus here is on the shape and form of the part, which influences the process by which they are fabricated.

Another advantage of dispersions is their capability for accepting larger amounts of fillers than PTFE powders. The process to incorporate the fillers is called co-coagulation. The main application of these

Table 9.1. Attributes of Polytetrafluoroethylene Impregnated/Coated Material

Attribute	Source
Good Sliding without adding lubricants	PTFE
Non-stick properties	PTFE
High service temperature	PTFE and substrate
Water repulsion	PTFE
Chemical resistance	PTFE
Greater mechanical strength than natural PTFE	Substrate

compounds is in the fabrication of special bearings. There are a number of other smaller uses of PTFE dispersions in fuel cells, batteries, de-dusting, and chloralkali processing.

Another approach to the classification of the applications of polytetrafluoroethylene dispersions is the nature of thermal treatment of the fabricated part. Some articles are sintered, some are not sintered but heated to remove the water and surfactant. In some application, the parts are neither sintered nor heated high enough to remove the surfactant. Table 9.3 summarizes the process-based categorization of dispersion applications.

Table 9.2. PTFE Dispersion Products and Applications

Products	Applications
Coated Woven Glass Cloth and Fiber	Architectural Fabrics, Gaskets and Laminates, Electrical Insulation, Release Sheets, Hoses
Impregnated Flax, Polyaramides and PTFE Yarn or Yarn Constructions (Asbestos in the past)	Packings, Seals and Gaskets
Dispersion Cast PTFE Film	Diaphragm and Dielectric Insulation in Small Capacitors, Composite Laminates
Coated Material Surfaces	Low Friction and Non-stick Surfaces
Fabric and Fiber Finishes	Yarns, Industrial Fabrics and Filter Cloth
Blends with Polymeric and Nonpolymeric Materials	Flame Non-drip Plastics

9.3 Storage and Handling

Most polytetrafluoroethylene dispersions should be stored at temperatures between 5°C to 20°C. Freezing the dispersion must be avoided due to its irreversible coagulating effect on PTFE particles. Maximum shelf life for dispersions is one year, although some dispersions may have shorter useful lives. Once a month, drums of stored dispersion should be rolled or gently agitated to rejuvenate them. Coagulation of PTFE particles can occur if the storage temperature is too high, if it is subjected to vigorous agitation or shearing, if shelf life is exceeded, monthly rejuvenation not done, and if chemicals are added to the dispersion.

Microscopic examination of PTFE dispersion can reveal coagulation of particles. Under magnification numerous white lumps indicate coagulation or spoilage of the dispersion. A normal dispersion, while it may contain an occasional coagulum, appears uniform and free of lumps.

Polytetrafluoroethylene dispersions may contain one or more surfactants (and other additives) such as perfluoroammoniumoctanoate, also known as C8 in the industry. There are health hazards associated with some of the surfactants such as C8. Thorough review of the material safety data sheet for each surfactant must be made and protection measures taken to avoid/minimize exposure and emission.

9.4 Principles of Coating Technology

The basic objective in coating a surface (substrate) is to attach a thin layer of a second substance to the substrate permanently or temporarily. An everyday example is painting house walls which is done by

Table 9.3. PTFE Dispersion Application Categories Based on Fabrication Processing

Sintered	Unsintered, Heated	Unsintered, Unheated
Coated Woven Glass Cloth	Filtration Cloth	Packings
PTFE Yarn	Batteries	Gaskets
Cast Films	Blends with Polymeric and Non-polymeric Materials	Batteries (sometimes heated)
Coated Metals		Dedusting
Co-coagulation Products		Paint Additives
Chloralkali Processing		
Fuel Cells		

applying a relatively thin layer of a well stirred paint using a roller, sprayer or a simple brush. A paint is required to adhere to the wall permanently which is accomplished by allowing it to dry. PTFE coatings are almost always required to adhere permanently. Properties of PTFE require that it be sintered well above its melting point (>342°C). A number of other considerations and requirements apply which are discussed in this section.

9.4.1 Coating Processes

Many different processes are used in the coating industry. One way of classifying the processes is by the number of layers that a method is capable of coating, and whether the method coats continuous or discrete units of substrates. PTFE dispersions are usually coated on continuous planar or fibril webs. Single and multiple layers can be coated on continuous webs. A great variety of techniques known as rod, dip, doctor blade, knife, gravure, reverse roll, air knife and forward roll are single layer methods. Slide coating techniques are multilayer techniques. Slot die coating and curtain coating can be tailored to coat single or multiple layers. Discrete methods include curtain, spray and dip coating techniques. Descriptions of coating methods can be found in a number of references including the review by Cohen and Gutoff,[1] and *Coating Technology Handbook* by Satas.[2][3]

There are three ways by which a liquid coating can be transferred to a substrate (web). First, the substrate can be dipped into the liquid coating, thus the term *dip-coating*, whereby an excess amount is entrained onto the surface of the substrate as it moves through the coating reservoir (trough). The excess material is removed usually by a blade or a bar and recycled to the reservoir, thereby metering the amount of liquid laid on the web. The second way to obtain a coating is by means of one or more rolls, which picks up the liquid from the trough and deposits it on the substrate. A doctor blade controls the thickness of the liquid on the roll. The metering action of the blade can be further improved by using rolls, which have engraved channels, with controlled volume. An example of this technique is *two-roll coating* (Fig. 9.1) and *reverse roll gravure coating*.

The third way to transfer a liquid coating to the web surface is by spraying. This technique atomizes the liquid into small droplets, which form discrete zones immediately after reaching the web surface. Surface tension forces drive the coalescence of these discrete zones towards forming a continuous layer. It is important that a sufficient number of droplets are sprayed on the web so that they would be in sufficiently close proximity to allow surface tension forces to work.

Dip coating, the most popular method of coating cloth and fibers with polytetrafluoroethylene dispersions, is described later in this chapter. Hard surfaces are coated by roll-coating techniques or spraying.

Each coating process operates according to a set of common variables with other processes and a few unique variables. It is essential to have a basic understanding of rheology (liquid flow) and surface chemistry (and physics) in order to understand coating processes. Flow of the liquid coating material and the interaction between the liquid and the substrate surface determine the state of the coating. A mostly qualitative review of these two concepts will be made here which should provide the reader with sufficient background to develop a working knowledge of polytetrafluoroethylene dispersion coating. More detailed treatment of these subjects can be found in works by Cohen & Gutoff and Satas (see Refs. 1, 2, and 3)

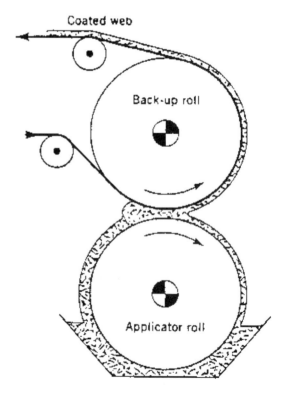

Figure 9.1 Two-roll forward roll coater.[1]

9.4.2 Rheology

Rheology is a general term referring to the science of studying flow and deformation of materials. *Viscosity* is defined as the resistance of a liquid to flow. For example, ketchup has a higher viscosity than water. An understanding of rheology and viscosity is essential to processing of polymers and coating. Polytetrafluoroethylene is an exception among thermoplastics where rheology does not enter the picture in most of its processing techniques. This is simply due to its ultra high viscosity, well above its melting point (10^{11}–10^{12} Pa·sec at 380°C) where no flow occurs. This is not true of PTFE coatings, which are subject to almost all considerations of coatings.

The subjects of rheology, viscosity and behavior of materials have been covered by numerous references. Plastics and coatings are subdivided into a number of categories based on their flow behavior when they are subjected to an external force. Examples of water, honey and ketchup which are encountered in everyday life serve to illustrate the differences among the flows. Ketchup requires more force than honey and honey more than water to flow. Viscosity as a measure of flow is defined by the ratio of shear stress and shear rate:

Eq. (9.1) $\text{Viscosity} = \dfrac{\text{Shear Stress}}{\text{Shear Rate}}$ (dynes·sec/cm^2)

Eq. (9.2) $\text{Shear Stress} = \dfrac{\text{Force}}{\text{Area}}$ (dynes/cm^2)

Eq. (9.3) $\text{Shear Rate} = \dfrac{\text{Velocity}}{\text{Thickness}}$ (1/sec)

The reader can learn a great deal by referring to a number of excellent works available about rheology and flow of materials.[4]-[7]

An applicable case to coating is *thixotropic* (or shear thinning) liquids, which become thinner when shear rate is increased. Viscosity increases after shear rate is reduced, although it does not immediately reach its original level. Figure 9.2 shows a general viscosity vs shear rate behavior for a Newtonian and thixotropic liquid. The area between the two graphs is called *hysteresis loop*. Eventually viscosity recovers to its original value, after the particle structure has set up in the liquid.

A simple example is household paint, which appears thinner during rapid stirring. Here, raising the speed of stirring is the method of increasing the shear rate. Thixotropy is the basis for the so-called *dripless* house paints. These paints undergo a reduction in viscosity while being brushed. When brushing is slowed, they return to higher viscosity and do not drip or sag. Both observations indicate a reduction in viscosity of paint as rapid motion (shear rate) is applied.

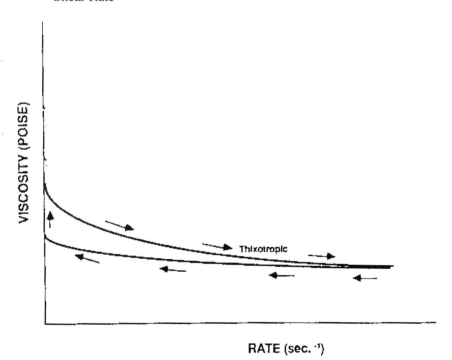

Figure 9.2 Shear stress shear rate curves: hysteresis loop.[3]

For the most part, polytetrafluoroethylene dispersions are not thixotropic. It is difficult to measure a viscosity decrease for them due to their fairly low values. PTFE dispersions exhibit near Newtonian behavior at the room temperature, that is, their viscosity is independent of shear rate. Organics, such as acrylic resins, and inorganic thickeners, such as barium nitrate, are added to PTFE dispersions, when a higher viscosity is desired, as in glass fabric coating, to allow the use of fillers without separation of the filler. The thickened PTFE dispersions exhibit shear-thinning behavior. Figure 9.3 is an example of viscosity behavior of a polytetrafluoroethylene thickened with barium nitrate. Measurements were made with a Brookfield viscometer using a #2 spindle.

Viscosity of PTFE dispersions is strongly affected by temperature. A typical thermoplastic polymer exhibits lower viscosity as temperature is increased above its melting point. Viscosity of PTFE dispersions actually increases at above room temperature. Figure 9.4 shows the effect of temperature on two PTFE dispersions. The increase is due to the reduced solubility of surfactant in water with temperature. This releases some of the PTFE particles from the micellar structures. Viscosity, then, begins to partly obey the law of slurries in which viscosity has an exponential relationship with the volume fraction of solid at higher concentrations. Polytetrafluoroethylene particles act as a filler due to their insolubility and inertness.

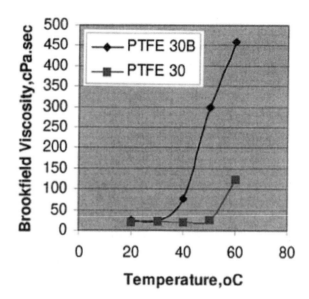

Figure 9.4 Viscosity behavior of PTFE dispersions as a function of temperature. (*Courtesy DuPont.*)

Viscosity of a slurry increases slowly at low concentrations. The increase becomes exponential as solids content increases. Physically, it can be explained that the solid particles set up structures in the liquid phase, which present a resistance to flow. There are a number of equations for the relationship between solids content and viscosity. Einstein's equation (9.4) can be used to calculate the viscosity of dilute solids in a Newtonian liquid. Marson and Pierce equation (9.5) is a simple expression for calculating viscosity throughout the range of solids concentration.[7]

Eq. (9.4) $$\eta = \eta_f (1 + 2.5\phi)$$

Eq. (9.5) $$\eta = \frac{1}{\left[1 - \left(\frac{\phi}{A}\right)^2\right]}$$

where ϕ is the volume fraction of solids, η is the viscosity of the slurry, η_f is the viscosity of the liquid and A is a constant with a value of 0.64–0.68 for spherical particles and smaller values for other shapes.

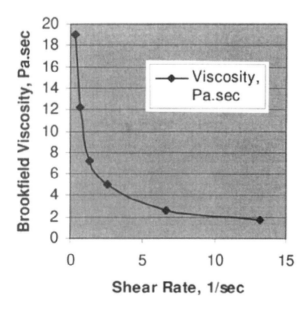

Figure 9.3 Shear thinning behavior of a thickened PTFE dispersion. (*Courtesy DuPont.*)

9.4.3 Surface Science

This branch of science studies the interfacial forces between materials. An important concept is surface tension defined as the sum of all forces exerted on a molecule by the surrounding molecules. At the interface of a solid and a liquid, surface tension can be used to predict wetting behavior of the liquid. In practice, surface tension between air and a liquid, and air and a solid, are neglected due to their relatively small value compared to the solid-liquid interaction.

A liquid is said to wet the surface of a solid when it spreads and flow out, and forms a thin film. A good example is the wetting of a metal pan surface by water. Non-wetting behavior is the opposite and can be best illustrated by observing a drop of oil on a PTFE-coated frying pan. The oil droplet basically retains its spherical shape and remains on the surface unchanged even after an extended period of time. The angle that the liquid meniscus forms with the solid surface is called *contact angle* (Fig. 9.5), which has a unique value for each pair of liquid-solid combinations at a given temperature. A wetting liquid forms a contact angle of 0°. An angle larger than zero indicates a non-wetting condition.

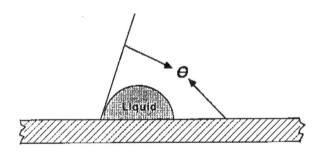

Figure 9.5 Contact angle.[3]

Surface tension and contact angle can be measured for combinations of liquids and solids. PTFE dispersions and solid surfaces have some of the lowest surface tensions of all materials against all liquids, values in the range of 15–18 dynes/cm. Although, there are very few liquids that wet the surface of polytetrafluoroethylene, most surfaces are wetted by PTFE dispersions due to the surfactant present in these dispersions. This is due to the criteria requiring a higher solid than liquid surface tension for wetting to take place (Eq. 9.6). This equation is the most simplified form of the wetting criteria in which other small interfacial forces have been neglected. Readers can refer to Ref. 21 in Ch. 2 for in-depth study of wetting criteria.

Eq. (9.6) $\quad \gamma_l < \gamma_s$

where γ_s is the specific surface tension of the solid (dynes/cm) and γ_l is the surface tension of the liquid (dynes/cm). Most metals and inorganic materials have significantly higher specific surface tensions than polytetrafluoroethylene dispersions and coatings.

Finally, leveling of the coating after application to the surface has to take place for a uniform film thickness to form. Leveling is actually a complex phenomenon and hard to control. Leveling is improved by longer time, lower surface tension of coating, lower viscosity, increased coating thickness, and smaller distance between ridges in the non-level areas.

9.5 Dispersion Formulation and Characteristics

This section reviews properties of dispersions and characteristics, which are important to the formulation and application of dispersions. They include solids content, pH, stability, and critical cracking thickness.

Dispersions of polytetrafluoroethylene are colloidal emulsions of small polymer particles (<0.25 µm) in water which are negatively charged. They contain relatively high concentrations of PTFE, which increases their specific gravity. An estimate of solids content can be obtained from the specific gravity (Table 9.4).

Table 9.4. Specific Gravity of PTFE Dispersions

Solids Concentration, %	Specific Gravity	Density of Solids, g/l
35	1.24	436
40	1.29	515
45	1.34	601
50	1.39	695
60	1.51	906

PTFE dispersions are supplied with a basic pH to prevent bacterial growth during storage, particularly when it is hot and humid. Bacteria feed on surfactant in the dispersion. Breakdown of the surfactants generates a rancid odor and brown discoloration in the dispersion. The pH can be reduced by adding acids, if necessary. It is important to control the amount of acid added to the dispersion because the increase in ionic strength of the dispersion can lead to coagulation of PTFE particles. As a matter of fact, increasing the ionic strength of polymer dispersion is one of the common methods for coagulation of dispersion to remove and dry the resin.

Ionic strength of a PTFE dispersion affects its conductivity. Conductivity is a very important characteristic of PTFE dispersions and can be a good indicator of its shelf life. It can be measured by a conductivity meter quickly and easily. Conductivity can also influence the viscosity and shear stability of the dispersion. Very high conductivity can destabilize the dispersion.

Stability of dispersion is important to its storage. It will partially settle during extended storage and when it is exposed to elevated temperatures (>60°C). A softly settled dispersion can be re-dispersed by gentle agitation. Freezing of the dispersion will lead to irreversible coagulation. Addition of water-soluble organic solvents or water-soluble inorganic salts and other compounds will also destabilize the PTFE dispersion and polymer coagulation will occur irreversibly.

The thickness of the wet coating affects the quality of the final sintered coating. An excessively thick layer will result in cracking after the polymer is dried. A *critical cracking thickness* is defined as the maximum thickness which can be coated in a single layer without formation of cracks. Layers in the thickness range of 5 µm to 25 µm can usually be cast without cracking concerns. The exact thickness is dependent on the formulation and type of dispersion, application process parameters, and the geometry of the article being coated. Multiple passes may be used to obtain higher thickness.

9.5.1 Formulation

Polytetrafluoroethylene dispersions usually contain nonionic surfactants which promote the wetting tendency of the dispersion. Triton® X-100 is an example of a common surfactant (supplied by Union Carbide.)[8] This type of surfactant does not survive the sintering temperatures of PTFE and decomposes. The products of surfactant degradation are mostly gaseous and evolve during the sintering of the coating, leaving little residue behind.

Many applications of dispersions require a number of properties in the end use which can be achieved by the addition of fillers, pigments, leveling enhancement additives, flow improvement additives, and other additives. For example, cold flow (creep) properties of the coating can be reduced by the addition of fillers such as fiberglass. Additives should be mixed only by mild stirring to avoid coagulation of PTFE.

There are applications where the viscosity of the dispersion must be increased to maintain uniform wet thickness in the process. Addition of water-soluble thickeners such as acrylic polymers is one way of increasing the viscosity of the dispersion. For example, the addition of 1% of Carbopol 934 (supplied by B. F. Goodrich Co.) can increase the viscosity of a dispersion containing 60% solids by 30 times to about 6 poise. (See "Teflon® PTFE Fluorocarbon Resin, Dispersion Properties and Processing Techniques.")[9] Other examples of thickeners include Acrysol® ASE acrylic polymers by Rohm & Haas and Natrosol® hydroxyethyl cellulose polymers by Hercules Corporation.

Another method is the addition of nonionic surfactants (Fig. 9.6), which can be added without increasing the viscosity to unacceptably high levels. Anionic surfactants are less desirable and the cationic type is unacceptable due to their coagulating or flocculating effects. Thickeners and surfactants degrade and evolve off during the sintering of the coating.

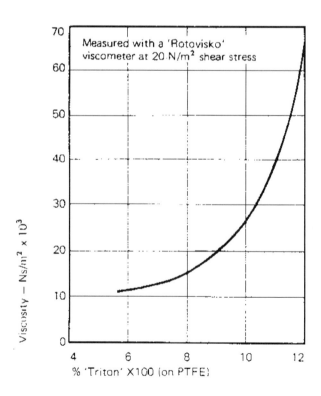

Figure 9.6 Viscosity measurements of a Fluon dispersion (60% PTFE) at 20°C (68°F) for various percentages of Triton X-100.[8]

9.6 Glass Cloth Coating

In this process, glass cloth is coated with PTFE dispersion, which is subsequently sintered in an oven. Typically, the glass fabric is supplied from a payoff roll and is passed through a trough filled with dispersion, followed by a drying and sintering oven. The glass cloth is then collected by a take-up roll. In some instances, the sintering step is omitted on the first few passes. The fabric is calendared to press any broken filaments into the soft PTFE coating and to "heal" mud cracks. The PTFE is then sintered in the remaining passes.

Glass cloth is woven from glass fibers. The fibers are coated with a "sizing" agent to act as a lubricant during the weaving process to prevent the fiber bundles from fraying. The sizing agent degrades and chars during the sintering process and leaves a color ranging from crème to brown. The color can be prevented by removal of the sizing chemically or, more economically, by heat.

Glass cloth has a smooth surface and is porous. It does not ionize in water or absorb the polytetrafluoroethylene dispersion. It picks up a small amount of dispersion per pass, therefore requiring multiple passes (up to a dozen) to obtain a smooth surface, if that is required.

9.6.1 Equipment

The equipment for glass cloth coating is shown in Fig. 9.7 which includes dip tank, oven for drying and sintering (tower), and payoff and take-up rolls. The dip tank should be made of stainless steel and equipped with a submerged roll or slide rod to allow advancing of the glass cloth through the tank. Multiple rolls help improve uniformity of side-to-side dispersion pickup. A partially submerged roll can help increase pickup by thick fabrics by forcing the dispersion through the cloth.

Some considerations for the equipment are listed below:

(1) To minimize foaming the tank should be shielded from air current and filled from the bottom, gravitationally.

(2) The equipment must be capable of constant speed to insure uniform dispersion pickup.

(3) A dip tank water jacket is sometimes necessary to maintain the dispersion temperature at 20 to 25°C.

(4) The dip tank should be designed with a minimum exposure to ambient air to avoid evaporation of water which would change the resin concentration, thus wet coating thickness.

(5) The oven should have three zones consisting of drying, baking and sintering.

(6) Drying zone should be capable of 100°C.

(7) Baking zone should reach 250°C.

(8) Sintering zone should be capable of 400°C.

(9) An annealing chamber is sometimes installed at the exit of the sintering oven to prevent the coated glass from cooling too rapidly to avoid wetting difficulty during the second pass coating.

(10) Finally, it is vitally important to exhaust the fumes of the ovens properly, thus avoiding exposure to the by products of the decomposition of surfactants, additives and polytetrafluoroethylene.

Equipment can vary in sophistication depending on the process needed to produce coated glass which meets the end-use requirements. A simple configuration equipment with a single stage oven is shown in Fig. 9.8. The design complications include multiplicity of rolls and increasing the number of independently controlled oven zones. Two designs can be seen in Fig. 9.9 and 9.10.

9.6.2 Processing

The processing steps include immersion in the dip tank, removal of excess dispersion, drying, baking, calendaring (sometimes) and sintering.

The dispersion should be gently stirred by an agitator for several minutes or its drums rolled, then filtered through a fine (5 µm to 20 mesh opening, depending on the application) filter. It should be next loaded into the dip tank and allowed to reach a constant temperature before beginning to coat the glass cloth. The coating speed is limited by some practical parameters such as the rate of return of excess dispersion to the dip tank, foam formation in the dispersion, and oven length and capacity.

Excessive coatings may be wiped from the glass cloth by applicators. They include, in the order of decreasing effectiveness, sharp-edged knives (*doctor blade*), round-edged knives, wire-wound rods, and fixed gap horizontal metering rolls. Coating thickness

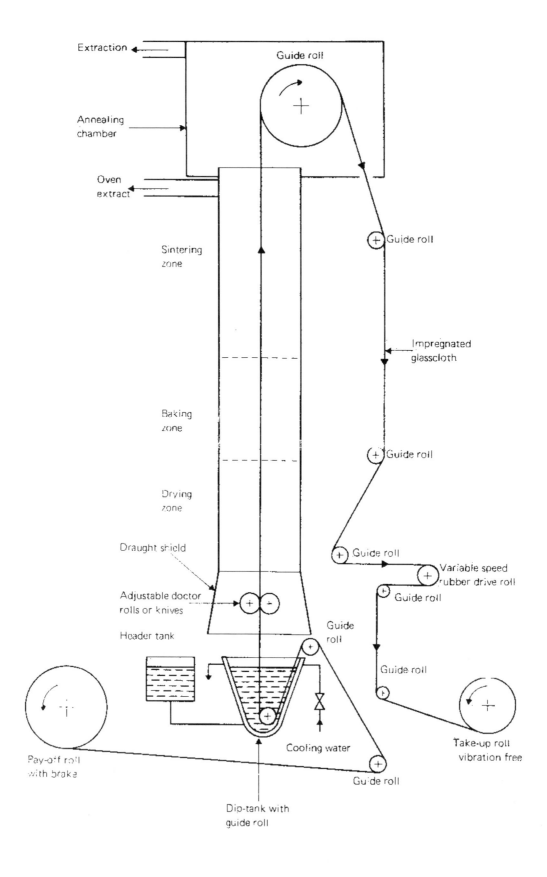

Figure 9.7 Glass cloth coating equipment.[8]

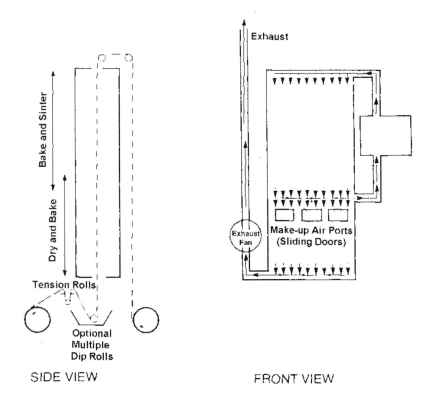

Figure 9.8 Simple counter-current tower.[9]

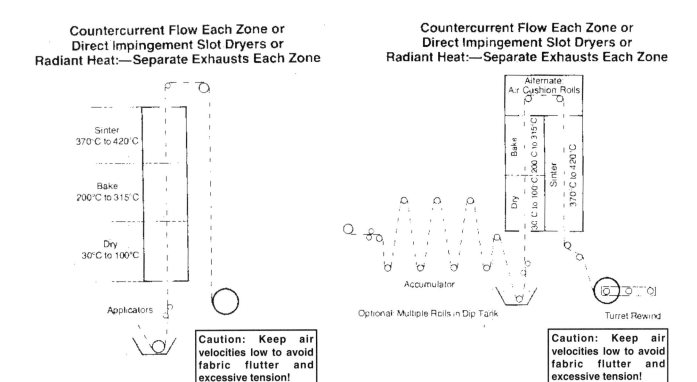

Figure 9.9 Modern coating towers.[9]

Figure 9.10 Modern coating towers.[9]

should be below critical cracking thickness. Sometimes multiple unsintered layers are coated on the glass cloth. The coated web must be first heated to remove the surfactant and then calendared before sintering to cure the cracks. Calendaring has the added benefit of flattening the fabric and tucking in the broken glass filaments in the coating. Broken filaments create defect point and wick moisture into the glass cloth which may affect its electrical properties.

The thickness and quality of the final coating, the type of glass fabric and the formulation will determine the number of passes that must be made. Typical process conditions for dispersions containing 45–60% PTFE are shown in Table 9.5. Glass fabric must reach the temperatures shown. Water is removed in the drying zone and the surfactant in the baking zone. It is preferable to extend the baking zone to complete the removal of surfactant as opposed to increasing the sintering temperature due to the reduction in the mechanical properties of glass fabric.

9.7 Impregnation of Flax and Polyaramide

Flax and polyaramide are used to fabricate packings and gaskets. They are available in a number of forms, which can be coated. The process described below was used in the past to impregnate asbestos and maybe used to process flax and polyaramides. The common forms of asbestos included yarn, cord, braid and cloth. Spun yarn could contain up to 5% cotton for production industrial packing. Extruded asbestos yarn had a smoother surface than spun yarn. Packing was produced by braiding and cloth by weaving from spun or extruded yarn. Asbestos packings have two major drawbacks: permeability and high coefficient of friction. Impregnation of asbestos with polytetrafluoroethylene reduced permeability and friction coefficient.

9.7.1 Processing

Impregnation of asbestos begins by dipping it in the PTFE dispersion followed by drying and baking. In general, each step is similar to the corresponding step for coating glass cloth. Asbestos is partly ionized in water and assumes positive charge, which promotes coagulation of the negatively charged dispersion on the surface of asbestos yarn. This inhibits the penetration of resin into the interior bulk of asbestos article.

Equipment for impregnating asbestos is conceptually similar to those used for glass cloth coating. Asbestos readily picks up PTFE dispersions and the amount of uptake depends on the concentration of PTFE in the dip tank. The dispersion may be diluted moderately with deionized water or substantially using a dilute aqueous solution of a nonionic surfactant such as Triton® X-100. The speed of the asbestos yarn movement through the dip tank has little effect on the amount of uptake. The length and capacity of the oven determines the maximum throughput of the process.

After dipping, the asbestos is dried in an oven (80–90°C) followed by calendaring to smooth the surface.

9.7.2 Impregnation of Porous Metals and Graphite

The applications of impregnated metal and graphite parts are in heat exchangers and bearings where impervious properties are required. The key issue in impregnation of graphite and porous metals is penetration of dispersion into the pores. This is accomplished by placing the dipping operation in a vacuum chamber. After pulling vacuum and removing the air from the pores of the article, air is gently allowed back into chamber. Pressure cycling drives the dispersion into the pores. More than one cycle can be used to increase impregnation. Sintering follows slow drying and baking.

Table 9.5. Process Conditions for Glass Cloth Coating [8]

Number of Passes	Dispersion Concentration, % PTFE	Added Surfactant	Web Speed, m/min	Web Temperature,°C		
				Drying	Baking	Sintering
1,2,3	45-50	No	1-2	90-100	200-250	380-400
4,5,6	55-60	Yes	1-2	90-100	200-250	380-400

9.8 Coating Metal and Hard Surfaces

Metal and ceramic surfaces are coated with polytetrafluoroethylene dispersions to protect them from corrosion and create nonstick surfaces. These objectives can be accomplished by using filled or unfilled coatings. Examples of applications include household and commercial cookware and industrial equipment. Each coating type has advantages and disadvantages. Unfilled PTFE forms a surface entirely made of polytetrafluoroethylene which is smoother and less porous than the coated filled resin. The filled formulation generates a harder surface, which wears at a slower rate than the unfilled coating. Unfilled PTFE coatings generally adhere to the surface mechanically, while pigmented dispersions adhere by priming the surface chemically. The middle layers of the coating usually contain the pigment.

9.8.1 Unfilled PTFE Coatings

Unfilled PTFE coatings are only suitable for adhesion to aluminum and few of its alloys such as aluminum/magnesium with a typical sintered coating thickness of < 25 µm. The preparation of the surface of aluminum is of paramount importance to obtain good adhesion of polytetrafluoroethylene coatings. The surface of aluminum must be roughened to maximize its mechanical engagement with the coated polymer. This can be best achieved by mechanically or chemically etching the surface of aluminum. Primers which do not need special surface preparations are also available.

The aluminum surface has to be degreased before it is etched to assure uniform etching. Grease and oil can mask the surface of metals and prevent etching. Organic solvents or solutions of phosphates can remove grease. Aqueous solutions of phosphates are preferable due to environmental emissions caused by organic solvents. The surface is simply washed with the phosphate solution followed by rinsing with tap water and deionized water.

Etching can be easily accomplished immersing the part in a dilute solution (20–30% by weight) of hydrochloric acid. An acid bath can be used to etch the surface of the article. A warm bath (30–40°C) requires several minutes for the etching to be completed. The duration of etching should be extended if the bath temperature is decreased. A longer etching time is required as the aluminum content of an alloy is decreased. Hydrogen is a by-product of metals reaction with acid, which must be removed to avoid explosion.

Following the rinsing of the part with water, it is immersed for a few minutes in a dilute solution of nitric acid at the room temperature. It is then washed thoroughly with tap water and deionized water and dried. After drying, the part must not be handled or contaminated with any foreign substance until the coating process has been completed. Adhesion of PTFE to contaminated areas of the surface will be weak.

The final step is coating the aluminum surface with the polytetrafluoroethylene. The coating can be applied by one of the methods described in Sec. 9.4. It is helpful to adjust the viscosity of the dispersion to 300–400 centipoise to improve its penetration into the pores of the roughened areas. This can be achieved by adding deionized water for small dilutions or a 3–5% aqueous solution of an appropriate surfactant, such as Triton® X-100, for large dilutions. The added surfactant preserves the stability of the dispersion, which is reduced by water addition.

After application to the part, the coating is dried and sintered, usually in a single oven comprised of two zones. The initial zone should heat the part and remove the moisture slowly to prevent rapid evaporation. A temperature setting below the boiling point of water around 90°C will remove the water in a few minutes. No water should be left in the coating when it is being sintered. The excessively rapid evaporation of water tends to form defects in the PTFE coating. Sintering of the coating can be accomplished in an oven at a temperature well above the melting point of the polymer (at least 380°C). The length of time depends on a number of variables such as the heat capacity of the part because the whole part must reach the sintering temperature. Sintering and drying oven must be exhausted properly to remove off gases, which could contain surfactant and PTFE degradation products.

9.8.2 Filled PTFE Coatings

This type of coating is desirable for applications where both good decorative and wear resistance qualities are important, such as household and commercial cookware and architectural finishes. In addition to improving properties such as wear rate, a pigment filled (usually inorganic) coating has the advantage of completely covering the substrate at a fairly low coating thickness (10–15 µm). These coatings do not adhere directly to metallic substrates and require chemical priming. Typical primers are mixtures of phosphoric and chromic acids.

Surface preparation prior to application of the pigmented coating consists of degreasing and sandblasting. Degreasing can be done chemically (Sec. 9.81) or by baking the article at a sufficiently high temperature to degrade the organic contaminants. After removal of grease, the article must be protected from handling and contamination to assure good adhesion bonding. Sandblasting is effective when a roughness height of 5–10 μm has been achieved.[10]

The filled PTFE dispersion is mixed with the priming agent before application. It is usually applied to the article by spray coating techniques. The choice of equipment material should be made to accommodate the highly corrosive priming agent. After spraying, the article is dried and sintered in three steps. Drying or water removal is carried out at 90–100°C, slowly. In the second step, the article is heated to 250–300°C and held to remove the surfactant. Finally, the part is sintered at > 380°C which completes the process.

9.9 PTFE Yarn Manufacturing

This process is designed to produce yarns of polytetrafluoroethylene by overcoming non-melt processibility of this polymer. Fibers of PTFE are used as staple, flock or multifilament yarn in the fabrication of bearings, filter bags and packings for valves, agitators and pumps.

Manufacturing of PTFE yarn takes advantage of the processes for conversion of cellulosic material, mainly wood pulp, into fiber. The basic technology has been described in the US patent 2,772,444 by Burrows and Jordan.[11] In this process, wood pulp is treated with a basic solution which converts the hydroxyl groups of cellulose to salt. The treated cellulose is blended with carbon disulfide, which converts the alkoxy salt groups to a thiocarboxy group called *xanthate*. The CS_2 converted material is a viscous dope which is thoroughly mixed, after further processing including filtration, with PTFE dispersion.

The mixture of xanthate and PTFE is spun into an acid bath. The acid converts the xanthate to carbon disulfide and cellulose. Carbon disulfide is recovered and recycled. The PTFE fiber is rinsed with water to remove the acid and other accompanying impurities. The fiber is then sintered and drawn to increase its tensile strength. Typical values of the ultimate strength (280–350 MPa) are about ten times higher than the strength of molded powders.

A specific example of PTFE yarn spinning process is described here.[11] A viscose solution containing 7% cellulose and 6% sodium hydroxide was prepared by using 30% by weight of carbon disulfide. The viscose solution was filtered and allowed to age before addition of a 60% PTFE dispersion containing 10% Triton® X-100. The spinning dope contained 40% polytetrafluoroethylene and 2.3% cellulose. This mixture was filtered and spun at 20°C. Each spinneret had 60 holes of 125 μm diameter. The spinning rate was 18 m/min.

The filaments entered an aqueous bath containing 10% sulfuric acid, 16% sodium sulfate and 10% zinc sulfate. After traveling for over one meter in the coagulating acid bath, PTFE yarn entered into a warm water (79°C) bath. The filament bundles were dried by taking a dozen wraps over a heated roll at 190°C. A this point the strength of the yarn is about 0.04 grams per denier. The cellulose was removed from the fiber by during the sintering of PTFE using a heated roll at 389°C. The outgoing tension was raised to 0.075 grams per denier resulting in seven times stretch over its original length. The resulting yarn had 60 filaments and 375 denier.

The fiber produced by this process has a black color due to the residual carbon left from the degradation of cellulose during the sintering. A bleaching process is necessary to obtain a white yarn. This can be accomplished by heating the yarn at a temperature over 300°C for a period of five days. The drawback of thermal bleaching is that half of the tenacity of the yarn is lost. Chemical bleaching can be accomplished by immersing the yarn in boiling sulfuric acid and adding a small amount of nitric acid. The disadvantage of this process is generation of acid waste and its disposal.

A number of other sources describe alternative processing techniques and refinement ideas for PTFE yarn spinning.[12]-[15]

9.10 Film Casting

Film casting is a process in which a continuous PTFE film is obtained by coating the dispersion on a carrier web followed by sintering. The cast product can be used after it has been stripped, or as a composite with the web. Typical applications include insulation of wire with the composite of PTFE and Kapton® carrier primarily for aerospace use.[16] Stripped PTFE film can be laminated to different substrates such as glass fabric. This process is also capable of producing film from thermoplastic fluoropolymers.

A casting process has been described in U. S. Patent No. 2,852,811.[17] This process involves coating a layer of PTFE dispersion on a metal carrier, followed by drying and sintering the coating. The low critical cracking thickness of PTFE dispersion limits the thickness which can be achieved in a single pass. Multiple coating passes are made until the desired thickness is obtained. The film is stripped and removed from the carrier in the final step. The nature of the carrier belt is very important. Highly polished corrosion resistant metal belts are used in casting processes.

PTFE cast films have highly desirable properties such as an absence of anisotropy (dependence of properties on direction) in their mechanical properties (see Table 9.6). The drawback of the casting process is its poor economics which has prevented the widespread use and acceptance of this manufacturing technique. The unfavorable casting economics are caused by the heavy carrier belts, which require special tracking mechanisms. The width of the belt is constant which limits the flexibility in the width of the cast film.

Film casting process has the advantage of allowing composite constructions. For example, PTFE films can be further coated with dispersions of other fluoropolymers such as FEP and PFA which are both thermoplastics. Alloys of fluoropolymers containing elastomers or fillers can be cast into film.

Another process [19] involves dip-coating a metal carrier with PTFE dispersion, which covers both sides of the belt and doubles the productivity (Fig. 9.11). After dipping in the dispersion, the belt is passed through a metering zone where metering bars remove the excess dispersion from the belt. The belt enters a drying oven for water removal followed by sintering zones to consolidate the PTFE and to form a film. Good thermal conductivity of the metal belt shortens the sintering time, thus, maximizing tensile strength and elongation. The coating is then peeled away from the web.

Multilayer films can be manufactured with good peel adhesion properties.[20][21] For example, polyimide films can be coated with a "tie layer" of FEP or PFA before being coated with PTFE dispersion. Other polymers such as ETFE and vinylidene fluoride can also be used as the tie layer. These films have good scrape abrasion resistance and are used for insulation of wires for aerospace and similar critical applications.

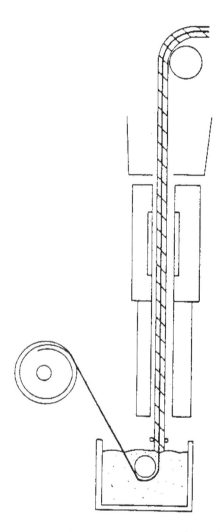

Figure 9.11 Film casting process schematic.[19]

Table 9.6. A Comparison of Cast and Skived Film Properties [18]

Film Type	Thickness, μm	Tensile Strength, MPa		Break Elongation, %		Elastic Modulus, MPa	
		MD[1]	TD[2]	MD	TD	MD	TD
Skived	76	52.3	40.4	450	360	469	517
Cast	68	35.8	34.5	530	510	434.5	434.5

[1] MD = Machine Direction   [2] TD = Transverse Direction

9.11 Anti-Drip Applications

Polytetrafluoroethylene has been added to other thermoplastic polymers to improve fire performance by suppressing dripping, for some time. Fibrillation of PTFE and formation of a structure which retains high viscosity in molten state, thus preventing dripping of the molten host polymer, has been credited for performance enhancement. Burn time is usually reduced by the incorporation of flame-retardants (Table 9.7).

Both fine powder and dispersion grade PTFE are effective in drip suppression. Small quantities (<1 % by weight) of PTFE are added to the host polymer, which must be distributed uniformly for maximum effectiveness. Dispersion can be easily added to the host material uniformly. The mixture is dried and water is removed, sometimes under vacuum, prior to extrusion compounding, usually in a twin screw extruder. Examples of common polymers for this application include polycarbonate[23][24] and polyethylene terphthalate,[25] PBT and ABS.

Table 9.7. Effect of PTFE on Drip-Suppression of Polycarbonate (PC)*

PTFE Content, wt %	MFR of PC, g/10min	Time to Flameout, sec	Number of Drips	UL-94 Rating
0	15	10.3	5/5	V-2
0.3	15	1.2	0/5	V-0
0.3	20	1.0	0/5	V-0

*Contained tetrabromobisphenol-A oligomer as flame-retardant.

A high molecular weight polytetrafluoroethylene is preferred because the fibrillation enhancement brought about by an increase in molecular weight improves drip suppression. Compounding difficulty increases with molecular weight due to premature fibrillation of PTFE during the mixing process. Dispersions of PTFE are easier to uniformly blend with host polymers.[26][27]

9.12 Filled Bearings

A special process has been used for decades to produce filled bearings by coagulation of PTFE in the presence of fillers. The main goal is to reduce wear and cold flow while taking advantage of low friction coefficient of PTFE. The original process (known as *DU Bearing Process*) was developed by Glacier Company and is described in US Patent 2,689,380. This technology is widely practiced today to fabricate filled compositions.

This process is also known as *co-coagulation*. PTFE dispersion is mixed with the filler, and a salt such as aluminum nitrate is added to convert the surfactant to an insoluble species, leading to coagulation of PTFE particles. The viscosity of the resulting paste ("mush" in the industry parlance) is adjusted by the addition of an organic solvent such as toluene. This paste is calendared onto the surface of a steel-backed porous bronze strip. The strip is manufactured by sintering bronze powder onto the steel strip. After sintering, the steel strip is rolled into bearing such that the filled PTFE would form its inside surface. Automotive applications are the main end use of these bearings, for example in shock absorbers.

The initial filler of choice was lead but lead-free formulations have been developed over the years. Graphite, bronze and zinc powders are used to fill PTFE for DU Bearings.[28][29]

9.13 De-Dusting Powders

One of the unusual applications of PTFE dispersion is de-dusting of powders. The objective is to prevent the dust in products such as cat litter from becoming airborne. Powder particles are not coated by PTFE. Most of the dust is trapped in the webs created by microscopic fibrils that are generated by shearing of the resin particles. The rest of the particles adhere to the web through contact. A popular commercial process for de-dusting powders is the *Harshaw Process for Dustless Powders*.[30]

The de-dusting process works with as little as 0.005% by weight of polytetrafluoroethylene in the powder but a concentration of 0.1% is common. This is equivalent to one kilogram of PTFE in one metric ton of the powder. It is necessary to work the PTFE to initiate fibrillation. This can be achieved by working (shearing) the mixture of PTFE and the powder in a blender or a slow turning mixer (like a cement mixer) at an elevated temperature. Room temperature is not effective and a shearing temperature around 100°C is needed. The time required to complete the shearing is dependent on the concentration of PTFE in the mixture and varies between a few seconds to a few minutes.

9.14 Other Applications

There are many other applications for PTFE dispersions. In automobile gaskets, a thick coating of dispersion is applied to the metal surface. The gasket is heated to remove the surfactant and water but not sintered. The resin undergoes cold flow under load when the gasket is tightened. This is helpful to insure a complete seal in addition to the other beneficial attributes of polytetrafluoroethylene.

Another interesting and unusual application is the fabrication of insect traps. Surfaces of substrates are coated with PTFE to prevent crawling insects from climbing on these surfaces. Low friction creates a slippery surface, which prevents the insects from leaving the trap.[31]

References

1. Cohen, E. D., and Gutoff, E. B., "Coating Processes," *Kirk-Othmer Ency. of Chem. Tech.*, 6:606–635, John Wiley & Sons, 1993.
2. *Coating Technology Handbook,* ed. D. Satas, Marcel Dekker, New York, 1991.
3. *Web Processing and Converting Technology and Equipment,* ed. D. Satas, Van Nostrand Reinhold, New York, 1984.
4. Bird, R. B., Stewart, W. E., and Lightfoot, E. N., *Transport Phenomena*, John Wiley & Sons, New York, 1960.
5. Schlichting, H., *Boundary layer Theory*, Translated by J. Kestin, *McGraw-Hill Series in Mechanical Engineering,* New York, 1968.
6. Patton, T. C., *Paint Flow and Pigment Dispersion,* 2nd ed., John Wiley & Sons, New York, 1979.
7. Dealy, J. M., and Wissbrun, K. F., *Melt Rheology and its Role in Plastics Processing: Theory and Applications*, Van Nostrand Reinhold, New York, 1990.
8. Ref. FLUON® Polytetrafluoroethylene, *Impregnation with PTFE Aqueous Dispersions*, Technical Service Note F6 – Third Division, ICI Petrochemicals and Plastics Division, Published 1984.
9. Teflon® PTFE, Dispersion Properties and Processing, *Bulletin No. X-50G*, DuPont Co., 1983.
10. Hoechst Plastics – Hostaflon®, Hoechst Aktiengesellschaft, Frankfurt, Germany, 1984.
11. Burrows, L. A., and Jordan, W. E., US Patent 2,772,444, assigned to DuPont, Dec. 4, 1956.
12. Boyer, C., US Patent 3,147,323, assigned to DuPont, Sep. 1, 1964.
13. Steuber, W., US Patent 3,242,120, assigned to DuPont, Mar. 22, 1966.
14. Kitagawa, H., Kinoshita, S., and Uchiyama, H., US Patent 3,397,944, assigned to Tokyo Rayon Kabushiki Kaisha, Aug. 20, 1968.
15. Gallup, A. R., US Patent 3,655,853, assigned to DuPont, Apr. 11, 1972.
16. Sahatjian, R. A., Ribbins, R. C., Steckel, M. G., US Patent 4,943,473, assigned to Chemical Fabrics Corporation, Sep. 24, 1990.
17. Petriello, J. V., U. S. Patent No. 2,852,811, Sep. 23, 1958.
18. Effenberger, J. A., Koerber, K. G., Latorrs, M. N., and Petriello, J. V., US Patent 5,075,065, assigned to Chemical Fabrics Corporation, Dec. 24, 1991.
19. Effenberger, J. A., Koerber, K. G., Latorrs, M. N., and Petriello, J. V., US Patent 4,883,716, assigned to Chemical Fabrics Corporation, Nov. 28, 1989.
20. Effenberger, J. A., and Koerber, K. G., US Patent 5,106,673, assigned to Chemical Fabrics Corporation, Apr. 21, 1992.
21. Effenberger, J. A., Koerber, K. G., and Lupton, E. C., US Patent 5,238,748, assigned to Chemical Fabrics Corporation, Aug. 24, 1993.

22. Ogoe, S. A., US Patent 5,041,479, assigned to The Dow Chemical Co., Aug. 20, 1991.
23. Kress, H. J., Bottenbruch, L., Witman, M., Kircher, K., Lindner, C., and Ott, K. H., US Patent 4,649,168, assigned to Bayer Aktiengensellschaft, Sep. 8, 1987.
24. Kress, H. J., Muller, F., Lindner, C., Horst, P., Buekers, J., US Patent 4,692,488, assigned to Bayer Aktiengensellschaft, Mar. 10, 1987.
25. Brink, T., and deGraaf, S. A. G., US Patent 4,356,281, assigned to Akzo NV, Oct. 7, 1980.
26. Pan, W. H., and Reed, R. A., US Patent 5,102,696, assigned to General Electric Co., Apr. 7, 1992.
27. Huang, J., Chen, F. S., and Brasser, J. J. M., US Patent 5,681,875, assigned to General Electric Co., Oct. 28, 1997.
28. Pratt, G. C., and Montpetit, M. C., US Patent 4,615,854, assigned to Federal Mogul Corp., Oct. 7, 1986.
29. Davies, G., et al., EP Patent 546070, assigned to Glacier Metal Co., Ltd, June, 16, 1993.
30. Product Information DuPont PTFE K-20, Fluoropolymer Resin, Aqueous Dispersion, DuPont Co. Wilmington, Delaware, Dec., 1996.
31. Long, R. H., US Patent 5,566,500, assigned to Consep, Inc., Oct. 22, 1996.

10 Processing of Polychlorotrifluoroethylene

10.1 Introduction

This chapter is about ways to process polychlorotrifluoroethylene (PCTFE) plastics into useful objects. Polychlorotrifluoroethylene consists of homopolymers and modified copolymers of chlorotrifluoroethylene. These thermoplastics are utilized due to their chemical resistance and mechanical and electrical properties. PCTFE polymers can be fabricated by conventional melt processing techniques like extrusion and injection molding. Compression molding is the most suitable technique for PCTFE because of the fit between its properties and the parameters of this process. These three processing techniques, ways to reduce residual stress in parts, and machining and bonding methods for polychlorotrifluoroethylene have been reviewed. In a separate section, application procedures of PCTFE dispersions have been covered.

10.2 Processing Considerations

Properties of polychlorotrifluoroethylene vary substantially with the processing technique, particularly molecular weight and mechanical properties. Three factors affect the mechanical properties and performance of PCTFE fabricated by melting the resin. They include molecular weight, crystallinity, and the amount of stress placed on the polymer during the processing. Molecular weight is represented by zero strength time (ZST) which is defined as the time required to hold a PCTFE specimen at 250°C under conditions specified by ASTM Method D 1430 (see Sec. 5.9). Each factor is discussed separately.

10.2.1 Zero Strength Time

In this discussion the reader should keep in mind that zero strength time (ZST) is equivalent to molecular weight. A high molecular weight is desirable for good physical properties. The generally accepted minimum value is a ZST of 100 seconds. Consequently, very high melt viscosity PCTFE must be processed to fabricate parts. The only way to reduce the viscosity of an existing polymer is to increase its processing temperature. Degradation temperature of CTFE polymers is about 296°C which means that processing at higher temperatures results in the degradation of the polymer and a lower molecular weight. A lower molecular weight means lower physical properties and a lower ZST. Table 10.1 provides a comparison of the thermal stability of PTFE and PCTFE in vacuum and in oxygen. Polychlorotrifluoroethylene is thermally less stable than polytetrafluoroethylene.

The objective should be to obtain the maximum ZST possible while processing the polymer. Design of a part has a great influence on the ZST because each part requires a unique set of processing variables. Injection molding and extrusion invariably lead to a lower ZST than compression molding, primarily due to higher process temperatures. There is ZST loss in a compression-molded part while the drop can be minimized in injection molding and extrusion. Figure 10.1 shows the effect of temperature and time on the zero strength time of PCTFE. Clearly, molecular weight of the polymer decreases rapidly at temperatures above 277°C even when it is held for a short time (~2 minutes).

Table 10.1. Temperature (°C) for 25% Weight Loss in Two Hours[1]

Polymer	In Vacuum	In Oxygen
PTFE	494	482
PCTFE	349	355

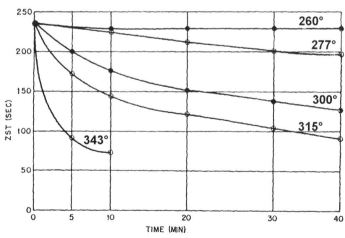

Figure 10.1 ZST vs time and temperature for polychlorotrifluoroethylene.[2]

10.2.2 Crystallinity

Crystallinity can be accurately measured by X-ray diffraction and less accurately but simply by specific gravity. The properties of polychlorotrifluoroethylene vary with the degree of crystallinity. Table 10.2 contains properties of two resins with different specific gravities. The crystalline resin, one with the higher specific gravity, has higher tensile yield strength and flexural modulus than the amorphous plastic. The amorphous polymer has higher deformation-under-load and impact strength than the crystalline resin.

How does processing alter the crystallinity of PCTFE? The degree or extent of crystallinity is determined by the rate of crystallization and the thermal history of the polymer. The rate increases with a decrease in ZST or increase in temperatures due to a reduction in the size of the molecules as a result of degradation. A reduction in the cooling rate leads to an increase in crystallinity at any ZST. The starting temperature for crystallization is 52°C and a maximum is reached in the range 150–190°C. The crystals completely melt upon reaching 204–218°C. Specific gravity change as a function of zero strength time can be seen in Table 10.3.

Table 10.2. Effect of Crystallinity on the Properties of PCTFE (ZST = 250, Compression-Molded Sheet)[2]

Property	Amorphous	Medium Crystallinity
Specific Gravity	2.105	2.131
Tensile Yield Strength, MPa		
Break Elongation, %	175	105
Flexural Modulus, MPa		
25°C	1,311	1,753
70°C	373	1,028
125°C	90	22
Deformation-under-load, % 24 hours at 6.9 MPa		
25°C	0.4	0.2
70°C	7.3	0.4
125°C	>25	3.6
Impact Strength, Notched Izod, J/m	267	80

Table 10.3. Crystallinity Change as a Function of ZST[2]

Specific Gravity	ZST = 111 sec	ZST = 225 sec	ZST = 300 sec
Initial	2.1240	2.1193	2.1186
After 24 hours at 149°C	2.1469	2.1368	2.1344
After 1,344 hours at 149°C	2.1556	2.1452	2.1422

10.2.3 Stress

Internal stress develops in all plastics during the processing. The more polymer flow is involved in the operations, the more residual stress is generated in the part. This means that injection molding and extrusion develop more stress than compression molding. High viscosity resins like PCTFE develop more stress than the low viscosity polymers. Fast cooling of a part "locks in" the internal stresses.

In injection molding of polychlorotrifluoroethylene, a high viscosity polymer is forced through a small opening which subjects the melt to high shear rates, and then it is allowed to cool rapidly. The shrinkage during the solidification also contributes to the trapping of the stresses. This is in contrast to compression molding where the flow is slow and cooling is usually gradual. Excessive residual stresses in a part can manifest themselves in the form of distortion or catastrophic failure, as cracks in the part. Distortion may not be seen until the temperature of the part is elevated.

In compression molding, very low shear rates are involved, typically in the range of 1 sec^{-1}.[3] Transition from the rubbery to melt state takes place at low temperatures which means very little rubbery component is present at the maximum process temperature. Consequently, a great relaxation of normal stresses takes place during the compression molding process. This is why parts made by compression molding contain minimal residual stresses and have excellent dimensional stability.

There are ways to minimize or overcome the problems arising from the residual stress of parts. Some preventive measures include designing the process equipment to reduce stress buildup in the part by: eliminating many restrictions, reducing shear rates, slowing the cooling rate, and cooling under pressure. These provisions are not always practical. Annealing is an efficient method of reducing stress in fabricated parts.

This technique consists of slow heating of the part in an oven to about 150°C, followed by slow cool down to room temperature. To combine annealing and an increase in crystallinity, the part can be heated to the maximum crystallization rate of PCTFE. In the case of extruded films, stress is relieved by heating the rolls while the film is being manufactured.

10.3 Compression Molding

How does compression molding work? To mold flat sheets of PCTFE, a press is required which is equipped with platens capable of reaching the process heating and cooling temperatures and pressures. The construction material must resist corrosion at elevated temperatures. Mirror-polished stainless steel plates can be used in conjunction with a thermally stable mold release agent such as silicone. For other shapes, an oven is required. The amount of resin is 2.15 g/ml of the final shape.

The resin is placed between the press platens using a polished plate and spacers on the lower platen. The second plate is placed on top before closing the platens. The assembly containing the resin can be alternatively prepared outside the press in which case the platens can be preheated. After closure of the platens and slight pressurization, the resin is allowed to heat to the required temperature. The pressure is slowly increased until it reaches the desired value and is maintained for the specified dwell. After 3–5 minutes have passed, the sheet can be cooled slowly in a press or quenched by removal of the assembly and immersion in water.

This is the fabrication technique to obtain a PCTFE part with the best physical properties due to its relatively low temperature. Thick sheets and thick wall tubes and rods are prepared by this method when thickness exceeds the capability of the extrusion process. The lengths produced by compression molding are limited to less than 0.5 m. High ZST (300–400 sec) resins are converted by compression molding at pressures of 13.8–20.7 MPa and temperatures of 246–274°C.[4] Molding time for a 3 mm thick sheet at 275°C is 7–14 minutes.

Compression molding large tubes and rods is done in two stages: heating and pressurization.[2] To make a rod, PCTFE pellets are placed in the mold and put in an oven for 48 minutes per cm of diameter and allowed to reach a temperature of 245–250°C. The assembly is then transferred to a press and pressurized at 1.7 MPa/cm of mold diameter. Pressure should be increased at a low rate to allow air to escape. Pressure must be maintained during the cooling cycle until the center of the rod is below 90°C to avoid cracking of the part. Compression molding processes for tubes and rods are similar. For a tube, wall thickness is the major dimension in the selection of process conditions versus diameter for a rod.

Compression molded parts should be annealed above the maximum service temperature prior to machining. This will increase the part's dimensional stability in the application.

10.4 Injection Molding
(This section has been adapted from Ref. 5)

We begin with a general review of the injection molding process. The principle of injection molding is simple. The plastic material is heated until it becomes a melt. It is then forced into a closed mold that defines the shape of the article to be fabricated. The material is allowed to cool to solidification in the mold, which is then opened to eject the part. Although the principle is simple, the actual process is anything but simple. This is the consequence of the complex behavior of molten polymers and the complexity of the parts made by this process. The essential elements of injection molding are heat transfer and forced melt flow. The equipment for injection molding (Fig. 10.2) consists of a machine, sometimes referred to as *press*, and a mold also known as *tool* or *die*.

There are a few varieties of injection molding machines, but they all perform the same basic functions. These include melting the resin, injecting it into the mold, holding the mold closed, and cooling the injected plastic. It is convenient to think of an injection molding machine as consisting of two units. The melting and injection functions are performed in the injection unit while the mold handling is conducted in the clamp unit. The two units are mounted on a common base and are connected by power and control systems. The main advantages of injection molding are high productivity/low production cost, ability to produce complex parts and reproducibility of the parts.

Injection molding polychlorotrifluoroethylene is a controlled reduction in the molecular weight of the resin to decrease the viscosity so that the melt would fill the mold at practical temperatures and pressures. Degradation of PCTFE produces corrosive by-products such as hydrochloric acid. Consequently, the surfaces in contact with the melted polymer should be made of corrosion resistant alloys such as Hastelloy C, chrome-

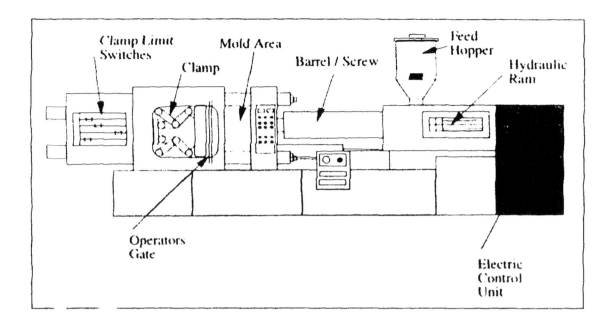

Figure 10.2 Typical injection molding machine.[5]

plated steel, or Xaloy 306.[6] Zero strength time decrease during injection molding depends on the thinnest section of the part. Minimum ZST with useful part properties is 100 sec. The maximum ZST possible must be achieved without leaving voids in the part.

Some of the requirements to process PCTFE include:

1. A minimum nozzle pressure of 210 MPa.
2. Clamping pressures to counteract the injection pressure.
3. Maximum melt inventory of 15 min.
4. Excellent heat and cycle controls.

Table 10.4 presents examples of process conditions for injection molding PCTFE. The independent process variable is the thickness of the thinnest section of the intended part. The actual conditions may vary based on the part design. The best indication of whether a part has been processed properly is its ZST. Specific gravity (measure of crystallinity), physical strength and appearance are the other measures of the part quality.

It is important to remember that PCTFE degrades during the processing and produces corrosive and toxic by-products requiring good ventilation. Very low molecular weight polychlorotrifluoroethylene products are oily and build up on the equipment surfaces. Their residues should be removed by an organic solvent such as trichloroethylene.

Table 10.4. Process Conditions for Injection Molding PCTFE[2]

Process Condition	Thickness, mm		
	<3	3-6	>6
Temperature, °C			
Rear Cylinder	260-280	260-280	260-280
Mid Cylinder	270-295	270-295	270-295
Forward Cylinder	280-315	280-315	280-315
Nozzle	315-350	315-350	315-350
Mold Face (coolant)	90-160	90-160	90-160
Melt leaving nozzle	280-305	280-305	280-305
Melt pressure at nozzle, MPa	207-414	172-314	138-276
Timing			
Total cycle, sec	40-90	60-120	70-180
Ram forward time, sec	10	15	25
Volume of mold cavity, cm^3 (assumed)	16	64	256
Recommended gate dimensions, mm			
Diameter (round), mm	1.6-2.7	2.7-6.2	4.7-9.4
Land length, mm	1.6	1.6-4.7	3.1-4.7

10.5 Extrusion

Rod stock, tubing, and film shapes of polychlorotrifluoroethylene can be produced by extrusion. Similar to injection molding, the thickness of the article being manufactured is the determinant of its ZST (molecular weight). Objects made by extrusion are not as strong as those produced by compression molding, but much longer lengths are possible. Wire insulation can also be produced by melt extrusion.

PCTFE can be extruded using most common extruders.[4] Sizes from 37.5–62.5 mm are adequate and larger extruders can be used to fabricate thick cross sections. The extruder must be capable of reaching a temperature of 340°C. Minimum extruder length to diameter ratio (L/D) is 16:1, but the preferred range is 20:1 to 24:1. Caution must be exercised in the design of the extruder screw because of the capability of high viscosity PCTFE melt to generate extremely high pressures, even in excess of barrel strength. The use of a breaker plate is not recommended due to high viscosity of this polymer and its thermal degradation. The die and transition pieces of the extruder should be designed to avoid stagnant areas to prevent degradation.

Thin (~25 µm) films of PCTFE are produced by extrusion. ZST values of the extruded films are usually in the range of 90–130 sec indicating fairly low molecular weight. These films are not very strong and prone to embrittlement when subjected to heat. Examples of extrusion conditions for a film and a heavy wall shape are given in Table 10.5

Typical properties of PCTFE parts can be found in Part III of the book.

10.6 Machining and Bonding

This plastic is easy to machine and behaves similarly to a soft metal while cutting on a lathe. Annealing before precision machining is helpful to insuring dimensional stability. It is recommended to use a coolant to maintain a constant temperature during the machining. High speed sharp cutters yield the best results.

PCTFE can be bonded by thermal, ultrasonic and dielectric methods. Commercial ultrasonic and thermal impulse sealing equipment can be used. It is easy to heat-seal the films of this polymer by heating them to 200–260°C. Adhesive bonding of PCTFE parts is possible using epoxy adhesives after treatment of the adhesion area with a sodium etching agent or another method. Surface treatment of fluoropolymers for adhesion is discussed in Ch. 16.

10.7 Processing PCTFE Dispersions

Commercial polychlorotrifluoroethylene dispersions are low viscosity aqueous anionic liquids. These dispersions are stabilized by the addition of surfactants and thickeners. A coalescing agent such as ethylene glycol (2–4% of polymer solids) is added to the PCTFE dispersion to reduce the minimum film formation temperature. The amount of surfactant should be minimized (<1%) to obtain high barrier properties. A number of other additives such as pigments, antifoamers, antifreeze chemicals and others. More description of formulation guidelines can be found in *Aclon® 400 Series Dispersions, Formulation Guidelines*, published by AlliedSignal Chemicals.

Application methods of PCTFE dispersions are similar to those of PTFE dispersions reviewed in Ch. 9. The major difference is the minimum film forming temperatures of polychlorotrifluoroethylene are lower than those of PTFE emulsions. PCTFE dispersions can easily form films at a temperature of 100–250°C.

Table 10.5. Extrusion Conditions for PCTFE (L/D = 20:1, rpm = 10)[2]

	Barrel Temperature,°C			Die Temperature,°C	
Part	Rear	Middle	Front	Body	Tip
Thin Film	200	304	321	318	349
Heavy Wall Shape	177	254	265	260	324

References

1. Critchley, J. P., Knight, G. J., and Wright, W. W., *Heat Resistant Polymers,* Plenum Press, New York, 1983.
2. Bringer, R. P. and Moreneau, G. A., "Processing of CTFE Plastics," *Plastics Design and Processing,* pp. 17–24, August 1968.
3. Mascia, L., *Thermoplastics – Material Engineering,* 2nd Ed., Elsevier Applied Science, 1989.
4. Bringer, R. P., "Chlorotrifluoroethylene Polymers," *EPST*, 7:204–210, 1967
5. Maier, C., and Calafut, T., *Polypropylene: The Definitive User's Guide and Databook,* 1st ed., *Plastics Design Library*, Norwich, NY, 1998.
6. Chandrasekaran, S., "Chlorotrifluoroethylene Polymers," *EPSE*, 3:465–480, 1989.

11 Fluoroadditives

11.1 Introduction

Fluoroadditives or micropowders are finely divided low molecular weight polytetrafluoroethylene powders, which are nearly entirely consumed by industries outside fluoropolymers. Micropowders are added to a great number of other products and compounded to enhance the properties of these products. The addition of a fluoroadditive imparts some of the properties of fluoropolymers to a host system. There are few applications, such as dry film lubricants, where a micropowder is used by itself.

In general, fluoroadditives have small particle size of the order of a few microns, hence the word *micropowders*. These powders are either granular (suspension polymerized) or fine powder based (dispersion polymerized), which have different particle morphologies, therefore, different properties and incorporation manner in the host material. Their molecular weight is in the range of a few ten thousand to a few hundred thousand compared to several million for the molding (granular and fine powder) resins (Table 11.1). This chapter describes the properties, methods of production, and some of the applications of micropowders.

Table 11.1. Estimated Molecular Weight of Polytetrafluoroethylene Resins

Resin	Molecular Weight	Melt Viscosity, Poise
Molding Resins	10^6-10^7	10^9-10^{11}
Fluoroadditives	10^4-10^5	10^2-10^5

11.2 Feedstock

Fluoroadditives are primarily produced by reducing the particle size of granular and fine powder polytetrafluoroethylene particles. Sometimes micropowders are produced by direct polymerization of tetrafluoroethylene but these are limited to dispersion-polymerized material.

High molecular weight of granular PTFE prevents grinding the particles below 20–25 µm. Reduction in molecular weight makes the resin brittle and easier to grind. Molecular weight has to be decreased to reduce the size of the particles to a few microns, which is required in most micropowder applications.

Fine powder particles are agglomerates with average particle size of 500 µm of small (<0.25 µm) primary particles. It is impossible to directly deagglomerate these particles because of fibrillation (see Ch. 8) and entanglement of the particles. The molecular weight has to be reduced to prevent fibrillation during the shearing to which the particles have to be subjected during deagglomeration or grinding. Sufficiently low molecular weight fine powder resins do not fibrillate.

The feedstock for micropowder production can come from a variety of sources. Micropowders were first developed as an outlet for the disposal of scrap resin. The highest quality micropowders are obtained by irradiation of first quality high molecular weight PTFE. Second quality resins could make high quality feedstock depending on the reason they are second quality. Contamination in the second quality feedstock would preclude production of high quality micropowder.

11.3 Degradation of Polytetrafluoroethylene

In this section, thermal and radiation degradation of polytetrafluoroethylene are discussed, which are the basis for the production of fluoroadditives. Heat or radiation are capable of degrading polytetrafluoroethylene if they are supplied in sufficient quantities to reach the temperature or radiation dose at which the polymer chain degrades.

There is a big difference between thermal decomposition of PTFE under vacuum or in the presence of inert gases compared to decomposition in an atmosphere of oxygen or air. Various studies have reported somewhat different degradation temperatures, which are attributable to the differences in the experimental conditions and polymer type. One common conclusion of all the studies is that PTFE degrades more at a lower temperature and more rapidly in the presence of oxygen or air. Another difference is in the type of products of degradation. Under vacuum or inert gases, the product of decomposition is mostly TFE and other small molecules. Under oxygen or air, smaller polymer chains are the product.

In one of the first studies, degradation of PTFE was delayed under vacuum until temperature reached 500°C where nearly pure monomer, tetrafluoroethylene, was produced.[1] Table 11.2 shows a comparison of degradation temperatures for PTFE and other polymers. In all cases, oxygen appears to promote the polymer decomposition. Polymer structure and monomer type influence thermal degradation.

It has been known for some time that exposure of polytetrafluoroethylene to high energy radiation such as x-ray, gamma ray and electron beam breaks down carbon-carbon bonds in the molecule's chain.[4] The cleavage of these bonds in vacuum or in an inert atmosphere (N_2) produces highly stable radicals, which prevent rapid degradation of PTFE. Some of these radicals recombine. When radiation is conducted in the air, the radicals react with oxygen leading to smaller molecular weight PTFE chains fairly quickly.

The data in Table 11.3 show that irradiation degrades PTFE. Subatmospheric pressure has little effect on the number-average molecular weight. Surface oxygen content (oxygen to carbon ratio in Table 11.3) is impacted by pressure and increases with pressure.

PTFE radicals react with oxygen the same way regardless of whether they have been produced by thermal decomposition or irradiation. Figure 11.1 shows the x-ray photoelectron spectroscopy analysis of the surfaces of irradiated and un-irradiated PTFE. It indicates significant oxygen content as a result of irradiation in air. The following reaction scheme has been widely accepted:[4][6][7]

$$-CF_2-CF_2- + \text{Heat or Irradiation} \rightarrow -CF_2-CF_2\cdot + -CF_2-\overset{\cdot}{C}F_2-CF_2-$$

$$-CF_2-CF_2\cdot + O_2 \rightarrow -CF_2CFO$$

$$-CF_2-CF_2-CF_2- + O_2 \rightarrow -CF_2-CF_2-CF_2- \rightarrow -CF_2-CF_2\cdot + -CF_2CFO$$
$$\underset{\cdot}{O}$$

The end group of degraded polytetrafluoroethylene is acyl fluoride (-CFO) which reacts with water and forms carboxylic acid group (-COOH) and evolves hydrofluoric acid (HF). End groups are usually identified by infrared spectroscopy. The number of end groups is too low, in most cases, to have a significant effect on the final properties of micropowders. In some applications, the end groups can have an effect.

Table 11.2. Degradation and Weight Loss in Two Hours at the Given Temperature

Atmosphere	Vacuum[1]	Oxygen[1]	Nitrogen[2]
Weight Loss, %	25	25	3
Polytetrafluoroethylene	494°C	482°C	460°C
Copolymer of tetrafluoroethylene and hexafluoropropylene	481°C	417°C	-
Polyvinylidenefluoride	403°C	354°C	-

[1]Ref. 2 [2]Ref. 3

Table 11.3. PTFE Irradiation in Air at Various Pressures[5]

Pressure, Pa	Irradiation Dose, Mrad	Surface Oxygen to Carbon ratio by XPS[1]	Number-average Molecular Weight[2] $\times 10^{-5}$
Atmospheric	0	0.024	30
0.2	5.3	0.032	0.75
2	5.3	0.088	0.73
8	5.3	0.139	0.74
20	5.3	0.134	0.79
Atmospheric	5.3	0.268	0.36

[1]XPS stands for x-ray photoelectron spectroscopy.
[2]Calculated form the heat of crystallization measured by differential scanning calorimetry.

For example, the end groups can promote adhesion of micropowders to metals.

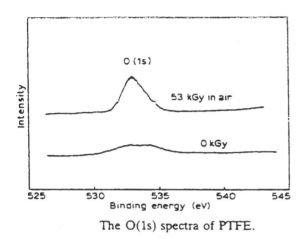

The O(1s) spectra of PTFE.

Figure 11.1 Surface analysis of irradiated and un-irradiated PTFE.[5]

11.4 Production Methods

PTFE molecules can be broken down by two techniques to produce lower molecular weight material; by thermal cracking and by irradiation. In the thermal method, the polymer is exposed to high temperatures beyond its degradation point, whereby covalent carbon-carbon bonds in its backbone are broken. The cleavage of a long chain results in several smaller molecules.

The same result can be achieved by exposing PTFE to gamma rays, x-rays, or electron beam. The most common commercial method to produce fluoroadditives is by electron beam exposure. The thermal cracking method is used to a much lesser extent. Both methods allow consumption of sintered and unsintered PTFE scraps which are produced during the various stages of processing (see Ch. 7 and 8). This section reviews the methods used for commercial production of micropowders.

11.4.1 Production of Fluoroadditives by Thermal Cracking (*Pyrolysis*)

A number of processes for thermal cracking of polytetrafluoroethylene into lower molecular weight material have been reported.[7]-[9] Typically, PTFE is heated for several hours from 400°C to 500°C producing low yields (10–60%) of low molecular weight material. The yield loss occurs due to the conversion of polytetrafluoroethylene to gaseous species. Significant improvements in the yield have had to be made for commercial viability of thermal cracking process.

High molecular weight polytetrafluoroethylene can be converted to lower molecular weight species, with high yield, by heating to temperatures up to 700°C under high pressure of an inert gas, like nitrogen.[10][11] Small quantities (0.1–5% by weight of PTFE) of oxides of nitrogen (NO, NO_2) and sulfur (SO_2, SO_3) and thionyl compounds can be added to accelerate the pyrolysis.[10] In this process, the time required for the thermal cracking was substantially reduced from the 2–8 hours by adding shear to the process.

For example, polytetrafluoroethylene scraps were loaded into a kneader which was equipped with two powerful sigma-type blades and with electric heating.[10] The vessel was flushed with nitrogen several times. The kneader was quickly heated to 500°C and the blades started, which help raise the temperature to 520–530°C due to frictional forces. After a reaction time of 10 minutes, excluding the time required for the polymer to reach temperature, the heat supply is shut off and the kneader is air-cooled. The pressure reaches 34–39 MPa/g in the pyrolysis zone. The reaction products are discharged by nitrogen pressure after temperature reaches 400°C. A brittle product is obtained at 97% yield with a melt temperature range of 321–323°C and a viscosity of 165 poise at 370°C. This indicates a major drop in the molecular weight of PTFE as a result of thermal cracking, considering that the molecular weight of molding grade polymer is of the order of 10^{10} Poise at 380°C.

Thermal cracking reduces the molecular weight of PTFE, which causes the reaction products to be brittle to the extent that the material can be broken up by mortar and pestle to achieve significant particle size reduction. Molecular weight decreases from 1–10 million to 2–200 thousand.

11.4.2 Production of Fluoroadditives by Electron Beam Irradiation

Electron beam is the most common commercial method of converting PTFE to fluoroadditives. The procedure for electron beam irradiation is relatively simple and can be found in a number of references.[12][13] In practice, a continuous process is used to improve the economics of the process. The resin is spread on a conveyor belt at a specific thickness and is passed under the electron beam at such a speed that the polymer is subjected to the desired dose (common unit is Mrads). Typically, multiple passes are made to expose the PTFE to a higher dose. After the total dose has been received, the irradiated material is removed and transferred to the grinding operation.

The conveyor belt can be shaped circularly or like a figure eight to carry the resin under the electron beam. The advantage of multiple pass irradiation is that it allows the polymer to be cooled after each pass. Dissipation of electron beam irradiation in polytetrafluoroethylene heats up the resin. Without removal of heat, PTFE will get very hot. It may even melt which will lead to sticking of the individual particles to each other, complicating the grinding process. Cleavage of bonds generates off-gases such as hydrofluoric acid during the irradiation, which must be removed by means of adequate ventilation from the processing areas. The stack effluents might have to be treated to remove the entrained particles and the evolved species prior to venting to the atmosphere. The nature of the stack gas treatment depends on its contents and the governing emission rules.

An electron beam is produced by an accelerator, which converts the electrical energy into accelerated filaments released from a tungsten filament source, typically 40 µm. The air space between the filament and the surface of the resin is of the order of 10 cm. An accelerator consists (Fig. 11.2) of four major systems: voltage generator, acceleration tube and electron gun, scan chamber and scan horn, and a control system. Each segment plays a key role and is briefly described.

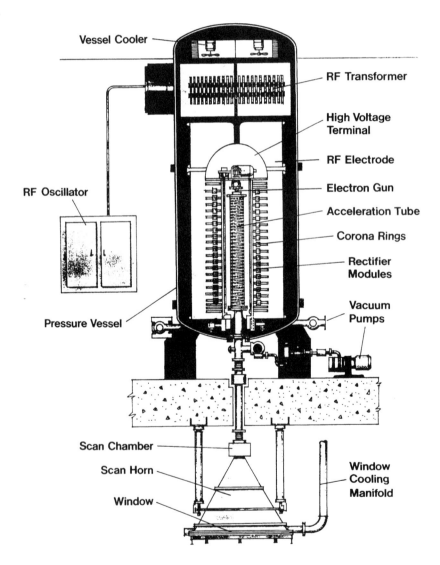

Figure 11.2 Schematic diagram of an electron-beam accelerator. *(Courtesy E-BEAM Services, Inc.)*

Low voltage power, typically 3-phase, 440 V AC, is converted to high frequency (100 kHz) radio frequency (RF) power by an oscillator. A number of rectifiers in series along with a radio frequency resonant transformer convert the RF power into ultra high voltage (2–5 MM V) direct current (DC) power. The DC voltage is sent to the acceleration tube. This tube is a hollow vacuum cylinder made from glass and metal. Glass rings are positioned as insulation rings in between metal rings, called *dynodes*. An electron gun is positioned at the entry point of the tube. A tungsten filament is the source of electrons which are emitted by heating the filament. The continuous electron stream is accelerated through a vacuum tube, supplying a high energy beam with a diameter of 2–3 cm.

The accelerated beam is transmitted through an oscillating magnetic filed with a frequency of 200 Hz which converts the tight beam cone into a wide curtain of electrons with a width of 1–2.5 m. A thin titanium window acts as a window or a filter before the beam reaches the product to be irradiated. A control system monitors all the functions of the various systems of accelerator to insure proper and consistent operation.

An accelerator operates based on a number of factors including voltage (electron energy), beam current and beam power. The voltage is calculated from the thickness and density of the material to be irradiated, PTFE in this case. The beam current is calculated from the surface dose and area flow rate. The beam power is calculated from mass flow rate and the average value of the dose.[14]

Other parameters in the process affect the irradiation of the product. They include the atomic composition, shape and size of the irradiated material and irradiation distribution. *Irradiation distribution* refers to the actual dose to which the resin at different thickness, is exposed. The maximum dose does not occur at the surface but somewhere within the body of the material. The higher the atomic weights of the irradiated material, the closer to the surface the maximum doses occur. Penetration of the electron beam increases with increasing accelerator voltage. Whenever the minimum dose is important, the thickness of the resin is reduced to allow the exiting beam dose to equal the entrance dose.

The selection of all operating variables for the accelerator shapes into a complex optimization task, which also depends on mechanical factors like the width, speed, and loading of the conveyor. The main objective of the optimization of the accelerator is to maximize the productivity of the equipment in order to produce high quality material at the lowest cost. This is particularly important in light of the substantial capital cost of the accelerator. The cost of irradiation is about 10–30% of the cost of the fluoroadditive, increasing with the dose. A detailed analysis of optimization has been presented by Cleland and Farrel [15] The specific beam power requirement can be obtained from simple equations such as Eq. (11.1), which assumes optimum product thickness requires an accelerator voltage of at least 2.0 MeV.[14]

Eq. (11.1) $$P = D_0 \frac{m}{3}$$

P = Beam power, kW
D_0 = Average dose, Mrad
m = Mass flow rate of PTFE in kg/hr

Equation 11.1 contains other assumptions such as energy utilization coefficient of 0.65 and beam current utilization of 0.85. The more general form of this equation requires determination of these values for which procedures can be found in the above references.

The effectiveness of electron beam irradiation increases with dosage; an example can be seen in Table 11.4. In this example, both unsintered and sintered PTFE were subjected to different doses and ground the same way. Temperature of the resin is held below 121°C during the irradiation. Particle size decreases rapidly as irradiation doses increase from 5 to 25 Mrad. The ground micropowder has an apparent density of 400 g/l and a melting point of 321–327°C.

Table 11.4. Effect of Dose on PTFE Micropowder Particle Size[12]

Test Case	PTFE Type	Irradiation Time (sec)	Dose (Mrads)	Average Particle Size (µm)
1	Unsintered	2.5	5	11.1
2	Unsintered	5.0	10	5.3
3	Unsintered	7.5	15	2.5
4	Unsintered	10	20	1.5
5	Sintered Scrap	12.5	25	0.9

In another method, irradiation and heat are combined to degrade PTFE.[13] A lower dose was reported to reduce the molecular weight, as determined by melt viscosity, than irradiation by itself in this method because of heat. No commercial use of this method has been disclosed.

A number of other devices have been reported for electron beam irradiation and grinding of polytetrafluoroethylene.[16][17] Sometimes irradiation and grinding have been combined to improve the operational efficiency of the process and the uniformity of the product.[18][19]

11.4.3 Grinding Irradiated PTFE

After the resin has received the sufficient dose of irradiation, it is ground to the desired particle size. The most popular method of grinding is in fluid energy mills, more commonly known as *jet mills* in which a compressible fluid such as air is used as a source of energy. The pulverization takes place by attrition between the solid particles being ground, in this case irradiated polytetrafluoroethylene. This type of mill can comminute solids such as PTFE which are difficult (or impossible) to process in media mills (e.g., ball mill) or roller mills (e.g., two-roll mill) and hammer mills.

Grinding in fluid energy mills results from the attrition between the particles of PTFE with the energy supplied by the compressed air (or other compressible fluid) entering the milling chamber at high velocities through jets, hence the name *jet mill*. A schematic diagram of the basic elements of a fluid energy mill is shown in Fig. 11.3.

The basic components of a jet mill include the grinding chamber into which the resin particles and air are introduced, a solids feeder, nozzles to inject the compressed air into the grinding chamber, and one or more cyclones to collect the ground particles. Most commercial operations require compressed air around 0.7 MPa gauge at the room temperature.

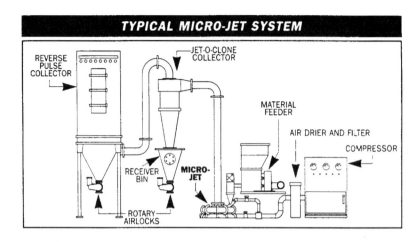
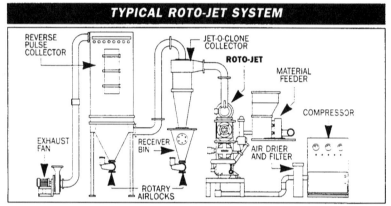
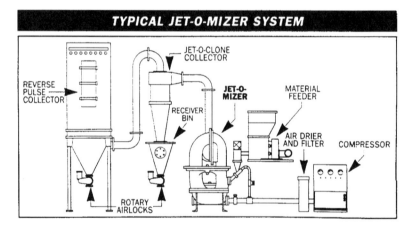

Figure 11.3 Three different types of mill supplied by Fluid Energy Processing and Equipment Co.[20]

Air and solids are entered separately into the shallow grinding chamber. Air emerges from nozzles that are spaced around the periphery of the chamber. The resin is fed by an aspirating venturi into the chamber where it circulates under the force of pressurized air. The finely pulverized product and low pressure (spent) air exit the chamber through a circular concentric outlet in the top (or bottom) plate. The collection system removes the resin particles by taking advantage of the rotational speed of the air in a cyclone. Better than 98% of the ground PTFE particles can be removed in a high efficiency collection system. The exhaust is normally cleaned by a bag filter.

Jet mills are rated according to the inside diameter of the grinding chamber, ranging from 5 cm to >100 cm. The height of a grinding chamber varies from 10 mm to >60 mm. The capacity of the mill depends on its size, type of solids and the desired fineness of grind. Jet mills with a capacity of a few hundred grams to a few tons per hour are commercially available.

A fluid energy mill requires considerable energy, primarily in the form of compressed air, which is the major cost component of the operation. To improve the economics of the fluid energy grinding, especially for hard solids, it is helpful to pre-grind in a conventional mill and use the product as feed stock for the jet mill.

11.5 Commercial Products

Polytetrafluoroethylene fluoroadditives (micropowders) are produced by irradiation of high molecular weight PTFE or by direct polymerization (dispersion). They have finely divided particles that are smaller than the particles of other PTFE types. Micropowders are mainly intended for use as a minor constituent of mixtures with other solids or liquids. They can impart some of the properties of fluoropolymers to the host systems. Fluoropolymer manufacturers offer a variety of virgin micropowders. Other companies supply fluoroadditives made from the irradiation of scrap and second grade PTFE. Tables 11.5–11.10 provide information about the grades and properties of fluoroadditives offered by fluoropolymer manufacturers. Fluoroadditive powders are specified by ASTM D5675 and aqueous dispersions are specified by ASTM D4441.

Table 11.5. Typical Properties of DuPont PTFE Fluoroadditive Powders[21]-[28]

Zonyl®	PTFE Polymer Type	Production Method	Specific Surface Area, m^2/g	Particle Size Distribution, μm	Average Particle Size, μm	Melting Peak Temperature, °C	Bulk Density, g/l
MP 1000	Emulsion	I	5-10	10%<3 90%<30	12	325	500
MP 1100	Emulsion	I	5-10	10%<0.3 90%<8	4	325	300
MP 1150	Emulsion	I	5-10	10%<2.0 90%<25	10	325	450
MP 1200	Suspension	I	1.5-3	10%<1 90%<9	3	325	450
MP 1300	Suspension	I	1.5-3	10%<3 90%<25	12	325	425
MP 1400	Suspension	I	1.5-3	10%<3 90%<20	10	325	425
MP 1500J	Emulsion	D	8-12	10%<10 90%<35	20	330	425
MP 1600N	Emulsion	D	8-12	-	12	325	350

Table 11.6. Typical Properties of DuPont PTFE Aqueous Dispersion Fluoroadditives[29]

Zonyl®	TE-3667
Solids Content, % by Weight	60
Dispersion Particle Size, μm	0.22
Specific Gravity at 20°C	1.5
Brookfield Viscosity at 25°C, Pa·Sec	0.02
pH (minimum)	9
Conductivity, μS/cm	700
Melting Peak Temperature, °C	325

Table 11.7. Typical Properties of Imperial Chemical Industry PTFE Fluoroadditive Powders[30]

FLUON®	PTFE Polymer Type	Specific Surface Area, m²/g	Average Particle Size, μm	Bulk Density, g/l
FL1680	Suspension	0.8	13	450
FL1690	Suspension	1.0	21	480
FL1700	Suspension	3.1	-	530
FL1710	Suspension	2.3	9	400

Table 11.8. Typical Properties of DYNEON PTFE Fluoroadditive Powders [31]

HOSTAFLON®	PTFE Polymer Type	Specific Surface Area, m²/g	Melt Flow Index (MFI) @ 372°C, g/10min		Average Particle Size, μm	Bulk Density, g/l
			Die diameter, mm (force, kg)	MFI		
TF-9202	Emulsion	15	2.08 (2.16)	<10	3	250
TF-9203	Emulsion	15	2.08 (1.0)	<2	6	350
TF-9205	Suspension	5	1.0 (3.0)	<20	10	400

Table 11.9. Typical Properties of Daikin PTFE Fluoroadditive Powders[32]

POLYFLON®	PTFE Polymer Type	Specific Surface Area, m²/g	Specific Gravity	Melting Point, °C	Average Particle Size, μm	Bulk Density, g/l	Volatiles Content*, %
L-2	Emulsion	9	2.2	330	4	400	400
L-5	Emulsion	11	2.2	327	7	400	400
L-5F	Emulsion	11	2.2	327	5	400	400

*Determined from weight loss after 1 hr at 300°C

Table 11.10. Typical Properties of Ausimont PTFE Fluoroadditive Powders[33]

Polymist®	PTFE Polymer Type	Specific Surface Area, m²/g	Particle Size Distribution, % <20 μm	Average Particle Size, μm	Melting Temperature, °C	Bulk Density, g/l	Specific Gravity
F5	Suspension	3	<5	<10	320-325	450	2.28
F5A	Suspension	2.5	<3	<6	320-325	300	2.28
F5A EX	Suspension	2.5	<4	<7	325-330	400	2.28
F510	Suspension	3	-	<25	325-330	475	2.28
XPH-284	Suspension	3	<5	<10	327-333	425	2.22

11.6 Applications

Polytetrafluoroethylene micropowders improve the properties of the host systems to which they are added in numerous ways. These benefits originate from the basic properties of PTFE, which include chemical resistance, low coefficient of friction, inertness to moisture and weather, and thermal stability. Major applications of fluoroadditives include thermoplastics and elastomers, paints and coatings, printing inks and oils and greases. Summaries of the important benefits that are obtained in commercial uses are listed in Table 11.11.

Major applications of PTFE micropowders are discussed in the following sections. A list of specific parts and host systems is shown in Table 11.12.

Table 11.11. Application and Benefits of PTFE Micropowders

Benefit	Example of Application
Inert to almost all chemicals and solvents	Coatings for chemical tanks
Reduce coefficient of friction	Engineering polymer parts
Improve wear characteristics	Plastic pulleys and gears
Ultraviolet and weather resistance	Outdoor applications
Improvement in non-stick and release properties	Paints
Increase rub resistance	Inks
Improve corrosion resistance	Coating
Improved thermal stability	Plastics
Excellent electrical insulator	Wire insulation
No moisture absorption	Coatings and plastics

Table 11.12. Applications of Fluoroadditives[32]

Application	Base material	Notes
Bearing parts, gears	Polyacetal Polyamide Polyphenylene sulphide (US: sulfide) Phenolic resin Polyester Thermosetting resins	Automotive parts Mechanical parts such as cameras and cinematographs Bearing pads
Brake pads	Phenolic resins/asbestos	Automobile brakes
Computer key board	Polycarbonate	Keys for telephones and computers
Sealing and packaging	Silicone rubber Urethane rubber	Oil seals Automobile parts
Car wax	Wax	Car wax, wax
Painting	Epoxy resin paints Acrylic resin paints	
Food containers	Polypropylene	For sticky food
Binders	Carbon powder	For batteries
Release agents	Mold release agents Copier toner Surface treating agents Printing ink	Copy paper
Plastic working aid	Individual use	For metal straightening
Stranding aid	Individual use	For stranding yarn

11.6.1 Thermoplastic Modification

Fluoroadditives are added to improve lubricity, reduce friction and increase wear resistance of high performance plastics such as polyacetals, nylon, polycarbonates, polyamides, polyimides, polyesters, polysulfides, polysulfones, and other resins. Micropowders are added to enhance the tribological characteristics of the polymer. These improvements aid in replacement of lubricated metal parts by plastics to provide corrosion resistance. Fluoroadditives increase the PV limits (Fig. 11.4) and reduce slipstick response of polymeric hosts in dynamic load bearing applications such as pulleys, slides, bearings, cams and gears.

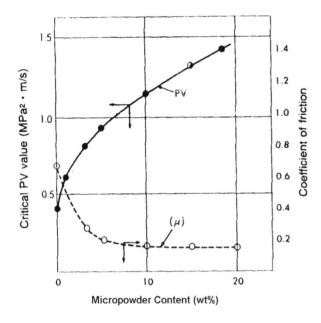

Note: Data shown above are average values for Daikin Polyflon® L-5 Micropowder
Test equipment: Thrust washer test apparatus (ringshaped contact surface)
Test conditions: Partner material: SUS-27
Load: 8 MPa (1.4 psi)
The revolution speed was increased in steps of 0.1 m/sec. The test was conducted at each of the speeds for ten minutes. Load (P) × speed (V) at the point where the friction becomes extremely strong constitutes the critical PV value. The coefficient of friction (μ) is taken from the value immediately preceding the critical PV value.

Figure 11.4 Critical PV value of polycarbonate.[32]

Table 11.13 shows the impact of adding 15% by weight of PTFE micropowders to four thermoplastics, which are widely used in mechanical applications. In all four cases the reduction in wear rate and wear factor are substantial. In the case of polycarbonate, the reduction is over 270 times, which has resulted in making this plastic viable for dynamic mechanical applications. The popularly accepted mechanism for these improvements is the formation of a dry film of polytetrafluoroethylene between the plastic part and the mating surface.[34] This film is formed by shearing of dispersed PTFE particles during the break-in period of the plastic. The consequence of this mechanism is, in effect, conversion of the interface of the mating surfaces to PTFE, thus allowing its properties to come into play.

11.6.2 Elastomers

Micropowders added to elastomers enhance the surface and bulk properties of parts. Examples of surface properties include mold release, friction, wear, and lubricity. Tear strength, abrasion resistance, and flex life are examples of bulk properties (Table 11.14) improved by fluoroadditives addition. A number of high performance materials such as silicone rubbers, fluoroelastomers and fluorosilicones have poor mechanical properties at elevated temperatures which can be detrimental to their processing. For instance, some of these elastomers tear easily during removal from mold at 177–191°C. Addition of fluoroadditives can reduce damage from sticking or tearing by increasing the mold release of the parts. Certain grades of PTFE fluoroadditives can increase the tear strength of elastomers as seen in Fig. 11.5, both at the room temperature and at 177°C.[35]

Fluoroadditives can be blended using the customary methods of incorporation of additives into elastomers, using Banbury or two-roll mixers. High concentrations (up to 30–50% by weight of base elastomer) of PTFE micropowders can be formulated in the elastomers without compromising elastic behavior. Use of a concentrated batch of micropowder is preferable for compounding low viscosity elastomers such as silicone or fluorosilicone elastomers. Additional shearing action is required during the "letting down" of the concentrate in the elastomer in order to disperse the PTFE uniformly.

11.6.3 Printing Inks

The main attribute imparted in printing inks by the addition of micropowders is scuff-resistance. Improvement in rub resistance, slip, and surface smoothness are achieved in addition to reduction in blocking and gloss. Lithographic, flexographic, and gravure inks can be all formulated to include 1–3% by weight of PTFE fluoroadditives. Formulations containing PTFE facilitate the production of high gloss inks by allowing a reduction in wax.

Lithographic formulations typically contain 1% by weight micropowder but higher performance inks for magazine covers contain higher levels of PTFE. Examples of the enhancement achieved by the addition of micropowders can be seen in Fig. 11.6. PTFE containing inks resist abrasion and rub appreciably better than inks without the additive.

11.6.4 Paints and Coatings

Fluoroadditives are used in coatings to improve release, mar resistance, slip, chemical resistance, weatherability and moisture repulsion. They also modify the texture of painted surfaces and help reduce their gloss. PTFE micropowders are sometimes incorporated in coatings on cookware and industrial products which may operate at moderate or high temperatures. Formulation of polytetrafluoroethylene additives requires consideration of low surface tension of this polymer (18 dynes/cm. Low surface tension solvents (hydrocarbons) and surfactants are the common means of producing stable coatings containing micropowders.

Interior latex semigloss and flat paints have been formulated with micropowders to improve stain and scrub resistance without affecting properties of the paint such as viscosity, fineness grind or blocking resistance. Wood stains containing fluoroadditives, for exterior applications, exhibit increased water repulsion without a reduction in viscosity. Typically, less than 5% of the total formulation solids need to be PTFE. Antifouling marine paints have been developed which contain up to 30–50% by weight of PTFE micropowders.

Industrial coatings containing fluoroadditives have enhanced mold release, wear or friction coefficient. These coatings are applied to the surface of complicated shapes by electrostatic or fluidized bed methods. They contain 5–15% by weight of PTFE powder.

11.6.5 Oils and Greases

Oils not only lubricate but also cool, transfer energy and insulate electrical equipment. Polytetrafluoroethylene micropowders have been formulated at small concentrations in additives for engine lubricating oils for many years. The inclusion of PTFE in motor oils has been rather controversial. The debate has been focused over the effectiveness of PTFE in internal combustion engines.

Fluoroadditives are also added to high performance greases as stable thickeners, which enhance their long-term performance reliability in hostile environments. Reductions in wear, friction coefficient, slipstick action and torque requirements of systems can be achieved by the addition of these PTFE powders to grease. Examples of applications include camera parts, aerospace parts, plastic gears, valves, bicycles and ski equipment.

Table 11.13. Thrust Bearing Wear Test Results (ASTM D3702)[34]

Thermoplastic Resin	TEFLON® Fluoroadditive, Wt. %	Wear Rate, cm/hr ($\times 10^{-6}$)	Wear Factor, $\frac{cm^3 \cdot min \, (\times 10^{-10})}{KJ \cdot sec}$	Dynamic Friction Coefficient
Polycarbonate	0	33,770	219,432	0.51
	15	124	807	0.21
Nylon 66	0	474	3,083	0.50
	15	123	804	0.29
Polyester (PBT)	0	300	1,948	0.37
	15	63.5	413	0.18
Acetal	0	1,719	11,184	0.38
	15	28	184	0.19

Table 11.14. Effect of TEFLON, Fluoroadditives on Typical Formulation of VITON, A-401C for O Rings (Press Cure 10 min at 177°C)[34]

Property	ASTM Method	Amounts of Fluoroadditive as Percent by Weight of Viton® Fluoroelastomer*					
			MP-1		MP-2		MP-3
		Reference	10%	20%	10%	20%	20%
Tear Resistance (Press Cure, Test with Grain)	D624 (Die B)						0.51
at 23°C kN/m		36.5	33.8	32.2	37.4	38.6	52.2
at 177°C kN/m		9.1	8.9	8.8	11.4	10.7	16
Abrasion Resistance Wt. Loss/Rev., mg (H-18 Wheel, 1 g load)	D3389 (Taber)	0.42	0.26	0.09	0.26	0.11	0.1
Hardness (Room Temperature)	D2240	77	78	78	78	82	82
Compression Set	D395 Method B						
22 h at 200°C		8.8	11.7	14.7	11.7	16.1	15.6
70 h at 200°C		14.2	17.6	17.6	20.5	22	20
Other Properties at R.T.							
50% Modulus, MPa	D412	2.9	2.8	2.7	3.4	3.9	5.4
100% Modulus, MPa	D412	6.5	5.6	4.9	6.4	6.6	8.6
Tensile Strength, MPa	D412	12.6	11.3	10.6	13.5	10.9	13.4
Elongation at Break, %	D412	177	193	205	203	172	198

*MP-1 is Zonyl® MP-1000 by DuPont Co.; MP-2 is Zonyl® MP-1400 by DuPont Co.; MP-3 is Zonyl® MP-1500 by DuPont Co.; Viton® is a registered trademark of DuPont Co.

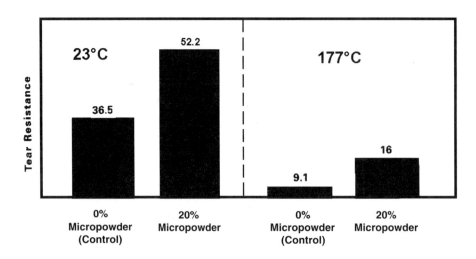

Figure 11.5 Resistance (in MPa) to tear strength for Viton® A-401C as a function of temperature.[34]

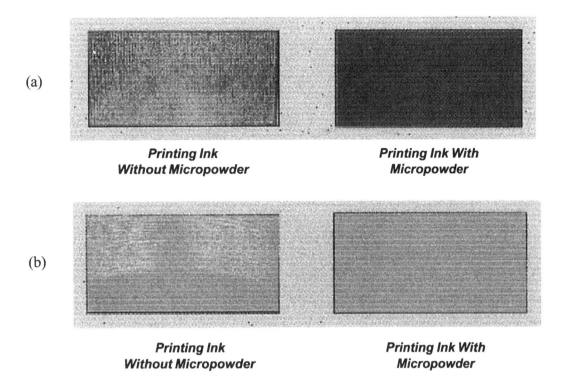

Figure 11.6 *(a)* The Graphic Arts Comprehensive Abrasion Test can simulate the relative motion of Surfaces during shipping. The ink used on the right contained 1% micropowder which largely eliminated ink abrasion as well as pickup on the blank sheet. The colors are heat set inks often used on magazine covers or inserts. *(b)* The Sutherland Rub Tester moves a weighted, test strip against a printed test specimen. After 50 strokes at four psi, ink containing 1% micropowder was far superior to the sample containing no additives.[34]

11.6.6 Dry Lubricant

PTFE micropowders can be applied directly as a dry powder or spray to surfaces such as slides, cables, glass cloth, window slides and locks. Fluoroadditives have lower coefficient of friction (Table 11.15) than graphite and molybdenum disulfide, which are popular, dry lubricants and are cleaner than them. Silicones tend to spread easily and have a strong propensity for contamination.

Table 11.15. Coefficient of Friction of Dry Lubricants[32]

Dry Lubricant	Static Coefficient of Friction
PTFE Micropowders	0.06–.07
Molybdenum Disulfide	0.15
Graphite	0.15–.16

11.7 Regulatory Compliance

Majority of fluoroadditives are produced by irradiation of high molecular weight polytetrafluoroethylene to facilitate their grinding into small particles. Food and Drug Administration (FDA) rule 21CFR177.1550, paragraph (c) specifies maximum allowable doses of radiation and maximum particle size processed by irradiation. This rule restricts the application of components containing irradiated fluoroadditives intended for repeated use in contact with food. Anyone planning to produce articles which come in contact with food should be sure of FDA compliance beforehand. Fluoroadditive manufacturers can usually supply FDA compliance information.

There are a number of other regulatory agencies including FDA, US Department of Agriculture and US Pharmacopia. It is important to investigate compliance issues when planning to formulate fluoroadditives into articles for use in applications where food, produce, and pharmaceutical contact may occur.

References

1. Lewis, R. F., and Naylor, A., *J. American Chem. Society*, 69:1968, 1947.
2. Heat Resistant Polymers, Critchley, J. P., Knight, G. J., and Wright, W. W., Plenum Press, New York, 1983.
3. Wall, L. A., Ed., *Fluoropolymers*, Wiley-Interscience, New York, 1972.
4. Golden, J. H., "The Degradation of Polytetrafluoroethylene by Ionizing Radiation," *J. of Polymer Science*, XLV(146):531–536, 1960.
5. Zhong, X., Yu, L., Sun, J., and Zhang, Y., "Radiation Stability of PTFE Irradiated under Various Conditions," *Polymer Degradation & Stability,* 39:187–191, 1993.
6. Fisher, W. K., and Corelli, J. C., "Effect of Ionizing Radiation on the Chemical Composition, Crystalline Content and Structure, and Flow Properties of Polytetrafluoroethylene," *J. of Polymer Science: Polymer Chemistry Ed.,* 19:2465–2493, 1981.
7. US Patent 2,496,978, Berry, K. L., assigned to DuPont, Feb. 7, 1950.
8. US Patent 2,978,519, Fischer, E., assigned to Hoechst Aktiengesellschaft, Apr. 4, 1961.
9. British Patent 1,074,768, assigned to Dow Chemical Co., Aug. 25, 1965.
10. US Patent 3,813,449, Hartwimmer, R. and Weiss, E., assigned to Farbewerke Hoechst Aktiengesellschaft, May 28, 1974.
11. US Patent 4,076,760, Hartwimmer, R., assigned to Farbewerke Hoechst Aktiengesellschaft, Feb. 28, 1978.
12. US Patent 3,766,031, Dillon, J. A., assigned to Garlock, Inc., Oct. 16, 1973.
13. US Patent 4,220,511, Darbyshire, R. L., assigned to Radiation Dynamics, Inc., Sept. 2, 1980.
14. Becker, R. C., Bly, J. H., Cleland, M. R and Farrell, J. P. "Accelerator Requirements for Electron Beam Processing," *Radiation Phys. Chemistry*, 14:353–375, 1979.
15. Cleland, M. R., and Farrell, J. P.; "Methods for Calculating Energy and Current Requirements for Industrial Electron Beam Processing," *Proceedings of the Fourth Conference on Application of Small Accelerators*, North Texas State University, Denton, TX, 1976.
16. US Patent 4,748,005, Neuberg, W. B., and Luniewski, R., assigned to Shamrock Chemicals Corporation, May 31, 1988.
17. US Patent 4,777,192, Neuberg, W. B., and Luniewski, R., assigned to Shamrock Chemicals Corporation, Oct. 11, 1988.
18. US Patent 5,149,727, Luniewski, R. S., assigned to Medical Sterilization, Inc., Sept. 22, 1992.
19. US Patent 5,296,113, Luniewski, R. S., assigned to Medical Sterilization, Inc., Mar. 22, 1994.
20. "Solutions for Fine Powder Processing," technical bulletins for Micro-Jet, Roto-Jet and Jet-O-Mizer size reduction systems, published by Fluid Energy Processing and Equipment Co., Plumsteadville, PA, 1999.
21. Technical Information Zonyl® MP 1000, Fluoroadditive, DuPont Co., Wilmington, DE, Sep., 1996.
22. Technical Information Zonyl® MP 1100, Fluoroadditive, DuPont Co., Wilmington, DE, Sep., 1996.
23. Technical Information Zonyl® MP 1150, Fluoroadditive, DuPont Co., Wilmington, DE, Sep., 1996.
24. Technical Information Zonyl® MP 1200, Fluoroadditive, DuPont Co., Wilmington, DE, Sep., 1996.
25. Technical Information Zonyl® MP 1300, Fluoroadditive, DuPont Co., Wilmington, DE, Sep., 1996.
26. Technical Information Zonyl® MP 1400, Fluoroadditive, DuPont Co., Wilmington, DE, Sep., 1996.
27. Technical Information Zonyl® MP 1500, Fluoroadditive, DuPont Co., Wilmington, DE, Sep., 1996.
28. Technical Information Zonyl® MP 1600, Fluoroadditive, DuPont Co., Wilmington, DE, Sep., 1996.

29. Technical Information Zonyl® PTFE TE 3667, Fluoroadditive, Aqueous Dispersion, DuPont Co. Wilmington, DE, Sep., 1996.
30. Fluon® Lubricant Polymers, Internet Website, http://www.fluoropolymers.com/, ICI Fluoropolymers, Business Unit of ICI Americas, Inc., Wilmington, DE, June, 1999.
31. Hostaflon® Micropowder Grades, in Hostaflon® Polytetrafluoroethylene, Product Comparison Guide, DYNEON, A 3M-Hoechst Enterprise, Minneapolis, MN., Feb., 1998.
32. Daikin-Polyflon® TFF Low Polymer, Daikin Industries, Ltd., Osaka, Jpn., Mar., 1992.
33. Polymist® PTFE LUBRICANT, in ALGOFLON® Polytetrafluoroethylene, Properties and Application Selection Guide, Ausimont Corp., Morristown, NJ.
34. Zonyl® Fluoroadditives, "A Minor Component, A Major Enhancement," *Properties and Applications Book,* DuPont Co. Publication, 1999.
35. US Patent 4,879,362, Morgan, R. A., assigned to DuPont, Nov. 7, 1989.

12 Filled Fluoropolymer Compounds

12.1 Introduction

Filled compounds of granular polytetrafluoroethylene are the most commercially significant commodities because of the large volume of consumption. Some companies produce their own compounds in-house while others purchase from specialty compounders who are skilled in formulation and production of various filled compounds. Compounds of dispersion polymerized PTFE are made in-house and fabricated into articles. PTFE dispersions are formulated into coatings by incorporation of a variety of additives and are not covered in this chapter. Polychlorotrifluoroethylene compounds are highly specialized and not very common.

The focus of this chapter is on compounds of polytetrafluoroethylene products. Pure or unfilled polytetrafluoroethylene is inadequate for a number of mechanically demanding engineering applications. Cold flow or creep would prevent the use of PTFE in many mechanical applications. In the 1960's, the addition of fillers was found to improve a number of physical properties of PTFE, particularly creep and wear rate. Filled granular (suspension polymerized PTFE) resins were found to be suitable for parts such as gaskets, shaft seals, bearings, bearing pads, and piston rings. Chemical resistance, low friction, and high temperature, combined with mechanical requirements, prompt specification of filled polytetrafluoroethylene as material of construction. This chapter describes the characteristics, properties and methods of production of filled polytetrafluoroethylene powders.

12.2 Granular-based Compounds

These compounds are made with polytetrafluoroethylene resin polymerized by suspension polymerization, known as *granular* powders. The choice and concentration of the filler depends on the desired properties of the final part. Glass fiber, bronze, steel, carbon, carbon fiber and graphite are among the common filler material. Up to 40% by volume of filler can be added to the resin without complete loss of physical properties. The impact of the additives below 5%, by volume, of filler on the properties of compound is insignificant. Above 40%, most physical properties of the compound drop sharply.

Close to one half of granular PTFE is consumed in the form of filled compounds. Many standard grades are offered by suppliers in the form of dry (low flow), free flow (granulated) and presintered powders. Applications of low and free powders are similar to those of unfilled polymer described in Ch. 7. Low flow resin is suitable for compression molding while free flow powder can be processed by isostatic and automatic molding and ram extrusion.

12.2.1 Fillers

Polytetrafluoroethylene is one of the more difficult polymers to compound. This is due to the extreme charge and functional neutrality of PTFE chains, which precludes any interaction with fillers. It does not have sufficiently low viscosity to flow upon melting and coat the surface of the fillers. Low coefficient of friction reduces mechanical interaction, rendering PTFE easily separable from the filler in physical mixtures. Compounding the filler with this polymer in the form of agglomerated granules is one method of locking in the uniformity of the mixture.

The only requirement for an additive to qualify as filler for PTFE is that it should be able to withstand the sintering temperatures of polytetrafluoroethylene. Sintering involves exposure to temperatures close to 400°C for several hours, which excludes a great many materials. Characteristics of the filler such as particle size and shape and the chemical composition of the filler affect the properties of compound. A list of most common fillers and descriptions of their important characteristics can be found in Table 12.1.

Glass fiber is the most common filler with a positive impact on creep performance of PTFE, which is reduced at low and high temperatures. This additive is chemically inert with the exception of its reactivity with hydrofluoric acid and strong bases. Glass filled compounds perform well in oxidizing environments. Wear characteristics of polytetrafluoroethylene are improved. Glass has little impact on the electrical properties of PTFE. Dielectric breakdown strength is somewhat adversely affected due to the increased porosity of parts. One drawback to glass is the discoloration of sintered parts, more prevalent at higher thicknesses. This problem can be solved by chemical treatment or baking of the glass fiber. In the latter case, the fiber is simply heated in an air oven for 24 hours at or above 260°C.

Table 12.1. Properties of Common Fillers[1]

Filler	Material Description	Particle Size (μm)	Particle Shape	Density, g/cm³
Glass	E glass	Diameter 13 mm Length 0.8 mm Aspect ratio>10	Milled fibers	2.5
Carbon	Amorphous petroleum coke	Diameter <75 μm	Roundish	1.8
Carbon Fiber	Pitch or PAN* based		Short fibers	
Graphite	>99% C, Synthetic or natural	<75 μm	Irregular shape	2.26
Bronze	9/1 Copper to tin ratio	<60 μm	Spherical or irregular shape	
Molybdenum Disulfide	Mineral (98% pure)	<65 μm		4.9

*PAN is an abbreviation for polyacrylonitrile. PAN fibers are thermally carbonized to obtain carbon fiber.

Carbon reduces creep, increases hardness and elevates thermal conductivity of polytetrafluoroethylene. Wear resistance of carbon filled compounds improves particularly when combined with graphite. Carbon-graphite compounds perform well in non-lubricated applications such as piston rings in compressor cylinders. Carbon filled PTFE has some electrical conductivity. Close tolerances can be achieved during machining by reducing wear rate of the tool, using a softer carbon powder.

Carbon fiber lowers creep, increases flex and compressive modulus, and raises hardness. These changes can be achieved with glass but less carbon fiber can achieve the same effects. Carbon fiber is inert to both hydrofluoric acid and strong bases which react with glass. Coefficient of thermal expansion is lowered and thermal conductivity is higher for compounds of carbon fiber PTFE. Carbon fiber parts are lubricated with water, that is, wear rate decreases, making them ideal for automotive applications in shock absorbers and water pumps.

Graphite filled polytetrafluoroethylene has an extremely low coefficient of friction due to the low friction characteristics of graphite. Graphite is chemically inert. It is also incorporated in combination with other additives such as carbon and glass. Graphite imparts excellent wear properties, especially against soft metals,[1] and high PV to PTFE.

Bronze is the most popular metallic filler, although steel powder is occasionally used. Large quantities (40–60% by weight) of bronze reduce deformation under load and raise thermal and electrical conductivity of PTFE compounds. These two characteristics are beneficial to applications where a part is subjected to load at extreme temperatures. Transmission and air-conditioner compressor seals are two examples of such parts. Bronze is an alloy of copper and tin and is attacked by acids and bases. It is oxidized and discolored during the sintering cycle with no impact on the quality. Nonoxidizing grades are available from suppliers such as US Bronze.

Molybdenum disulfide is an interesting additive. It increases the hardness of the surface while decreasing friction. Electrical properties of the compound are virtually unaffected. It is normally used in small proportions combined with other fillers such as glass. MoS_2 reacts with oxidizing acids and is inert towards most other chemicals.

A number of other fillers are incorporated in polytetrafluoroethylene to produce compounds for specialty applications. Calcium fluoride can replace glass in end-uses where glass is attacked by chemicals. Alumina (Al_2O_3) is an excellent electrical insulator, which also improves the mechanical properties of the compound for use in high voltage components. Inorganic pigments, able to withstand sintering temperatures, are added to impart color to PTFE compounds for customization or ease of identification of parts. Mica has a platelet structure and imparts desirable properties to polytetrafluoroethylene. Mica particles orient themselves perpendicularly to the direction of pressing. This orientation results in a significant reduction in shrinkage and the coefficient of thermal expansion in the direction of orientation. Physical properties of the compounds are severely lowered, thus rendering mica compounds only useful for compressive applications.

PTFE can be compounded with polymeric filler with adequate thermal stability. Examples of such polymers include polyarylates such as polyetherketone, polyphenylene sulfone and polyphenylene sulfide. Surface characteristics such as wear rate, coefficient of friction and surface tension can be modified with these additives.

12.2.2 Polytetrafluoroethylene Selection

Polymer selection for compounding suspension polymerized PTFE is relatively straightforward. Fine cut resins are used as a starting point to produce filled compounds. These powders have relatively small particle size and form the most uniform compounds. Examples of commercial grades of these resins are provided in Ch. 6. They include Teflon® 7A, 701N and 7C, Hostaflon® 1750 and Polyflon® M-12 and M-15. Typically, smaller particle size resins produce compounds with higher physical properties, e.g., Teflon® 7C and Polyflon® M-12.

12.2.3 Filled PTFE—Production Techniques

Granular polytetrafluoroethylene compounds containing fillers are converted into parts by the same molding techniques as those used for neat resin. These techniques have been described in Ch. 7. The compounding techniques aim at producing uniform blends of PTFE with fillers that can be processed in the same molding equipment. This section describes methods by which compounds can be made.

The first step in making a compound is blending the filler(s) with PTFE powder. The mixture is usually first tumbled by a drum-tumbler or another device to achieve a rough blend of the filler and PTFE. It is then milled to increase the uniformity of the polytetrafluoroethylene and the filler(s). Milling can be accomplished in a number of commercial mills. Different procedures and processing conditions are required for specific filled compounds. Commercial mills operate based on two different principles.

Hammer mills include commercial examples such as Tornado® mill (supplied by Tornado Engineering, Division of Solus Industries, Niagara Falls, New York) and Rietz mill (supplied by Bepex Corporation, Santa Rosa, California). The blend is subjected to a rotating set of small hammers while residing in a mesh screen basket. The particles pass through the mesh screen as they are dispersed and comminuted. The important process parameters are the size of the openings in the mesh screen and the number and speed of the hammers.

The second group of mills is V-shape devices, which have been described in Ch. 8. They consist of two connected cones and are usually equipped with an attrition bar. The mill rotates around a horizontal axis. The material is transferred from one cone to another during each rotation, thus mixing the filler and the resin powder. The attrition bar is parallel to the rotation axis and can be operated at different speeds. Various designs are available which subject the blend to different shear rates. Patterson Kelly Corporation (Division of HARSCO Corp., East Stroudsburg, Pennsylvania) is a major supplier of V-shape blenders.

The blends obtained from these mills are called *low flow* PTFE compounds and can be molded by billet and sheet molding techniques. A major concern is stratification or filler separation during transportation and handling. PTFE lacks affinity for most materials including common fillers as has been described in Ch. 2. There is virtually no interaction between PTFE molecules and the fillers, which can result in filler separation when the mixture is subjected to motion. The low coefficient of friction of polytetrafluoroethylene expedites the separation.

Low flow blends are also feed for the production of *free flow* compounds. Production of free flow compounds is based on granulation (agglomeration) of the particles of resin and filler. This is accomplished by mixing the ingredients (the low flow blend) with a water-immiscible organic solvent and/or water and sometimes a surfactant, and then heating the mixture and shearing it. Granules formed in this process are separated from the liquid and dried to remove the liquid residues. A *dry process* is one where no water is used.

Uniform mixing of PTFE powder with a hydrophilic or semi-hydrophilic filler like glass in a *wet process* (in water or a mixture of water and an organic solvent) is difficult because the filler tends to migrate into the aqueous phase. Consequently, some of the filler may remain in the water and not be incorporated in the agglomerates. This phenomenon, called *filler separation,* can be overcome by subjecting the surface of the filler to a hydrophobic surface treatment prior to lowering its surface energy closer to the surface energy of polytetrafluoroethylene. An alternative way is to add a compound to the wet process, which would be absorbed into the surface of the filler, to reduce its energy. Examples of compounds used for surface treatment include silicone oils, organosilanes and fatty acids. Specific processes for producing free flow compounds from patent literature are described below.

United States Patent 3,915,916[2] introduced a process for preparing agglomerated granules of PTFE containing glass. The compound was made in a stainless steel vessel (15.2 cm diameter and 20.3 cm deep) equipped with a stirrer and two baffles. The stirrer had four marine blades. In one example, 1600 ml of demineralized water, 50 ml of tetrachloroethylene and 200 g of a blend of fine-cut PTFE and filler was charged to the vessel. Small amounts of an aminofunctional organosilane (such as 0.5 ml gamma-aminopropyl triethoxisilane) and a silicon resin (such as 0.05 g phenyl methyl siloxane) were added to the mixture to improve the coupling of glass and PTFE. Glass fibers were added to the vessel in the range of 20–40% by weight on a dry basis in the compound. The mixture was heated and stirred to form a uniform mix of the ingredients and form agglomerates.

The vessel was agitated for 5 minutes at 1000 rpm after it had been heated to 40°C. The stirrer was slowed to 600 rpm and agitated for 25 minutes at this speed. The agglomerated product was recovered by filtering out the liquid and drying the granules in a vacuum oven at 130°C. The free flow compounds had apparent density of 850–900 g/l and average particle size of 600–750 µm. Typical properties of two compounds are summarized in Table 12.2. Increasing the glass fiber content of the compound reduces standard specific gravity (SSG) and shrinkage moderately, while physical properties decline more drastically.

In another patent,[3] glass fiber or bronze was compounded in a similar manner to the procedure in US Patent 3,915,916.[2] Silicon resin was replaced by a fatty acid (C_{12}-C_{18}) and an anionic surfactant such as ammonium perfluorooctanoate. Compounds with excellent flow and properties were obtained.

In another case,[4] a compound of an organic filler such as an aromatic polyester (Ekonol® from Norton Chemplast or Ryton from Phillips 66) and a fine-cut PTFE was made. The filler could be treated with a silane coupling agent, before or during the compounding operation. A low flow powder could be made by dissolving the silane in a ketone or isopropanol and adding the filler. PTFE was then blended dry with the dried filler in a mixer. In the second method, PTFE, the filler, and silane were all added to the solvent and agitated in a mixer such as a high speed disperser or a Henschel mixer. Excellent utilization of the filler was obtained and less than 0.005% by weight of the filler was not incorporated.

Honda et al.[5] reported an improvement over Leverett[2] by reducing the amount of silicon oil (phenyltrimethoxysilane) added to the compound by a factor of ten using a similar vessel and agglomeration process. They treated the filler surface with an organosilane which improved the color of molded parts due to the higher thermal stability of organisilanes than amino-functional organosilanes. The solvent was a combination of water and a halogenated hydrocarbon (C_2 or C_3) containing one fluorine and one hydrogen, such as 1,1-dichloro-2,2,2-trifluoroethane. Free flow compounds of PTFE with bronze, carbon powder, and glass beads were made by this method.

12.3 Fine Powder-Based Compounds

Dispersion polymerized polytetrafluoroethylene or fine powder is compounded to a much lesser extent than suspension polymerized resin. It is relatively difficult to mix solids with fine powder PTFE to form a uniform blend. This is due to the large average agglomerate size (several hundred microns) of fine powders. Large concentration of filler particles serves as points of stress concentration, thus deteriorating the physical properties of the compound. Excessive shearing will lead to fibrillation of polymer particles as opposed to deagglomeration. Fibrillation of PTFE

Table 12.2. Properties of Free Flow Glass Compound[2]

Glass Fiber, wt. %	Average Particle Size, µm	SSG	Apparent Density, g/l	Shrinkage, %	Tensile Strength, MPa	Break Elongation, %
25.2	610	2.193	877	1.33	13.8	179
38.8	670	2.174	862	0.53	8.3	79

during the mixing will render the mixture unusable for extrusion. See Ch. 8 for a detailed discussion of this subject.

The limitation in the shear that can be applied to the resin limits the filled volume fraction. Size and shape of the filler particles impact the maximum volume fraction; more of the smaller particles can be incorporated. Incorporation of additives including fillers and pigments into fine powder polymer is usually intended to accomplish one of the following objectives:

Achieve a color

Increase electrical conductivity

Increase abrasion resistance

Addition of pigments to fine powder polytetrafluoroethylene has been discussed in Ch. 8. This section focuses on the fillers that modify functional properties of the polymer.

12.3.1 Fillers and Compounding Methods

Some applications such as fuel transport hoses require static charge dissipation to avoid igniting the fuel. Conductive grades of carbon black can impart electrostatic charge dissipation to the polymer. A small amount of carbon black (1–2%) is compounded by adding a dispersion of the carbon black in the lubricant to polymer powder and tumbling them in a V-shape mixer like Patterson-Kelly machine (see Ch. 8). This compound forms the inner layer of the tubing by co-extrusion of the paste.

An important family of fillers is engineering polymers. These plastics are hard and, when combined with polytetrafluoroethylene, produce compounds that exhibit low coefficient of friction and a lower wear rate relative to PTFE. Flow under load or creep is also reduced. An application for such compounds is extruded tubing for push-pull cables (automotive) where low friction and wear rate is required. Examples of these polymers include polyarylene and polyamide-imide resins. Commercial polyarylene plastics include polyetheretherketone (PEEK), polyetherketone (PEK), polyetherketoneketone (PEKK), polyphenylenesulfide (PPS), and polyphenylene sulfone. A successful compound has been reported with Torlon® 400TF.[6] The powder form of these polymers is mixed with PTFE resin by briefly pre-mixing the dry powders below the transition point of polytetrafluoroethylene followed by adding lubricant and tumbling in a V-shaped mixer.

PTFE compounds may contain other additives in the form of fiber or bead, which are less expensive than engineering polymers. Some useful fillers include glass, and metals oxides[6] which also induce abrasion resistance. Abrasion resistance testing in a push-pull cable indicated that inclusion of 10 parts per hundred parts (phr) of PTFE of polyphenylene sulfide and 25 phr of glass beads increased the life cycle to more than 500,000 from 25,000 for an unfilled PTFE.[7] Similar results were obtained by the addition of 15 phr polyamide-imide to polytetrafluoroethylene.

Generally, the following steps are taken in the manufacturing of filled fine powder compounds.

1. Load the PTFE powder carefully into the V-blender to avoid shearing the resin.

2. Screen the filler(s) before adding it to the dry PTFE. After adding the filler to the resin, roll or tumble for 5–10 minutes to mix the contents thoroughly.

3. Add the lubricant evenly to the resin.

4. Set the blender to tumble at 24 rpm for 13 minutes for a 25 kg batch of resin. Longer rotation times may be required for larger batches of resin.

5. Screen the compound to break up loose lumps and separate those that do not break easily.

6. Empty the blender and store the lubricated resin in a jar or the original drum and make sure the lid is sealed tightly. Allow the blend to age for at least 12 hours at 35°C to allow complete diffusion of the lubricant into the polymer particles.

12.3.2 Fabrication of Reinforced Gasketing Material

Reinforced fine powder polytetrafluoroethylene material is primarily used for application as gaskets and seals in extreme temperature, pressure, and chemical environments. A gasket in this type of application must be resilient and resistant to corrosive chemicals and also maintain a high tensile strength and dimensional stability at high temperature and pressure. PTFE has the necessary corrosion resistance to the majority of industrial chemicals up to its melting point (327°C), but in its neat (without fillers or additives) form, it is not satisfactory in many applications because of the high cold flow (creep) that is inherent to PTFE. After a short while, an unfilled PTFE gasket will begin to "creep" under the pressure exerted by bolt loads that squeeze the gasket between flanges. The net result of cold flow is loss of gasket thickness and leaks. An increase in temperature both accelerates and increases creep. See Ch. 14 for a discussion of cold flow or deformation under load.

The reinforcement approach deals with the problem of cold flow by highly filling PTFE with a variety of fillers, as high as 90% by weight. Fillers are usually hard materials such as metal powders, ceramic, glass fiber, carbon, and others. The processing technique to distribute the fillers and the choice of the polymer type help retain good physical properties in the reinforced (filled) material. Fabrication of reinforced gasket material is accomplished by filling the fine powder PTFE using a somewhat unusual process which incorporates the fillers in the polymer structure. Typically, sheets of the material are made out of gaskets which can be stamped.

12.3.2.1 Process Background

Wire insulation and tubing made from fine powder PTFE often has unbalanced physical properties. Orientation of the polymer in the direction of the extrusion increases the tensile strength of the insulation or tubing in the extrusion direction above the strength in the transverse direction, as shown in the Fig. 12.1(a). This means that, when subjected to stress, the insulation could fail in the transverse direction.

The balance of tensile strength can be improved by quenching the molten insulation in cold water immediately after sintering (Fig. 12.1b). The rapid cooling freezes the polymer molecules in random positions and prevents preferential orientation in the extrusion direction. Figure 12.1(c) shows the stress-strain curve for reinforced gasketing material made by the process in this section without the need for cold water quenching. The biaxial balance is obtained by fibrillation of the resin in both directions.

12.3.2.2 Process Description

There are five steps in fabricating biaxially oriented sheeting or tape including slurrying the fine powder PTFE with a large excess of lubricant, filtration of the excess lubricant (see Ch. 8), rolling or calendaring the wet cake in transverse directions, drying the cake of the lubricant below the sintering temperatures, and sintering the dried cake. These steps are further described below. This process could be used to make biaxially oriented unfilled PTFE sheeting as well as filled material.

The slurry is typically made at a polymer plus filler to lubricant ratio of 1 to 4 or 1 to 5. Thorough mixing is achieved by vigorously stirring the mixture in a Waring Blender or other similar mixer. The aim is to wet the surface of the resin and distribute its particles uniformly in the lubricant. Anywhere from 1 to 10 minutes of mixing may be necessary to complete the process.

Filtration of the slurry should be done immediately after mixing has been completed. A suitable filter media is a 30 to 50 mesh cloth of nylon, cotton polyester, or Whatman No. 4 paper. The slurry should be transferred carefully to assure even distribution of the solids over the filter cloth. Vacuum or pressure aid could expedite the filtration, but care must be taken to avoid too much lubricant removal. After filtration, the desirable range of the lubricant is in the 18–22% range by weight. Filled compositions and more complex shape parts require more lubricant than unfilled PTFE and sheet form.

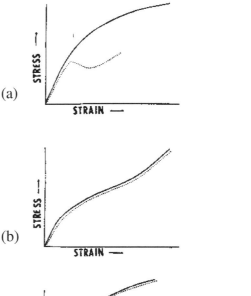
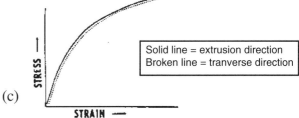

Figure 12.1 Stress-strain curve for various orientation states.[8] *(a)* Unbalanced orientation (short sintering and air quenching). *(b)* Random orientation (post sintering and water quenching). *(c)* Biaxial orientation.

The next step is rolling the wet cake, manually or mechanically, under pressures of 0.03–0.1 MPa in two perpendicular directions to induce biaxial orientation as shown in Fig. 12.2. Care must be taken to avoid reducing thickness by more than 75% in each pass. After reaching the desired thickness, the sheet is doubled up to continue effecting work. Calendaring rolls (Fig. 12.3) are more convenient in reducing thickness in a controlled manner.[9] Calendaring can be done in the transverse direction by collecting and folding the sheet after orientation in the machine direction. The sheet should be saturated with lubricant at all times during the calendaring operation. The color of the sheet becomes lighter when solvent content is reduced and can be recovered by soaking it in a bath of lubricant.

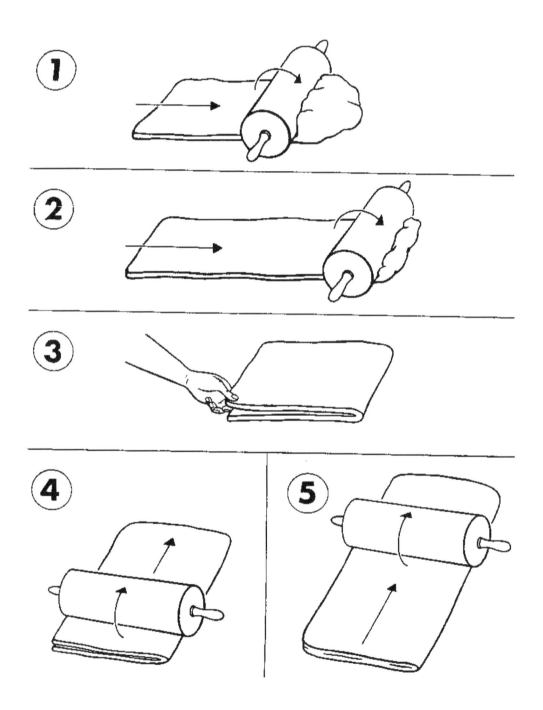

Figure 12.2 Procedure for biaxial orientation of sheeting.[8]

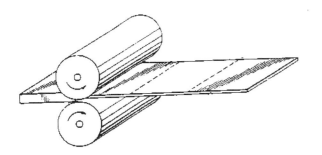

Figure 12.3 Oriented sheeting formation by calendaring.[9]

Drying of the sheet follows the calendaring step which completes the orientation. The lubricant is removed by drying the sheet in an oven prior to sintering. An optional step is to compact the sheet at 10–15 MPa at 150–200°C before sintering. Sintering of the sheet is carried out at 360–380°C for 2–30 minutes, depending on the thickness. This process can prepare sheets with a finished thickness of 6 mm.

Examples of tensile properties of biaxially oriented sheets, unfilled and filled, are given in Table 12.3. Comparison of biaxially oriented fine powder PTFE, at different temperatures, with filled granular demonstrates the advantages of the biaxial orientation process. Deformation under load data for biaxially oriented sheets and their comparison with filled granular are given in Table 12.4, clearly illustrating the advantage of biaxial orientation in reducing deformation under load.

Further improvement in deformation can be obtained by alternately inserting perforated stainless steel plates (Fig. 12.4) between the oriented sheets prior to sintering.[9] The sintering process is conducted under pressure to consolidate the composite and obtain a predetermined thickness. At the end of sintering the composite is cooled rapidly under pressure. PTFE actually bonds to stainless steel in this process. The decrease in creep is often substantial, for example by insertion of two steel plates, the creep relaxation of a silica-filled PTFE is decreased from 57.9% to 30.5% at 100°C.[9]

Table 12.3. Tensile Properties of Biaxially Oriented Sheeting[8]

	Biaxially Oriented Unfilled Resin	Unfilled Paste Extruded PTFE	Biaxially Oriented 40% Graphite-Filled Fine Powder PTFE	35% Graphite-Filled Granular PTFE	Biaxially Oriented 25% Glass Fiber-Filled Fine Powder PTFE	35% Glass Fiber-Filled Granular PTFE
Temperature,°C	-40	-51	-40	-54	-40	-54
Tensile Strength, MPa	56.8	39.4	32.8	8.6	32	10.9
Break Elongation, %	85	50	15	6	48	8
Temperature,°C	23	23	23	23	23	23
Tensile Strength, MPa	35.6	27.6	16.8	6.7	18.1	7.6
Break Elongation, %	450	340	130	4	280	79
Temperature,°C	100	100	100	100	100	100
Tensile Strength, MPa	19.3		9.8	4	9.1	3.7
Break Elongation, %	380		77	5	240	144
Temperature,°C	260	232	260	260	260	260
Tensile Strength, MPa	9	8	4.9	1.8	3.2	1.5
Break Elongation, %	340	320	42	11	200	108

Table 12.4. Deformation-Under-Load[8]

Compound	Deformation @ 23°C at 14 MPa, ASTM D621, %
Biaxially Oriented, 40% Bronze	1.9
Granular PTFE, 60% Bronze	5.8
Biaxially Oriented, 25% Glass Fiber	2.6
Granular PTFE, 25% Glass Fiber	6.4

12.4 Co-Coagulated Compounds

Co-coagulation is the method by which large quantities of fillers can be incorporated in dispersion polymerized polytetrafluoroethylene. The addition of fillers takes place prior to coagulation of the resin from its dispersion state. Polymerization of tetrafluoroethylene by dispersion method and coagulation to produce fine powder (*coagulated dispersion*) has been described in Ch. 5. PTFE dispersion contains primary particles that are in the range of 0.15–0.25 µm in size. Coagulated particles called *agglomerates* have an average size of several hundred microns, each consisting of thousands primary particles.

In the process of co-coagulation, the additives are added to the polytetrafluoroethylene dispersion and mixed. This dispersion is coagulated and the compound is recovered. The smaller the filler particles, the smaller the points of stress rise in the compound will be. Significantly larger quantities of filler can be compounded in PTFE by this technique.

The main steps in the co-coagulation process consist of the following.

1. Mixing the fillers with water and a small amount of the surfactant present in the polytetrafluoroethylene dispersion.
2. Adding polytetrafluoroethylene dispersion to the mixture in step one, while stirring gently.
3. Coagulating the filled dispersion by increasing the agitation rate resulting in a thickening of the mixture.
4. Filtering the coagulated material through cheese cloth.
5. Drying the wet cake in trays (made from stainless steel mesh) in oven at 120°C for 4–16 hours. This mixture is suitable for paste extrusion.

Coagulation is quite rapid and is completed in less than 3 minutes. Clear upper water indicates complete coagulation while the appearance of a milky layer is a sign of incomplete coagulation. Agitation of the dispersion should continue until coagulation is completed. Afterwards, minimal shear should be applied to the compound to avoid separation of the polymer and the fillers.

Co-coagulated compounds can also be molded into thick sections. In this case, the surfactant (Triton® X-100) should be removed by heating the material for 4 hours at 300°C. The lumps formed during this step can be broken by chilling and pulverizing.

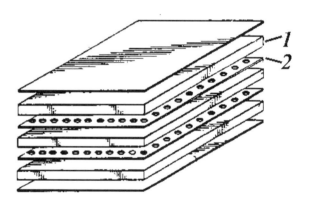

Figure 12.4 A composite of biaxially oriented sheeting with perforated steel plates.[9] *(1)* Biaxially oriented sheeting. *(2)* Perforated stainless-steel plate.

12.5 Processing Compounds

The principles of compression molding described in Ch. 7 are applicable to the molding of filled compounds of granular polytetrafluoroethylene. There are some differences in the processing conditions due to the presence of the fillers. Higher molding pressure is one requirement for successful molding of parts. Figures 12.5 and 12.6 provide preform pressure ranges for a few low flow and free flow compounds. Preforming pressure affects the mechanical and physical properties of parts molded from compounds. Figures 12.7 through 12.12 summarize the impact of pressure on properties for a few compounds. Specific gravity, tensile strength, and elongation at break rise and reach plateau values with increases in preform pressure. Additional pressure has a detrimental effect on the properties.

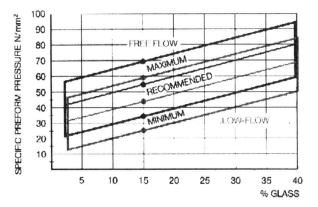

Figure 12.5 Recommended preform pressure ranges for glass-filled compounds.[1]

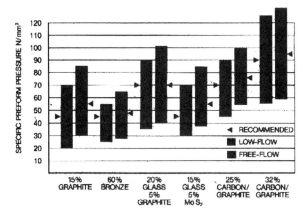

Figure 12.6 Recommended preform pressure ranges for various compounds.[1]

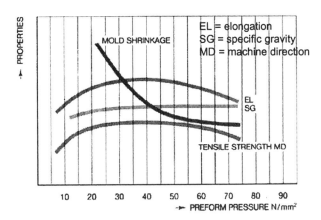

Figure 12.7 Influence of perform pressure on physical properties. Compound: 15/25% glass.[1]

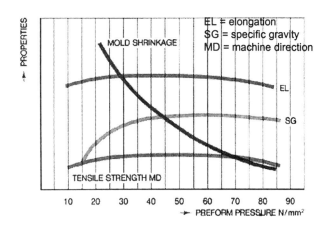

Figure 12.8 Influence of perform pressure on physical properties. Compound: 15% graphite.[1]

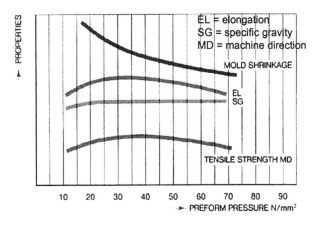

Figure 12.9 Influence of perform pressure on physical properties. Compound: 60% bronze.[1]

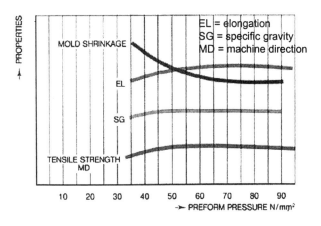

Figure 12.10 Influence of perform pressure on physical properties. Compound: 20%/5% glass/graphite.[1]

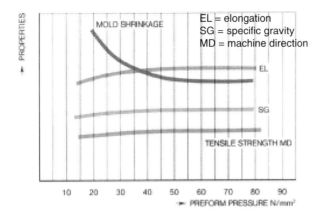

Figure 12.11 Influence of perform pressure on physical properties. Compound: 15%/5% glass/MoS$_2$.[1]

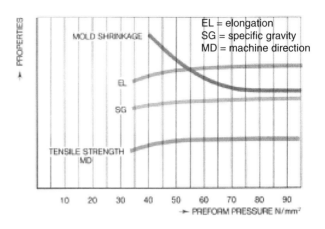

Figure 12.12 Influence of perform pressure on physical properties. Compound: 23/2/29 C/graphite.[1]

In comparison with unfilled PTFE, compounds can by sintered using of shorter sintering cycles due to the increase in thermal conductivity of the preform by incorporation of fillers (Table 12.5). Resin manufacturers often recommend conservative (i.e., long) sintering cycles. Trial and error is the quickest means of finding an economic sintering cycle. Filled PTFE compound parts can be sintered freely or in-mold under pressure similar to unfilled resin. Some recommended cycles are given in Table 12.6.

12.6 Typical Properties of Filled Fluoropolymers

Properties of granular PTFE compounds are presented in this section. Mechanical and tribological properties have been covered in separate sections. A few points should be noted about filled PTFE compounds. Shape and size of the filler and resin particles, type of molding process, and processing conditions all influence, to different extents, the properties of parts made from compounds. Properties of compounds made with non-spherical particles are usually anisotropic, i.e., they are dependent on the direction. Properties in the mold direction (MD) and perpendicularly cross direction (CD) must be measured to characterize fabricated articles. Reproducibility is only obtained by controlling all the variables.

The volume fraction of the fillers in the compound depends on its specific gravity. Table 12.7 lists the weight and volume fractions of the filler for a number of common filled compounds. A higher weight fraction of heavier fillers such as bronze can be incorporated than lighter additives like glass. Physical properties of the compound deteriorate with increase in the volume fraction of the filler as illustrated in Fig. 12.13.

Table 12.5. Thermal Conductivity of PTFE Compounds[1]

Compound Type	Unfilled PTFE	15% wt. Glass Fiber	15% wt. Glass Fiber	60% wt. Bronze	23% Carbon/ 2% Graphite (wt.)	23% Carbon/ 2% Graphite (wt.)
Thermal Conductivity, W/(m.K)	0.24	0.33	0.41	0.57	0.58	0.63

Table 12.6. Suggested Sintering Cycles for Filled PTFE Compounds[11]

Size of preform		Sintering cycle		
Size Dia. × length (mm)	Weight (kg)	Heating rate	Sintering period	Cooling rate
50 × 50	0.2	120°C/hr.	365°C for 5 hrs.	75°C/hr.
90 × 100	1.4	120°C/hr.	365°C for 8 hrs.	50°C/hr.
174/52 × 100	4.8	Room temp. to 300°C 50°C/hr. 300 to 340°C 20°C/hr. 340 to 365°C 50°C/hr.	365°C for 10 hrs.	365 to 300°C 50°C/hr. 340 to 300°C 20°C/hr. 300°C to room temp. 50°C/hr.
256/74 × 200	20.7	18°C/hr.	365°C for 12 hrs.	18°C/hr.
500 × 500 × 1t	0.55	100°C/hr.	365°C for 4 hrs.	30°C/hr.

Table 12.7. Weight and Volume Fractions of Filled PTFE Compounds

Filler Type	Weight Fraction of Filler, %	Volume Fraction of Filler, %
Glass	15	13.2
Glass	25	22.4
Glass	40	36.5
MoS$_2$	5	2.3
Graphite	15	14.5
Bronze	60	26.7
Glass/Graphite	10/10	18.5
Glass/ MoS$_2$	15/5	15.9
Carbon/Graphite	23/2	28.4

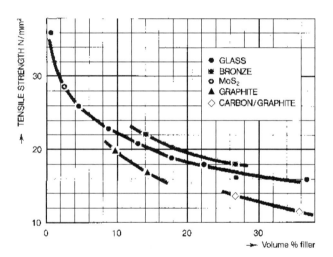

Figure 12.13 Tensile strength as a function of percent filler.[1]

12.6.1 Mechanical Properties

Polytetrafluoroethylene retains excellent properties at very low and high temperatures. Table 12.8 provides summary of some of the mechanical properties of three different compounds containing 65% bronze, 15% carbon and 25% glass fiber at different temperatures. Properties of unfilled PTFE have been listed for comparison.

Tensile strength and break elongation at elevated temperatures are given in Table 12.9. All the listed compounds retain excellent tensile properties at above room temperature.

Deformation-under-load of all filled polytetrafluoroethylene compounds decreases in comparison to unfilled resin, as seen in Table 12.10. Combinations of carbon and graphite reduce deformation the most at the room temperature and at elevated temperatures. The next effective filler in reducing deformation-under-load is bronze at 60% by weight. Hardness is increased by the addition of additives, particularly bronze, carbon and graphite (Table 12.11).

Compressive strength and flexural data are presented in Tables 12.12 and 12.13.

Table 12.8. Mechanical Properties of Filled PTFE Compounds[1]

Property	Test method	Units	Temperature °C	Unfilled[1]	Compound 65% Bronze	Compound 15% Carbon	Compound 25% Glass Fibre
Tensile yield strength	ASTM D 1708	N/mm²	−253 −196 −183 −129 − 79 + 23	123 92 (84) 53 36 18	– 47 (44) (35) (26) 12	– 44 (41) (32) (24) 8	30 24 (23) (18) (14) 10
Ultimate tensile strength	ASTM D 1708	N/mm²	−253 −196 −183 −129 − 79 + 23	124 103 (95) 63 42 36	– 47 (45) (35) (26) 15	– 47 (45) (35) (27) 16	31 25 (24) 21 20 19
Tensile modulus	ASTM D 1708	N/mm²	−253 −196 −183 −129 − 79 + 23	4300 3200 3100 2100 1400 750	– – – – – –	– – – – – –	3200 2600 (2500) (2000) (1600) 800
Elongation	ASTM D 1708	%	−253 −196 −183 −129 − 79 + 23	3 10 (12) 70 130 400	0 4 (5) (34) (70) 200	0 2 (4) (23) (50) 140	1 4 (6) (35) (80) 240
Flexural strength	ASTM D 790	N/mm²	−253 −196 −183 −129 − 79 + 23	Did not break	Did not break	Did not break	Did not break

Note: Parantheses indicate values estimated by interpolation

(1) Crystallinity 41–71%

(Cont'd.)

Table 12.8. *(Cont'd.)*

Property	Test method	Units	Temperature °C	Unfilled[1]	Compound 65% Bronze	Compound 15% Carbon	Compound 25% Glass Fibre
Flexural modulus	ASTM D 790	N/mm²	−253	5100	–	–	2800
			−196	4700	–	–	2500
			−183	4600	–	–	2300
			−129	3200	–	–	1900
			− 79	1600	–	–	1700
			+ 23	700	–	–	1000
Compressive strength	ASTM D 695	N/mm²	−253	220	–	–	188
			−196	171	–	–	152
			−183	(168)	–	–	(140)
			−129	158	–	–	(109)
			− 79	143	–	–	(79)
			+ 23	118	–	–	24
Compressive modulus	ASTM D 695	N/mm²	−253	6200	6200	6100	6800
			−196	5500	5700	6000	5900
			−183	(5400)	(5400)	(5500)	(5600)
			−129	4100	(4100)	(4300)	(4400)
			− 79	1900	(3000)	(3200)	(3200)
			+ 23	700	550	900	860
Torsional modulus of rigidity	ASTM D 1043	N/mm²	−253	2200	–	–	1700
			−196	1500	–	–	720
			−183	(1400)	–	–	(690)
			−129	1000	–	–	(520)
			− 79	500	–	–	(390)
			+ 23	160	–	–	110
Izod impact strength (notched)	ASTM D 256	J/m	−253	75	67	51	53
			−196	80	70	60	59
			−183	(85)	(73)	(64)	(63)
			−129	(97)	(86)	(103)	(98)
			− 79	133	(121)	(139)	(143)
			+ 23	160	161	169	173

Table 12.9. Tensile Properties of Filled PTFE Compounds Measured According to ASTM D 1708[1]

Temperature: 23°C	Unfilled PTFE	15% Glass	25% Glass	60% Bronze	23C/2Gr	29C/2Gr
Tensile strength, N/mm², MD	35,5	25,5	19,5	17,0	11,0	7,0
CD	36,5	24,5	18,5	17,0	15,0	10,0
Ultimate elongation, %, MD	400	285	235	250	65	35
CD	450	290	250	235	105	50
Temperature: 150°C	701-N	1103-N	1105-N	1146-N	1191-N	1192-N
Tensile strength, N/mm², MD	13,5	8,6	6,3	5,4	3,8	3,0
CD	14,0	8,4	5,8	6,4	5,6	3,8
Ultimate elongation, %, MD	440	350	305	305	120	60
CD	480	340	300	330	185	85

C = Carbon, Gr = Graphite

Table 12.10. Deformation-Under-Load of Filled PTFE Compounds Measured According to ASTM D 621A, Molding Direction[1]

	Unfilled PTFE	15% Glass	25% Glass	60% Bronze	23C/2Gr	29C/2Gr
% deformation under load 1 h, 23°C, 14,2 N/mm²	11,8	9,8	9,0	7,8	4,8	4,7
% deformation under load 24 h, 23°C, 14,2 N/mm²	14,3	12,1	12,4	11,1	6,6	6,1
% permanent deformation	7,9	6,3	6,4	5,8	2,9	2,6
% deformation under load 1 h, 150°C, 5 N/mm²	10,0	9,8	9,2	8,2	6,6	5,6

C = Carbon, Gr = Graphite

Table 12.11. Hardness of filled PTFE compounds.[1] Hardness (Shore D) is measured according to ASTM D 2240, 23°C. Indentation test is measured according to DIN 53456.

Compound	Hardness (Shore D)[1]	Indentation Hardness test[2]
Unfilled PTFE	57	28
15% glass	62	24
25% glass	65	27
5% moly	62	30
15% graphite	63	36
60% bronze	70	39
25% carbon/graphite	68	39
32% carbon/graphite	70	–

(1) Measured according to ASTM D2240 at 23°C   (2) Measured according to DIN 53456

Table 12.12. Compressive Strength of Filled PTFE Compounds Measured According to ASTM 695M, Molding Direction[1]

	15% Glass	25% Glass	60% Bronze	23C/2Gr	29C/3Gr
Compressive strength, N/mm^2, 0,2% offset, 23°C	6,8	7,2	8,2	8,8	9,0
Compressive modulus, N/mm^2, 23°C	725	860	1050	970	815
Compressive strength, N/mm^2, 0,2% offset, 150°C	1,6	1,8	2,1	2,3	2,8
Compressive modulus, N/mm^2, 150°C	160	174	219	236	224

C = Carbon, Gr = Graphite

Table 12.13. Flexural Properties of Filled PTFE Compounds Measured According to ASTM D 790 M1, Molding Direction[1]

	Unfilled PTFE	15% Glass	25% Glass	60% Bronze	23C/2Gr	29C/3Gr
Flexural yield strength, N/mm^2, 23°C, 0,2% offset	–	6,2	5,5	8,8	10,3	8,8
Flexural modulus, N/mm^2, 23°C	690	820	1000	1290	1090	1180

C = Carbon, Gr = Graphite

12.6.2 Thermal Properties

Fillers reduce the linear coefficient of thermal expansion and contraction of compounds. Tables 12.14 and 12.15 provide data for several compounds at different temperatures. Aluminum reduces the coefficient of thermal contraction the most due to its flat platelet structure; mica has a similar effect.

12.6.3 Electrical Properties

Fillers and additives significantly increase the porosity of polytetrafluoroethylene compounds. Electrical properties are affected by the void content as well as the filler characteristics. Dielectric strength drops while dielectric constant and dissipation factor rise. Metals, carbon and graphite increase the thermal conductivity of PTFE compounds. Tables 12.16 and 12.17 present electrical properties of a few common compounds.

12.6.4 Chemical Properties

Permeability of compounds increases due to the voids. The effect of preforming and sintering conditions on the extent of permeability increase are shown in Fig. 12.14.

Table 12.14. Coefficient of Linear Thermal Expansion of Filled PTFE Compounds[1]

Material	Linear thermal contraction from 15°C to: (% or µm/m x 10^4)				
	−79°C	−129°C	−183°C	−197°C	−253°C
SAE 1020 Steel	0.1	0.1	0.1	0.2	0.2
Copper	0.2	0.2	0.3	0.3	0.3
Aluminium	0.2	0.3	0.4	0.4	0.4
PTFE, unfilled	1.5	1.9	2.0	2.1	2.1
PTFE, 60% Bronze	0.8	1.0	1.2	1.2	1.4
PTFE, 25% Glass fibre (MD)	1.0	1.2	1.5	1.5	1.7
PTFE, 25% Glass fibre (CD)	0.6	0.8	0.9	0.9	0.9
PTFE, 15% Graphite	0.9	1.2	1.4	1.4	1.5

C = Carbon, Gr = Graphite

Table 12.15. Coefficient of Linear Thermal Expansion of Filled PTFE Compounds Measured according to ASTM E 831[1]

Temp. Range (°C)	Unfilled PTFE		15% Glass		25% Glass		60% Bronze		23C/2Gr		29C/3Gr	
	MD	CD	MD	CD	MD	CD	MD	CD	MD	CD	MD	CD
−150 to +15	103	96	88	74	83	61	70	66	79	57	67	50
−100 to +15	119	109	102	86	96	69	80	77	90	64	77	57
−50 to +15	131	117	111	93	106	74	87	84	95	67	84	61
+15 to +23	472	286	332	278	284	180	201	207	315	158	222	133
+23 to +100	125	129	135	123	109	66	117	110	114	70	108	80
+23 to +200	142	152	156	153	136	84	134	132	136	88	132	99
+23 to +250	159	176	181	179	159	102	155	152	158	107	152	115
Moulded at Preform Pressure N/mm²	21		45		55		45		70		90	

C = Carbon, Gr = Graphite

Table 12.16. Electrical properties of filled PTFE compounds measured according to ASTM D 149, D150. (Note that bronze was too conductive to be measured.)[1]

	Unfilled PTFE	15% Glass Fibre	25% Glass Fibre	60% Bronze	20% Glass 5% Graphite	15% Glass 5% MoS_2
Dielectric strength, kV/mm						
In air	59	17,6	12,9	x	2,48	27,2
In oil	–	36,2	34,2	x	7,36	36,7
Dielectric constant						
60 Hz	2,1	2,50	2,63	x	3,38	2,71
10^6 Hz	2,1	2,35	2,85	x	3,25	2,68
Dissipation factor						
60 Hz	<0,0003	0,0753	0,0718	x	0,0761	0,0464
10^6 Hz	<0,0003	0,0029	0,0028	x	0,0024	0,0061

Table 12.17. Resistivity of Filled PTFE Compounds Measured According to ASTM D 257[1]

	Unfilled PTFE	15% Glass	25% Glass	60% Bronze	23C/2Gr	29C/3Gr
Surface resistivity (Ω)	10^{17}	>10^{16}	>10^{16}	>10^{16}	10^7	10^7
Volume resistivity (Ω·cm)	10^{18}	>10^{17}	>10^{17}	10^{12}	10^5	10^6

C = Carbon, Gr = Graphite

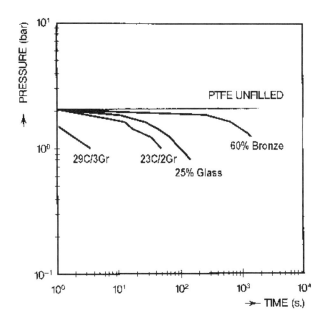

Figure 12.14 Effect of preforming conditions on the permeability of filled PTFE compounds.[1] (Nitrogen permeability of Teflon compounds measured on a membrane 50 mm diameter, 0.5 mm thickness.) C = carbon, Gr = graphite.

Polytetrafluoroethylene has excellent chemical resistance properties as discussed in Ch. 13. The effect of incorporation of additives on chemical properties depends on the type of the filler and the specific chemicals. In general, chemical properties of filled PTFE compounds are not as good as those of the unfilled resin. Table 12.18 shows the effect of a number of chemicals on carbon/graphite, glass and bronze compounds.

12.6.5 Tribological Properties of Filled PTFE

Wear and friction are two independent and mutually exclusive variables, that is, they are independent of each other. Unfilled polytetrafluoroethylene has an extremely low coefficient of friction and a high wear rate. Incorporation of fillers does not affect the coefficient of friction while the wear rate of the compound is reduced. The wear rate of the mating surface may be high as a result of *abrasion* while that of the PTFE compound is small. Abrasion is a strong function of

Table 12.18. Chemical Resistance of Filled PTFE Compounds[1]

CHEMICAL	FILLER		
	CARBON/ GRAPHITE	GLASS	BRONZE
Acetaldehyde	A	A	A
Acetone	A	A	A
Aluminium sulphate	A	A	B
Ammonium chloride	A	A	C
Ammonium hydroxide	A	B	C
Aniline	A	A	C
Benzene	A	A	A
Boric acid	A	A	A
Brine	A	A	A
Bromine (anhydrous)	C	B	C
Carbon disulphide	A	A	A
Chloroacetic acid	A	A	B
Chlorobenzene	A	A	A
Chloroform	A	A	A
Chromic acid	B	B	C
Citric acid	A	A	A
Diethyl ether	A	A	A
Ethylene glycol	A	A	A
Fatty acids	A	A	A
Ferric chloride	A	A	C
Ferric sulphate	A	A	C
Fluorosilicic acid	B	C	C
Formic acid	A	A	A
Freon* (liquid)	A	A	A
Hydroboric acid	A	B	C
Hydrochloric acid	A	B	C
Hydrocyanic acid	A	B	C
Hydrofluoric acid	A	C	C
Hydrogen sulphide (solution)	A	C	C
Lead acetate	A	A	C
Maleic acid	A	A	B
Mercury salts	A	A	C
Molasses	A	A	B

(Cont'd.)

Table 12.18. *(Cont'd.)*

CHEMICAL	FILLER		
	CARBON/ GRAPHITE	GLASS	BRONZE
Naphtha	A	A	B
Naphthalene	A	A	B
Nickel salts	A	A	A
Nitric acid (0-50%)	C	B	C
Nitro benzene	A	A	A
Phenol	A	B	A
Phosphoric acid	A	A	C
Phthalic acid	A	A	A
Picric acid	A	A	A
Pyridine	A	A	C
Salicylic acid	A	A	B
Silver nitrate	A	A	C
Sodium carbonate	A	A	A
Sodium hydroxide	A	B	A
Sodium nitrite	A	A	A
Sodium peroxide	B	A	C
Sodium silicate	A	C	A
Sodium sulphide	A	A	C
Starch	A	A	A
Sulphuric acid	B	A	C
Tallow	A	A	A
Tannic acid	A	A	A
Tartaric acid	A	A	A
Trichloroethylene	A	A	B
Zinc chloride	A	A	C

A = excellent
B = Fair
C = Unsatisfactory

*FREON is Du Pont's registered trademark for its fluorocarbon solvents

the type, morphology and concentration of the filler(s). Wear, friction and abrasion are in general a function of many variables including load, velocity, type of movement, degree of coverage, temperature, filler, finishing of the part, break-in conditions, material surface, lubrication, environment and the presence of wear debris. The influence of these variables is discussed.

The coefficient of friction is inversely proportional to pressure and proportional to velocity. Wear of compounds of polytetrafluoroethylene is proportional to load (P) and velocity (V). Combinations of pressure and velocity are defined where the material can be used, thus a *PV limit* is defined. Above this PV limit, the wear increases exponentially because of the heat generated as a result of motion. Figure 12.15 shows the useful area of application of polytetrafluoroethylene compounds. Other than PV limit, creep or deformation under load limits the use of the compounds at high pressures where material starts to creep. Generally, a PTFE compound can be characterized by PV limit, creep limit and wear factor. *Wear factor* or *specific wear rate* is defined as the volume of material worn away per unit of sliding distance and per unit of load.

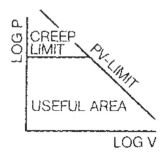

Figure 12.15 Useful area of material application defined by PV limit and creep.[1]

The characteristics of the filler and its content impact the wear behavior of compounds. Particle size, shape and structure are the key contributing filler variables. Coefficient of the friction of the compound is only mildly affected by the filler while wear factor is strong function of the filler. It is difficult to draw general conclusions from the performance of one filler under one set of conditions to others. For example, under moderate wear conditions, bronze does not perform as well as glass fiber. Under severe wear conditions (high velocity and load), bronze or graphite filled compounds perform better due to their ability to remove the excessive heat from the surface. Fibrous fillers such as glass fiber wear less rapidly than spherical ones such as glass beads. The lowest wear is achieved when filler particle size is close to that of the resin; very fine fillers have high wear rate.

Examples of the tribological characteristics of compounds are presented. Vespel® is included as a benchmark for excellent tribological performance. Wear factors for a number of materials are shown in Fig. 12.16 as a function load and velocity. Coefficient of friction for the same material is given in Fig. 12.17.

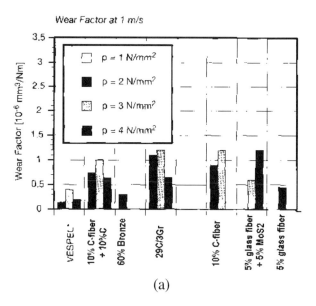

(a)

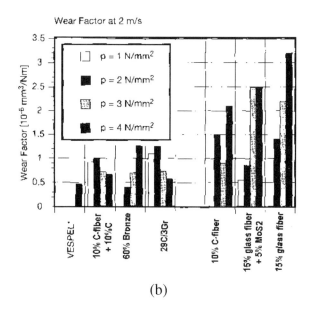

(b)

Figure 12.16 Wear factor of PTFE compounds vs velocity and load.[12] (C = carbon, Gr = graphite.) Vespel® is a trademark of DuPont, Co.

12.7 Commercial Products

A number of resin manufacturers and independent compounders supply filled compounds. Tables 12.19 through 12.28 provide information made available by these companies. These tables cover a few of the compounds offered by the suppliers. More complete information can be obtained by contacting the compound manufacturers.

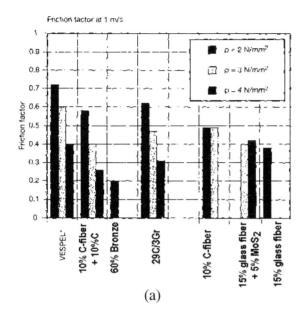

(a)

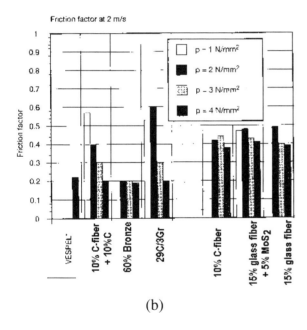

(b)

Figure 12.17 Coefficient of friction of PTFE compounds vs velocity and load.[12] (C = carbon, Gr = graphite.) Vespel® is a trademark of DuPont Co.

Table 12.19. Filled PTFE Compounds of Dyneon Custom Compounding[13]

Property	Test Method	CC 144/S (40% Bronze)	CC 174-N (15% glass fiber/ MoS$_2$)	CC 191-HE (25% Carbon/ Graphite)	CC 622 (10% Graphite)
Tensile Strength, MPa	ASTM D 4745	22.8	24.8	13.1	18.6
Tensile Elongation, %	ASTM D 4745	250	240	120	200
Hardness (initial), Shore D	ASTM D 2240	67	67	68	64
Compressive Strength, MPa	ASTM D 695	10.4	11	10.4	7
Deformation under Load, % Initial at 13.8 MPa and 21°C	ASTM D 621	5.3	4.4	4.1	6.5
Deformation under Load, % Permanent at 13.8 MPa and 21°C	ASTM D 621	3.1	2.5	2.5	3.7
Static Coefficient of Friction	ASTM D 1894	0.21	0.19	0.2	0.17
Dynamic Coefficient of Friction	ASTM D 3702	0.23	0.30	0.24	0.18
Coefficient of Thermal Expansion, (cm/°C)·10^{-4}		2.8	2.5	2.2	3
Wear Factor, [cm^3·min/kg·m·h]·10^{-8}	ASTM D 3702				

Table 12.20. Filled Compounds of Dyneon Custom Compounding[13]

Property	Test Method	CC 603 (15% glass fiber)	CC 605 (25% glass fiber)	CC 607 (35% glass fiber)
Tensile Strength, MPa	ASTM D 4745	20.7	17.2	13.1
Tensile Elongation, %	ASTM D 4745	280	230	180
Hardness (initial), Shore D	ASTM D 2240	63	63	63
Compressive Strength, MPa	ASTM D 695	7	9.1	7.8
Deformation under Load, % Initial at 13.8 MPa and 21°C	ASTM D 621	9.6	6.1	7.6
Deformation under Load, % Permanent at 13.8 MPa and 21°C	ASTM D 621	5.4	3.7	4.6
Static Coefficient of Friction	ASTM D 1894	0.23	0.24	0.27
Dynamic Coefficient of Friction	ASTM D 3702	0.31	0.33	0.36
Coefficient of Thermal Expansion, (cm/°C)·10^{-4}		2.8	2.5	2.5
Wear Factor, [cm^3·min/kg·m·h] 10^{-8}	ASTM D 3702			

Table 12.21. Filled PTFE Compounds of Ausimont[14]

Property	15GL (15% glass fiber)	25GL (25% glass fiber)	25CAR (25% Coke/Graphite)	35CAR (35% Coke/Graphite)
Bulk Density, g/l	500	510	430	410
Radial Shrinkage, %	2.0	1.6	2.2	2.0
Tensile Strength, MPa	24	20	19	17
Break Elongation, %	280	220	80	30
Hardness, Shore D	58	59	63	65
Permanent Deformation under Load, %	6	5.5	1.9	1.2
Wear Factor, [cm^3·min/kg·m·h] 10^{-8}	20	15	35	30
Coefficient of Linear Expansion, 10^{-5}/°C	16	15	15	13

Table 12.22. Filled PTFE Compounds of Ausimont[14]

Property	25CAR B (25% electrog. carbon)	35CAR B (35% electrog. carbon)	GR (15% Graphite)	60BZ (60% bronze)
Bulk Density, g/l	450	460	440	930
Radial Shrinkage, %	2.4	2.4	2.1	1.9
Tensile Strength, MPa	21	19	24	20
Tensile Elongation, %	220	70	230	230
Hardness, Shore D	61	64	57	65
Permanent Deformation under Load, %	1.8	1.6	5.0	1.9
Wear Factor, [cm^3·min/kg·m·h] 10^{-8}	150	150	1.750	10
Coefficient of Linear Expansion, 10^{-5}/°C	14	12	19	12

Table 12.23. Filled PTFE Compounds of Daikin[11]

Property	MG-1030 (MG-1030F[1]) (15% glass fiber)	MG-1040 (MG-1040F) (20% glass fiber)	MG-1050 (MG-1050F) (25% glass fiber)	MG-6030 (MG-6030F) (15% Carbon Fiber)
Bulk Density, g/l	390 (720)	390 (720)	400 (710)	370 (680)
Tensile Strength, MPa	33 (26)	27 (23)	25 (19)	21 (16.5)
Break Elongation, %	340 (330)	325 (310)	310 (280)	280 (120)
Hardness, Shore A	53-63	55-65	56-66	60-65
Deformation under Load, % Permanent at 13.8 Mpa, 24 hrs and 24°C	13.7	12.4	12	6.9
Static Coefficient of Friction	0.065	0.073	0.085	-
Dynamic Coefficient of Friction	0.24	0.24	0.24	0.28
Wear Factor, [cm^2·mm/kg$_f$km] 10^{-5}	1.2	1.1	1.0	1.0
Coefficient of Linear Expansion, 10^{-5}/°C	11.1	10.2	9.3	7.3

[1]Free flow

Table 12.24. Filled PTFE Compounds of Daikin[11]

Property	MG-2030 (15% Graphite)	MG-2060 (15% Graphite)	MG-3060 (60% Bronze)	MG-6050 (MG-6050F) (25% Carbon)
Bulk Density, g/l	390	400	870	360 (630)
Tensile Strength, MPa	22	13.5	18.5	19 (17)
Break Elongation, %	325	155	275	50 (30)
Hardness, Shore A	53-63	55-65	62-72	65-75
Deformation under Load, % Permanent at 13.8 Mpa, 24 hrs and 24°C	5.6	3.3	3.1	6.1
Static Coefficient of Friction	0.058	0.065	0.090	-
Dynamic Coefficient of Friction	0.23	0.25	0.24	0.29
Wear Factor, [cm^2·mm/kg$_f$km] 10^{-5}	6.7	2.0	0.7	2.0
Coefficient of Linear Expansion, 10^{-5}/°C	11.5	8.8	10.7	5.5

Table 12.25. Filled PTFE Compounds of ICI [15]

Property	FC-100-15 (15% glass fiber)	FC-100-25 (25% glass fiber)	FC-110-30 (30% glass fiber)	FC-140-15 (15% Graphite)
Bulk Density, g/l	500	515	530	460
Tensile Strength, MPa	21	15	15	14
Break Elongation, %				
Preforming Pressure, MPa	50	70	70	50
Fabrication Methods	General Molding	General Molding	General Molding	General Molding

Table 12.26. Filled PTFE Compounds of ICI [15]

Property	FC-150-25 (60% Powdered Coke)	FC-160-60 (60% Bronze)	FC-180-50 (50% Stainless Steel)	FC-200-25 (25% glass fiber)
Bulk Density, g/l	450	1050	850	530
Tensile Strength, MPa	11	14	14	10.3
Break Elongation, %				
Preforming Pressure, MPa	70	70	70	
Fabrication Methods	General Molding	General Molding	General Molding	Ram Extrusion

Table 12.27. Filled PTFE Compounds of PTFE Compounds, Inc. [16]

Compound	PTFE Compounds Description	Tensile Strength (PSI)	Tensile Elongation (%)	Specific Gravity
15% Glass Low Flow	15 FG LF	3100	300	2.20
15% Glass Free Flow	15 FG FF	2900	270	2.20
25% Glass Low Flow	25 FG LF	2700	260	2.22
25% Glass Free Flow	25 FG FF	2400	240	2.22
15% Glass 5% MoS_2 Low Flow	15FG 5M LF	2800	270	2.23
15% Glass 5% MoS_2 Free Flow	15FG 5M FF	2500	240	2.23
10% Graphite Low Flow	10 GR LF	2900	270	2.11
10% Graphite Free Flow	10 GR FF	2500	240	2.11
15% Graphite Low Flow	15 GR LF	2600	220	2.10
15% Graphite Free Flow	15 GR FF	2200	200	2.10
23% Carbon 2% Graphite Low Flow	23C 2GR LF	2100	70	2.08
23% Carbon 2% Graphite Free Flow	23C 2GR FF	1900	50	2.08
25% Carbon Low Flow	25 C LF	2200	100	2.04
25% Carbon Free Flow	25 C FF	2000	80	2.04
60% Bronze Low Flow	60 BZ LF	2500	150	3.85
60% Bronze Free Flow	60 BZ FF	2300	110	3.85
55% Bronze 5% MoS_2 Low Flow	55BZ 5M LF	2200	140	3.70
55% Bronze 5% MoS_2 Free Flow	55BZ 5M FF	2000	100	3.70

References

1. "Filled Compounds of Teflon® PTFE," DuPont Co. Publication, Oct., 1992.
2. US Patent 3,915,916, Leverett, G. L., assigned to DuPont, Oct. 28, 1975.
3. US Patent 4,370,436, Nakamura, Y., and Kawachi, S., assigned to Daikin Kogyo Co., Jan. 25, 1983.
4. US Patent 5,264,523, Honda, N., Hirata, T., and Yukawa, H., assigned to Daikin Industries Ltd., Nov. 23, 1993.
5. US Patent 5,321,059, Honda, N., Sawada, K., Idemori, K., and Yukawa, H., assigned to Daikin Industries Ltd., June 14, 1994.
6. US Patent 5,045,600, Giataras, J. L., Kray, K. K., and Marino, C. P., assigned to Markel Corporation, Sep. 3, 1991.
7. US Patent 4,362,069, Giataras, J. L., Golas, E. B., and Reed, W. F., assigned to Markel Corporation, Dec. 7, 1982.
8. "Fabrication of Sheeting and Tape by the HS-10 Process," DuPont Co. Bulletin, PIB#22.
9. US Patent 4,913,951, Pitolaj, S., Garlock Inc., Apr. 3, 1990.
10. "Teflon® 42 Aqueous Dispersion Properties and Processing Technique," Information Bulletin X-91d, DuPont Company.
11. "Daikin Filled PTFE Molding Powders," Technical Information, Daikin Fluorocarbon Polymers, Daikin Kogyo Co., Apr., 1981.
12. "A Tribological Characterization of Teflon® PTFE Compounds," Technical Information Bulletin No. H-38205, DuPont, S. A., June, 1992.
13. Technical Information, Custom Compounding, Division of Dyneon.
14. "ALGOFLON® Properties and Application Selection Guide," Ausimont Corp., Morristown, NJ.
15. "Imperial Chemical Industry, Fluon® Filled PTFE Granular Polymers," Technical Information, ICI Web Page, May, 1999.
16. "PTFE Compounds, Inc.," Filled Compounds Bulletin, New Castle, DE

PART III
13 Chemical Properties of Fluoropolymers

13.1 Introduction

A fundamental property of fluoropolymers is their resistance to organic and inorganic chemicals. Increased content of fluorine enhances the chemical resistance of the polymer. The overwhelming majority of the applications of fluoropolymers take advantage of their inertness to chemicals. Chemical properties of fluoropolymers are not affected by fabrication conditions. Another aspect of the interaction of these plastics with chemicals is permeation. Even though a reagent may not react with a fluoropolymer, it may be able to permeate through the polymer structure. The extent and rate of permeation is dependent on the structure and properties of the plastic article as well as the type and concentration of permeant. Temperature and pressure usually influence the permeation process. This chapter reviews chemical compatibility of fluoropolymers and their permeation behavior towards different chemicals.

13.2 Chemical Compatibility of Polytetrafluoroethylene

Polytetrafluoroethylene is by far the most chemically resistant polymer among thermoplastics. Few substances chemically interact with PTFE. The exceptions among those commercially encountered include molten alkali metals, gaseous fluorine at high temperatures and pressures, and few organic halogenated compounds such as chlorine trifluoride (ClF_3) and oxygen difluoride (OF_2). The inertness of PTFE arises from its molecular structure.[1]

A few other chemicals have been reported to have attacked PTFE at or near its upper service temperature (260°C).[2] PTFE reacts with 80% sodium or potassium hydroxide. It also reacts with some strong Lewis bases including metal hydrides such as boranes (B_2H_6), aluminum chloride, ammonia, and some amines (R-NH_2) and imines (R=NH). Slow oxidative attack may take place by 70% nitric acid at 250°C under pressure. It is important to test the effect of these reagents on PTFE under the specific application temperature to determine the material limitations.

Polytetrafluoroethylene derives its chemical resistance from an extremely strong carbon-fluorine bond and an impermeable sheath of fluorine atoms surrounding the carbon-carbon chain. High crystallinity renders PTFE insoluble in solvents. Tables 13.1-13.4 summarize the effect of a number of representative organic and inorganic compounds on PTFE.

Some halogenated solvents are absorbed by the polymer without any chemical interaction or degradation. The action is strictly physical and the removal of the absorbed species from them restores it back to its original state. Too much absorption by a PTFE sample can be an indication of excessive porosity. A highly porous sample may appear blistered due to the expansion of vapors in the surface pores. A properly fabricated part does not exhibit blistering.

Table 13.1. Chemical Resistance of PTFE to Common Solvents[3]

Solvent	Exposure Temp., °C	Exposure Time	Weight Gain, %
Acetone	20	12 mo.	0.3
	50	12 mo.	0.4
	70	2 wk.	0
Benzene	78	96 hr	0.5
	100	8 hr	0.6
	200	8 hr	1.0
Carbon Tetrachloride	25	12 mo.	0.6
	50	12 mo.	1.6
	70	2 wk.	1.9
	100	8 hr	2.5
	200	8 hr	3.7
Ethanol (95%)	25	12 mo.	0
	50	12 mo.	0
	70	2 wk.	0
	100	8 hr	0.1
	200	8 hr	0.3
Ethyl Acetate	25	12 mo.	0.5
	50	12 mo.	0.7
	70	2 wk.	0.7
Toluene	25	12 mo.	0.3
	50	12 mo.	0.6
	70	2 wk.	0.6

Table 13.2. Chemical Resistance of PTFE to Common Acids and Bases[3]

Reagent	Exposure Temperature, °C	Exposure Time	Weight Gain, %
Hydrochloric Acid			
10%	25	12 mo.	0
10%	50	12 mo.	0
10%	70	12 mo.	0
20%	100	8 hr	0
20%	200	8 hr	0
Nitric Acid			
10%	25	12 mo.	0
10%	70	12 mo.	0.1
Sulfuric Acid			
30%	25	12 mo.	0
30%	70	12 mo.	0
30%	100	8 hr	0
30%	200	8 hr	0.1
Sodium Hydroxide			
10%	25	12 mo.	0
10%	70	12 mo.	0.1
50%	100	8 hr	0
50%	200	8 hr	0
Ammonium Hydroxide			
10%	25	12 mo.	0
10%	70	12 mo.	0.1

Table 13.3. Chemical Compatibility of PTFE with Halogenated Chemicals[3]

Chemical	Effect on PTFE Sample
Chloroform	Wets, insoluble at boiling point
Ethylene Bromide	0.3% weight gain after 24 hr. at 100°C
Fluorinated Hydrocarbons	Wets, swelling occurs in boiling solvent
Fluoro-naphthalene	Insoluble at boiling point, some swelling
Fluoronitrobenzene	Insoluble at boiling point, some swelling
Pentachlorobenzamide	Insoluble
Perfluoroxylene	Insoluble at boiling point, slight swelling
Tetrabromoethane	Insoluble at boiling point
Tetrachloroethylene	Wets, some swelling after 2 hr at 120°C
Trichloroacetic Acid	Insoluble at boiling point
Trichloroethylene	Insoluble at boiling point after 1 hr.

Table 13.4. Chemical Compatibility of PTFE with Various Chemicals[3]

Chemical	Effect on PTFE Sample
Abietic Acid	Insoluble at boiling point
Acetic Acid	Wets
Acetophenone	Insoluble – 0.2% weight gain after 24 hr. at 150°C
Acrylic Anhydride	No effect at room temperature
Allyl Acetate	No effect at room temperature
Allyl Methacrylate	No effect at room temperature
Aluminum Chloride	Insoluble in solution with NaCl; 1-5% anhydrous $AlCl_3$ affects mechanical properties
Ammonium Chloride	Insoluble at boiling point
Aniline	Insoluble – 0.3% weight gain after 24 hr. at 150°C
Borax	No wetting or effect by 5% solution
Boric Acid	Insoluble at boiling point
Butyl Acetate	Insoluble at boiling point
Butyl Methacrylate	No effect at room temperature
Calcium Chloride	No effect by saturated solution in methanol
Carbon Disulfide	Insoluble at boiling point
Cetane	Wets, insoluble at boiling point
Chromic Acid	Insoluble at boiling point
Cyclohexanone	No effect observed
Dibutyl Phthalate	Wets, no effect at 250°C
Diethyl Carbonate	No effect at the room temperature
Dimethyl Ether	No effect observed
Dimethyl Formamide	No effect observed
Ethyl Ether	Wets
Ethylene Glycol	Insoluble at boiling point
Ferric Chloride	1-5% $FeCl_3.6H_2O$ reduces mechanical properties
Ferric Phosphate	No effect by 5% solution
Formaldehyde	Insoluble at boiling point after 2 hr.
Formic Acid	Insoluble at boiling point
Hexane	Wets
Hydrogen Fluoride	Wets, no effect by 100% HF at the room temperature
Lead	No effect
Magnesium Chloride	Insoluble at boiling point
Mercury	Insoluble at boiling point
Methacrylic Acid	No effect at the room temperature
Methanol	Wets
Methyl Methacrylate	Wets above melting point
Naphthalene	No effect
Nitrobenzene	No effect
2-Nitro-Butanol	No effect
Nitromethane	No effect
2-Nitro-2-Methyl Propanol	No effect
n-Octadecyl Alcohol	Wets
Phenol	Insoluble at boiling point
Phthalic Acid	Wets

13.2.1 Effect of Ozone with Polytetrafluoroethylene

Ozone is considered a reactive substance against plastics due to its ability to readily degrade into an atom and a molecule of oxygen. The atomic oxygen is highly reactive, due to its unpaired electrons in its last orbital, allowing it to attack and etch most polymers. Polytetrafluoroethylene has been reported to be very resistant to etching by ozone in low earth orbit environment, where atomic oxygen is the most abundant species.[4][5]

$$O_3 \rightarrow O_2 + O$$

Resistance of polymers to ozone attack is studied in space environments in "actual" applications. In the laboratory, glow discharge or plasma etching is the common method for laboratory study of ozone effect. Plasma and low earth orbit environments are not equivalent. For instance, oxygen plasma contains a variety of other particles including electrons and free radicals in addition to atomic oxygen. In contrast, atomic oxygen is the dominant constituent of low earth orbit.

In general, oxygen uptake was least for PTFE and most for polyethylene in experiments in which a series of fluorinated polyolefins were exposed to ozone "out of glow." This means that plasma or glow discharge was used to produce atomic oxygen that etched the sample placed outside the discharge zone. The results of "out of glow" plasma etching and low earth orbit, by and large, are in agreement.[6]

Maximum oxygen uptake decreases with an increase in the fluorine content of the polymers. For example, polyvinylidene fluoride took up less oxygen as a result of O_3 exposure than polyethylene, though more than PTFE, in the same experiment.[6] The exception to this trend is polyvinyl fluoride, which has a higher etch rate than polyethylene (Table 13.5). Why is PVF more susceptible to ozone attack than polyethylene?

Golub[7] has proposed an explanation attributing the high etch rate of PVF to the ease of fluorine formation from the decomposition of this polymer. Fluorine promotes degradation of molecular oxygen to its active atomic form ($O_2 \rightarrow 2O$) or by further reaction with PVF, thus enhancing the etch rate. Polyvinylidene Fluoride and FEP molecules have lower etch rates due to the higher stability of CF_2 group to oxygen attack than CFH in PVF or CH_2 in polyethylene. Both polyvinylidene Fluoride and copolymers of tetrafluoroethylene and ethylene (ETFE, Tefzel®) have low etch rates due to the protection provided to the CH_2 groups bonded to CF_2 groups. The ultimate stability is reached in the linear PTFE chain, which consists of all CF_2 groups, with the exception of the few end groups.

Table 13.5. Relative Mass Loss Rates for Polymer Films Exposed to Low Earth Orbit [6]

Polymer	Mass Loss Rate	Fluorine to Carbon Ratio
Polyimide (Kapton®)	1	0
Polyetherphthalate (Mylar®)	1.06	0
Polyethylene (low density)	0.43	0
Polyethylene (high density)	0.80	0
Polyvinyl Fluoride (Tedlar®)	1.16	0.5
Polyvinylidene Fluoride (Kynar®)	0.25	1
PTFE (Teflon®)	<0.03	2
Tetrafluoroethylene/ Hexafluoropropylene (Teflon® FEP)	<0.03	2

13.2.2 Oxygen Compatibility of PTFE

Oxygen is singled out due to its propensity to facilitate auto-ignition of organic material including plastics. Fluoropolymers are extensively used for oxygen services because of their low flammability. Oxygen does not interact with polytetrafluoroethylene under most circumstances.

Limiting oxygen index (LOI) of PTFE is greater than 95% under ambient conditions. This means that PTFE does not burn without an ignition source in an atmosphere containing less than 95% oxygen. LOI is not a complete predictor of all practical conditions in which oxygen and PTFE may interact. A number of considerations apply.

Increased pressure and temperature can accelerate auto-ignition of PTFE, that is, ignition can occur without initiation by an external source. Combinations of high temperatures and high flow rates through small orifices can initiate reaction with PTFE. Organic impurities such as grease, oil and dirt react violently with oxygen and generate heat, which can be sufficient for igniting PTFE. Surface is the most common location of contamination. Before use, all surfaces that come in contact with oxygen must be meticulously cleaned to assure that they are devoid of any organic contaminants.

Surface condition can affect the safety of liquid oxygen contact with PTFE. Despite its extremely low temperature, liquid oxygen can react with organic contamination and ignite PTFE. Thorough cleaning of the surfaces of all parts that come in contact with oxygen is required. ASTM Method D2512 is used to determine the compatibility of materials with liquid oxygen. This method determines the relative sensitivity of materials under impact conditions, by using an impact tester developed at the Army Ballistic Missile Agency. Materials that are sensitive to impact energy are usually likely to react with oxygen when subjected to other forms of energy such as heat.

13.3 Chemical Compatibility of Polychlorotrifluoroethylene

Polychlorotrifluoroethylene (PCTFE) resists attack by most industrial chemicals. The exceptions include alkali metal complexes and organic amines. Chlorine gas, nitrogen tetraoxide, and a number of halogenated solvents are absorbed by PCTFE. Most silicones induce stress cracking. Tables 13.6-13.7 summarize weight gain of low and high molecular weight grades after exposure to each chemical for a period of two weeks at the room temperature.

13.4 Permeation Fundamentals

Permeation can be defined as the passage of gases and liquids through a second material such as a solid. It is a significant consideration in the selection of plastic material for the construction of chemical processing equipment because process fluids may travel across the thickness of the polymer by permeation. Permeated species in sufficient quantities could cause one or more of corrosion, contamination and unacceptable environmental emission.

In its simplest form, permeation can be expressed as a product of the solubility and diffusion coefficient of the permeant in the polymer. Permeation of a gas can be calculated from Eq. 13.1. This equation is derived from Fick's First Law of mass transfer. Permeation concerns the movement of a species through the molecules of another species, e.g., a gas through a polymer. It does not take into account transport of material through cracks, voids and in general physical flaws in the structure of the second species such as the polymer. To be sure, both phenomenon result in the migration of chemicals through the structure. This means that after an appropriate plastic material has been selected to meet the permeation requirements of a process, the equipment must be fabricated carefully to avoid flaws in the polymer structure.

Eq. (13.1) $\quad P = D \cdot S$

P (cm^3/sec·cm·atm) is the permeability of the gas, D is the diffusion coefficient (cm^3/sec) and S (cm^3/cm^3·atm) is the solubility coefficient.

Table 13.6. Weight Gain of PCTFE after Two Weeks to Inorganic Reagents at the Ambient Temperature[8]

Reagent	Low Molecular Weight PCTFE, SG = 2.08	High Molecular Weight PCTFE, SG = 2.12
Ammonium Hydroxide	None	None
Aqua Regia	0.10	0.04
Bromine	0.15	0.1
Carbon Disulfide	0.4	0.2
Hydrochloric Acid (10%)	None	None
Hydrochloric Acid (36%)	None	None
Hydrofluoric Acid (60%)	None	None
Hydrogen Peroxide (30%)	0.23	None
Nitric Acid (10%)	None	None
Nitric Acid (70%)	None	None
Nitric Acid (Red Fuming))	0.07	0.04
Nitric Acid/Hydrofluoric Acid(50:50) (60%)	None	None
Sodium Hydroxide (50%)	None	None
Sodium Hypochlorite	None	None
Sulfuric Acid (30%)	None	None
Sulfuric Acid (Fuming 20%)	0.03	0.02

Table 13.7. Weight Gain of PCTFE after Two Weeks to Organic Chemicals at the Ambient Temperature[8]

Reagent	Low Molecular Weight PCTFE, SG = 2.08	High Molecular Weight PCTFE, SG = 2.12
Acetic Acid (3%)	None	None
Acetic Acid (Glacial)	0.09	0.03
Acetone	5.17	0.5
Acetophenone	None	None
Aniline	0.01	None
Benzaldhyde	0.02	None
Benzene	2.4	0.6
Benzoyl Chloride	0.14	None
Butyl Alcohol	-	None
Carbon Tetrachloride	4.1	1.6
Citric Acid (3%)	None	None
Cyclohexanone	0.35	None
1,2-Dichloroethane	0.11	0.03
2,4-Dichlorotoluene	0.15	0.06
Diethyl Phthalate	None	None
Dimethylhydrazine (Anhydrous)	3.9	1.8
Ethyl Acetate	7.65	6.0
Ethyl Alcohol (Anhydrous)	None	None
Ethyl Ether	5.6	5.2
Ethylene Oxide	5.8	4.0
Formic Acid	None	None
Furan (B.P. 31-32°C)	5.4	3.7
Premium Gasoline	0.83	0.2
Heptane	None	None
Hexachloroacetone	None	None
Hydraulic Fluid	None	None
Lactic Acid	None	None
Methanol	0.10	None
Methyl Ethyl Ketone	5.9	1.2
Motor Oil	0.01	0.01
2,4-Pentanedione	0.17	0.20
Pyridine	0.55	0.10
Toluene Diisocyanate	0.44	-
1,1,2-Trichloroethane	0.04	0.02
Trichloroethylene	10.9	7.8

Several factors affect the permeation rate of the polymer. Temperature increase raises the permeation rate for two reasons. First, solubility of the permeant increases in the polymer at higher temperatures. Second, polymer chain movements are more abundant which allow easier diffusion of the permeant. Permeation rate of gases increases at higher partial pressures. For liquids, permeation rates rise with an increase in the concentration of the permeant. Unless the permeant species are highly soluble in the polymer, permeation rate increases linearly with pressure, concentration and the area of permeation.

Permeation rate decreases at higher thickness, as illustrated in Fig. 13.1 for water transmission rate through a perfluorinated ethylene propylene copolymer (FEP). Effect of thickness is usually nonlinear. Permeation rate is very high at a low thickness and rapidly decreases with an increase in the in the thickness. After a critical thickness is reached, the effect of thickness is diminished and permeation rate reaches a plateau. At lower thicknesses, the effect of surface structure begins to play a significant role in the permeation. A more oriented (ordered) surface will serve to inhibit permeation.

Chemical and physical characteristics of the polymer have powerful effects of the rate of permeation, as much as four orders of magnitude.[10] Chemical affinity for the permeant, intermolecular forces such as van der Waals and hydrogen bonding forces, degree of crystallinity, and degree of cross-linking are the influential variables.

A similarity of chemical functional structures of the polymer and the permeant will promote its solubility and permeation rate. Higher intermolecular forces of the polymer result in less permeation because of the resistance that they present to the development of space between adjacent molecules required for the passage of the permeant. Crystallinity is an important factor, which can be controlled during the processing of the polymer. Crystalline phase can be considered impermeable by most species because of its orderly structure (packing) which usually minimizes its specific volume. This means that there is little or no free space among the polymer chains for the passage of permeant. Amorphous phase has the opposite construction and is disorderly with interchain space available for permeation. The specific volumes of the crystalline (0.43 cm^3/g) and amorphous (0.5 cm^3/g) phases provide evidence for the argument. Amorphous phase has a 13% higher specific volume, which translates in additional space for permeation. Finally, cross-linking acts somewhat similarly to crystallinity, though less effectively, to limit space for permeation. Cross-linking is size-dependent and smaller species may permeate.

The molecular size of the permeant, its chemical structure and condensation characteristics affect permeation. Diffusion of the permeant increases as its molecular size decreases, thus, contributing to an increase in permeation. Molecular structure is important. A polar chemical will normally have a lower permeation rate in a nonpolar polymer than a nonpolar species and vice versa. This is due to the ability of chemicals with similar structures to the polymer to swell the polymer, that is, to create space between the chains for permeation. A more easily condensed chemical will also be more effective in swelling of the polymer, resulting in higher rates of permeation.

13.5 Permeation Measurement and Data

A number of methods can be used to measure permeation rate through polymers including fluoropolymers. These methods are helpful for comparison of different materials. The extent of the information obtained is limited due to the inability of these techniques to account for real-world conditions. Typically, a film of fluoropolymer acts as a barrier to keep a gas or liquid in a reservoir. Figure 13.2 shows the schematic of a device for measuring liquid and vapor permeation. Permeation rate is calculated from the measured pressure or weight loss in the reservoir. Examples of techniques include ASTM Methods D813 and F-739-81. There are more complex means of measuring permeation, which approach the actual applications conditions. One example is a controlled recirculation of a fluid through a closed loop system, which contains commercially manufactured parts, made with the fluoropolymers, such as lined components. In these systems gas chromatography and mass spectroscopy are used to analyze the permeation.

A comparison of moisture vapor permeation through various polymers can be seen in Table 13.8. Notice that PCTFE is only second to FEP and both are among the most resistant plastics to water vapor permeation. Permeation data for gases and liquids through PTFE are presented in Figs. 13.3 and 13.4. Permeability data for PTFE can be found in Appendix I. Permeability and physical property data for films of PTFE can be found in Appendix II.

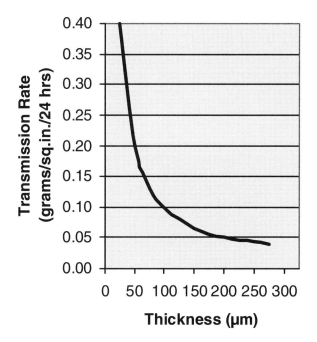

Figure 13.1 Water vapor transmission rate of Teflon® FEP resins at 40°C. Note: Values are averages only and not for specification purposes.

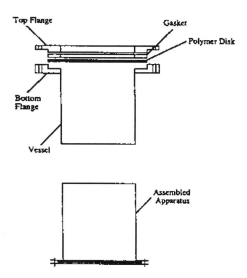

Figure 13.2. Schematic of a device for measuring liquid and vapor permeation.

Table 13.8. Permeation (Transmission) of Water and Gases Through Polymer Films[8]

	Water Vapor Transmission gm-mil/100 in²/24 hrs. @ 100°F @ 90% RH (gm-mm/m²/24 hrs. @ 37.8°C @ 90% RH)	Gas Transmission cc (STP) mil/100 in²/24 hr-ATM @ 77°F (cc [STP]-mm/m²/24 hr-ATM @ 25°C)		
		O_2	N_2	CO_2
ACLAR UltRx, SupRx, Rx	0.016 (0.006)	7 (2.8)	1 (0.4)	14 (5.5)
ACLAR 33C	0.020 (0.008)	7 (2.8)	1 (0.4)	16 (6.3)
ACLAR 22C	0.026 (0.010)	15 (5.9)	2.5 (1.0)	40 (15.7)
ACLAR 22A	0.027 (0.011)	12 (4.7)	2.5 (1.0)	30 (12.0)
PVC, PVdC Copolymer	0.20–0.6 (0.08–0.24)	0.8–6.9 (0.3–2.7)	0.12–1.5 (0.05–0.6)	38–44 (15–17)
Polyethylene Low Density	1.0–1.5 (0.39–0.59)	500 (195)	180 (71)	2700 (1060)
Medium Density	0.7 (0.28)	250–535 (100-210)	85–315 (35–125)	100–2500 (40–985)
High Density	0.3 (0.12)	185 (73)	42 (17)	580 (230)
CAPRAN® 77C (Nylon 6)	19–20 (7.5–7.9)	2.6 (1.0)	0.9 (0.35)	4.7 (1.9)
Fluorinated Ethylene Propylene	0.22 (0.08)	715 (281)	320 (125)	1670 (660)
Polyvinyl Fluoride	2.1 (0.81)	7.5 (3)	0.25 (0.10)	11 (4.3)
Polyvinylidene Fluoride	2.5 (1.0)	3.4 (1.34)	9 (3.5)	5.5 (2.2)
Polyester–PET Oriented	1.0–1.3 (0.39–0.51)	3.0–6.0 (1.2–2.4)	0.7–1.0 (0.28–0.39)	15–25 (5.9–9.8)

CAPRAN® is a register trademark of AlliedSignal Inc.

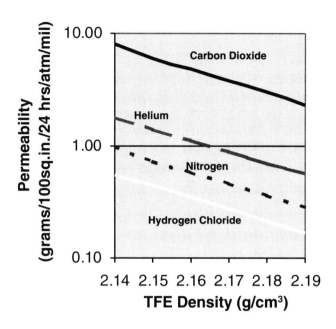

Figure 13.3 Effect of density of Teflon® FEP resins on their permeability to gases at 30°C.

13.6 Environmental Stress Cracking

A weakness of many polymers is their tendency to fail at fairly low stress levels due to the impact of some hostile environments. This phenomenon is known as *environmental stress cracking*. Cracking occurs when the polymer is stressed for a long time under loads that are relatively small compared to the yield point of the material. Well-known examples include failure of vulcanized natural rubber in the presence of ozone. It reacts with unsaturated hydrocarbons at the surface and even when the elastomers are subjected to low stresses cracks can develop and lead to failure. Another example is stress cracking of polyolefins such as high density polyethylene in the presence of surfactants. When polyethylene is held under stress in the presence of some detergents, its behavior changes from short time ductile failure at high stresses to brittle fracture at low stresses after relatively longer time periods with very small break elongations.[11].

Even though environmental stress cracking must be considered in designing parts from fluoropolymers, it is not considered an extensive problem for this family of plastics. Permeation variables have strong influence on stress cracking. Different fluoropolymers differ in their propensity to environmental stress

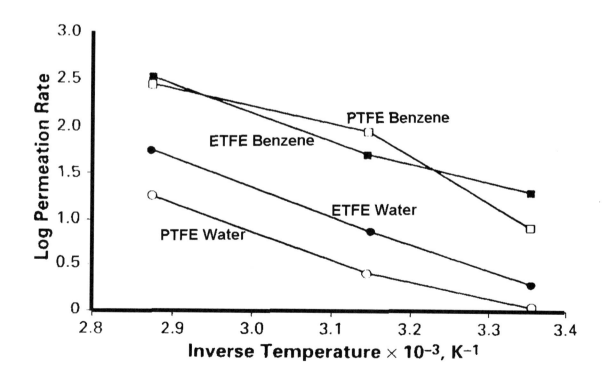

Figure 13.4 Permeability through PTFE and ETFE.[9]

cracking, primarily based on their degree of crystallinity. Crystallinity can be lowered by adding a comonomer and varying its concentration. Higher comonomer content decreases crystalline phase content of the polymer. Resin processing can affect crystallinity. Fast cooling (or quenching) at the end of the fabrication process serves to reduce crystalline content and increase amorphous content. Lowering the crystalline phase content of the part tends to increase resistance to stress cracking due to the increaseing break elongation. Increasing the molecular weight of the polymer reduces its crystallinity and enhances its stress crack resistance. Longer chains have higher tensile strength, i.e., load-bearing ability.

Chemicals with similar structure to the polymer tend to permeate and plasticize, thus, reducing its mechanical strength. Fluoropolymers can be permeated by small halogenated molecules, usually containing chlorine and fluorine due to their similarity of structures. Environmental stress cracking effect of chemicals on polymers can be measured by exposing the polymer to the chemical under the desired conditions. Tensile properties of the exposed sample can then be measured. Any loss of elongation and tensile strength would indicate environmental stress cracking.

Processing of the fluoropolymers plays an important role in minimizing the tendency to undergo stress crack resistance. Reducing crystallinity as much as possible and minimizing residual stress lower the tendency of a part to experience environmental stress cracking. These objectives can be achieved by reducing processing times and cool-down rates, and annealing.

Another issue is the stress that a part experiences in application. Tensile loads cause less tendency for stress cracking than compressive loads. Fluoropolymer-lined equipment and parts are examples of parts, which may contain residual stress left due to their design and or fabrication.

References

1. "Teflon® PTFE Fluoropolymer Resin," *Properties Handbook,* DuPont Co., July, 1996.
2. "Teflon® Fluorocarbon Resin," Performance Guide for the Chemical Processing Industry, DuPont, Technical Bulletin No. E-21623-3.
3. "Exploring the Chemical Resistance of Teflon® Resins," *The Journal of Teflon®,* 3(9), Nov./Dec., 1962.
4. Leger, L. J., and Visentine, J. T., *J. Spacecraft Rockets,* 23:505, 1986.
5. Leger, L. J., Visentine, T., and Santos-Mason, B., *International SAMPE Technical Conference,* 18:015, 1986.
6. Golub, M. A., "Reactions of Atomic Oxygen [O(^{3}P)] with Polymer Films," ol. Symp. 53:379–391, Huthig & Wepf Verlag. Basel, 1992.
7. Golub, M. A., and Wydeven, T., *Polymer Degrad. Stab.,* 22:325, 1988.
8. "ACLAR® Barrier Films," Technical Bulletin SFI-14, Allied Signal Advanced Materials, Morristown, New Jersey.
9. "Teflon® Fluorocarbon Resins in Chemical Service," *Journal of Teflon®,* Pub. by DuPont Co., Jan.–Feb., 1970.
10. Imbalzano, J. F., Washburn, D. N., and Mehta, P. M., *Basics of Permeation and Environmental Stress Cracking in Relation to Fluoropolymers,* Technical Information, DuPont Co., No. H-24240–1, Nov., 1993.
11. Young, R. J., and Lovell, P. A., *Introduction to Polymers,* Chapman & Hall, London, 2nd Ed, 1991.
12. Buxton, L. W., Goldsberry, D. R., and Henthorn, G. V., *Fluoropolymer-Lined Chemical Systems and Permeation,* DuPont Fluoroproducts Technical Information Bulletin No. H-54485, Sep., 1993.

14 Properties of Tetrafluoroethylene Homopolymers

14.1 Introduction

Carbon and fluorine are the major elements that form the tetrafluoroethylene homopolymers. They are the main reason that these plastics have many special properties which surpass those of most other polymers. These properties span mechanical, electrical, and thermal characteristics of the polymers in addition to chemical properties. This chapter concentrates on presenting the key properties of polytetrafluoroethylene (PTFE).

14.2 Influence of Processing

Polytetrafluoroethylene resins are semicrystalline and highly stable at temperatures up to 400°C. The degree of crystallinity in a fabricated part is dependent on processing conditions. Different processing conditions such as the length of exposure to sintering temperatures affect some of the properties of the resin. A group of PTFE properties remains relatively independent of the fabrication conditions (Table 14.1).

Table 14.1. Properties Unaffected by Fabrication Conditions[1]

Mechanical	Electrical	Chemical
1. Flexibility at low temperatures	1. Low dielectric constant	1. Insolubility
2. Low coefficient of friction	2. Low dissipation factor	2. Chemical resistance
3. Stability at high temperatures	3. High arc resistance	3. Weatherability
	4. High volume resistivity	4. Low energy surface
	5. High surface resistivity	5. Nonflammability

Three steps are common to fabrication conditions of PTFE. They include preforming, sintering, and cooling. All three steps can influence properties of the part such as flex life, permeability, stiffness, resiliency, and impact strength.

Process conditions affect five different variables, which influence the functional properties of any part.

1. Macroscopic flaws such as impurities, cracks, charge-to-charge adhesion, and bubbles.

2. Microscopic flaws or *microporosity* are voids which are created by poor coalescence of particles.

3. Crystallinity is a measure of the orderly packing of the molecules.

4. Molecular weight is an indication of the size of the polymer molecules.

5. Degree of orientation which is an indicator of the alignment of polymer molecules in a specific direction.

These five factors are difficult to measure and control routinely due to a need for special equipment and procedures. A number of indirect properties have been defined to measure the impact of the five parameters. They include *specific gravity, tensile strength, break elongation, dielectric strength* and *heat of fusion*. These properties are measured by relatively simple and quick methods.

Dielectric strength is a function of microporosity but not molecular weight or crystallinity. The number and size of the microvoids affect the dielectric strength. Table 14.2 shows the relationship of dielectric to microporosity.

Specific gravity (SG) is easily measured by gradient-tube or water displacement as described in ASTM methods D1505 or D792. The value of specific gravity of a sample is affected by microvoids. It is important to correct the measured value of specific gravity for the void content. The following formula (Eq. 14.1) can be used to calculate the inherent specific gravity of a part:

Eq. (14.1)

$$\text{Inherent SG} = \frac{\text{Measured SG}}{1 - 0.01 \times \% \text{ void content}}$$

Inherent specific gravity of a PTFE part free of voids can be calculated from Eq. 14.2 as a function of crystalline content.

Eq. (14.2)

$$\text{Inherent SG} = \frac{[2.302 \times \% \text{ Crystalline} + 2.00 \times \% \text{ Amorphous}]}{100}$$

Eq. (14.3)   % Amorphous = 100 - % Crystalline

After substituting from Eq. (14.3) into (14.2), the relationship between crystallinity and inherent specific gravity is reached.

Eq. (14.4)

$$\text{Inherent SG} = \frac{[2.302 \times \% \text{ Crystalline} + 2.00 \times (100 - \% \text{ Crystalline})]}{100}$$

Table 14.2. PTFE Granular Resin: Relation of Dielectric Strength to Degree of Microporosity[3]

Sample	Appearance of Cross Section in Microscope	Dielectric Strength, kV/mm
A	No visible voids at 100× magnification	30.4
B	Scattered 25 µm voids between particles	23
C	Scattered 125 µm voids	17.8
D	Numerous 125 µm voids	10

Figure 14.1 graphically illustrates the relationship between inherent specific gravity and crystallinity of a part. Equations 14.1 and 14.4 can be used to calculate specific gravity of a part at different void contents, as shown in the Fig. 14.1.

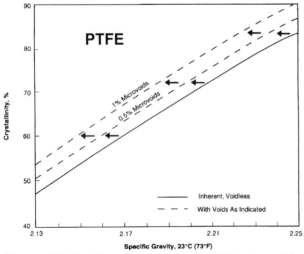

Figure 14.1 Relation of percent crystallinity to specific gravity.[3]

Break elongation and tensile strength are affected by all of the above five factors. Tensile properties are reduced due to microvoids. An increase in crystallinity results in a reduction in the tensile strength. Elongation rises with crystallinity up to a point and then falls. The impact of microvoids is particularly significant on high tensile strength (i.e., low crystallinity) samples. Table 14.3 shows the effect of microporosity on tensile strength and break elongation at four levels.

Tensile properties are also influenced by the degree of orientation. Tensile strength is greater in the oriented direction, but break elongation is lower than the unoriented specimen.

Table 14.3. PTFE Granular Resin: Effect of Microporosity on Tensile Strength and Elongation[3]

Sample*	Extent of Microporosity	Tensile Strength, Mpa	Ultimate Elongation, %
A	Negligible	24.8	390
B	Slight	17.4	350
C	Moderate	13.9	300
D	Severe	12.4	170

*Free-cooled 1/16 in specimens with relative crystallinity of 65–68% tested by ASTM D4894/4895.

14.2.1 Measurement of Flaws

It is clear that *macroscopic* and *microscopic* flaws affect the properties of PTFE parts. It is important to characterize their impact quantitatively. These flaws can be directly measured by a number of techniques. The measurement methods are complex and not suitable for routine quality control. Macroscopic flaws are frequently detectable visually or by a simple magnifying glass. Internal defects can be detected by x-ray techniques such as radiography. ASTM method E94 can be applied as a guideline in establishing procedures.

Excessive microporosity can be detected visually by a trained observer. Other detection methods do not depend on the experience of the observer. The skills needed for the visual technique are difficult to describe due to its heavy dependence on experience. A few steps can be taken to increase the likelihood of successful inspection of parts for microporosity. The best approach is to compare a series of samples, which have been processed by the same sintering and cooling cycles. These samples will have the same inherent specific gravity. A thin section should be obtained due to the ease of light passage. A powerful transmitted light source should be employed for the observation of the voids.

Another way to detect the microvoids is by dye penetration. In this technique, bright color dyes are placed on the surface of the PTFE part (thickness <1.5 cm) being tested. An "acceptable" sample is compared with the one in question in this method. A porous sample will have an appearance of fluorescent spots similar to stars in the sky. ARDROX P135E is one of the common fluorescent dyes and can be purchased from ARDROX, Inc., La Mirada, California. Another dye is formulated by mixing 1 g "Calco" Red A-1700 dye (supplied by Cyanamid, Division of American Home Products), 100 ml of reagent grade Turpentine and 5 g $CaCl_2$.

A quantitative way of measuring microporosity is by an indirect technique using specific gravity and crystalline content. Equation 14.1 can be rearranged to obtain Eq. 14.5. Inherent SG can be calculated from Eq. 14.4, provided that crystalline content of the sample could be measured. A few techniques are available for the measurement of crystallinity. They include infrared spectroscopy,[4] torsional damping,[5] x-ray diffraction,[6] and ultrasonics.[7]

Eq. (14.5) % Void content
$$= \frac{100 \times (\text{Inherent SG} - \text{Measured SG})}{\text{Inherent SG}}$$

Infrared spectroscopy and torsional dampening are the most reproducible techniques for the measurement of crystallinity. The former is most flexible in accommodating different sample shapes. Averages of duplicate measurements yield the best results. Orientation introduces significant errors into these methods, thus, prohibiting their application for paste extruded samples. However, void content of most paste extruded PTFE is usually low.

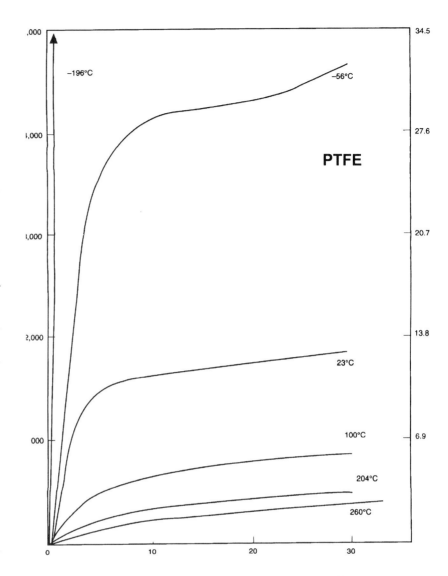

Figure 14.2 Tensile stress, based on original cross section.[3]

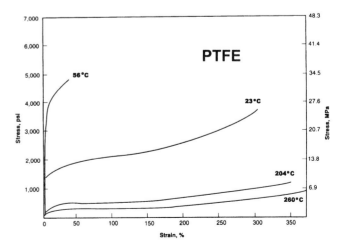

Figure 14.3 Stress vs. strain in tension.[3]

14.3 Mechanical Properties

These properties are among the most important factors for the design of plastic parts and systems. In this section data for stress-strain behavior of PTFE in different modes and conditions are presented. Stress-strain curves for polytetrafluoroethylene are shown in Figs. 14.2 and 14.3 at various temperatures.

Figure 14.2 indicates that yield occurs at fairly high strain levels. The yield deformation decreases as temperature increases. The elastic modulus, defined as the slope of the linear part of the stress-strain curve, increases significantly with decreasing temperature. Figure 14.3 shows the ultimate values of stress before failure by fracture occurs. Tensile strength decreases while break elongation increases with increasing temperature. Figures 14.4 through 14.7 provide stress-strain data for tension, compression and shear modes. Yield stress in compressive mode has a similar value to that in tensile mode.

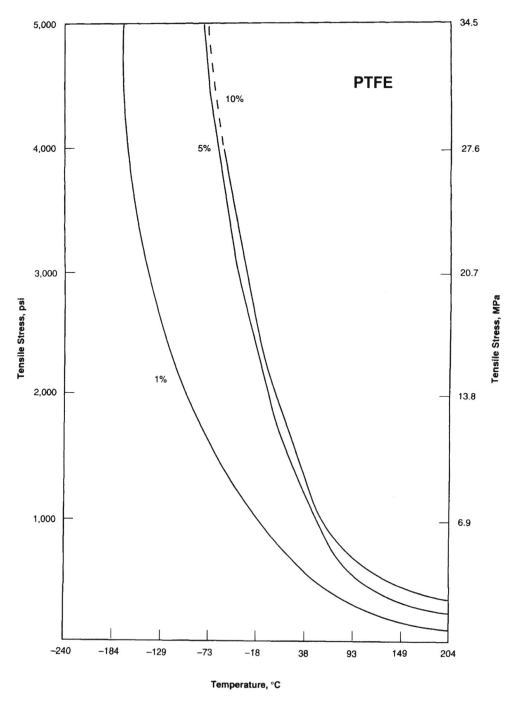

Figure 14.4 Tensile stress vs. temperature at constant strain.[3]

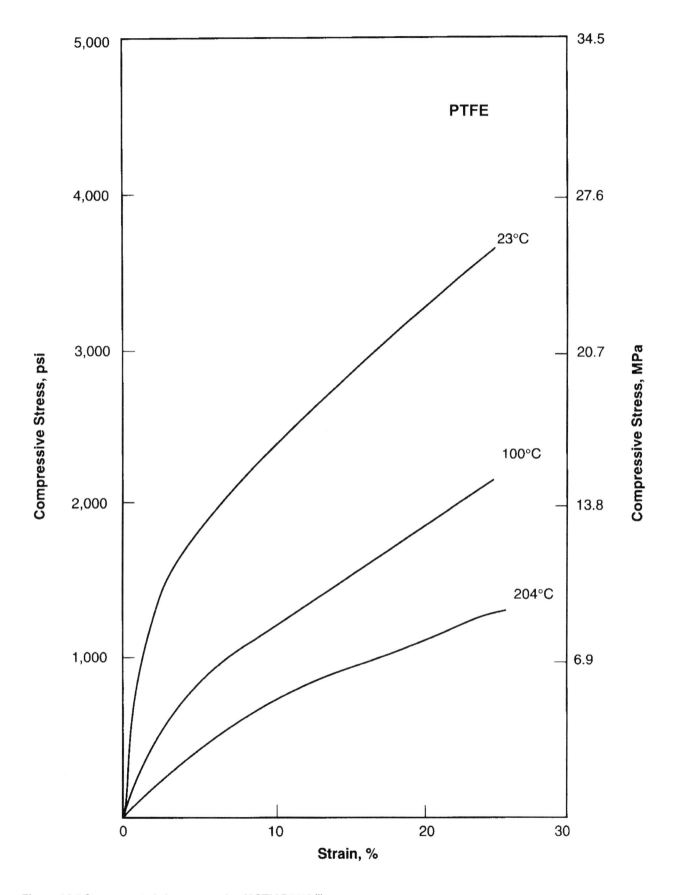

Figure 14.5 Stress vs. strain in compression (ASTM D695).[3]

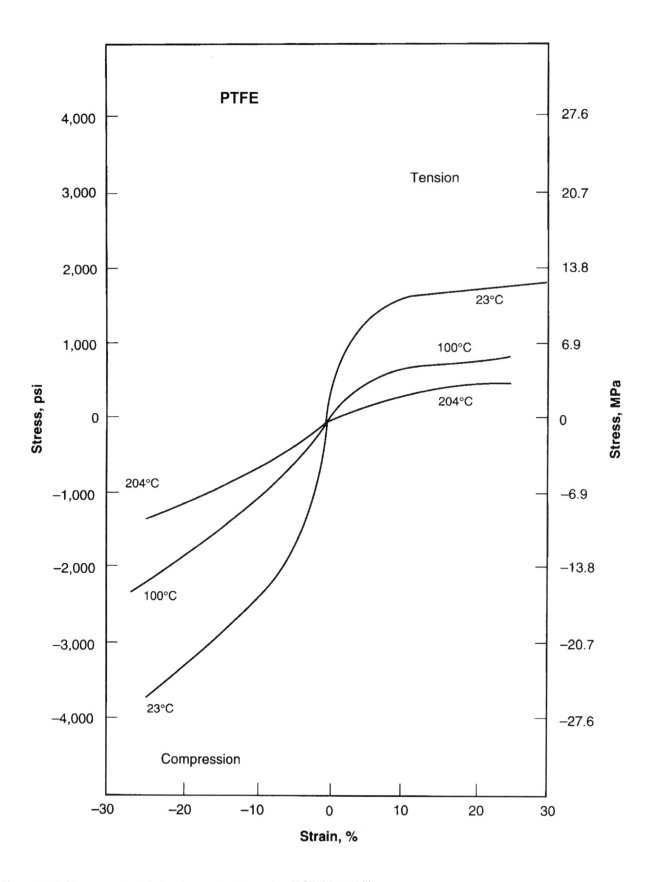

Figure 14.6 Stress vs. strain in tension and compression (ASTM D695).[3]

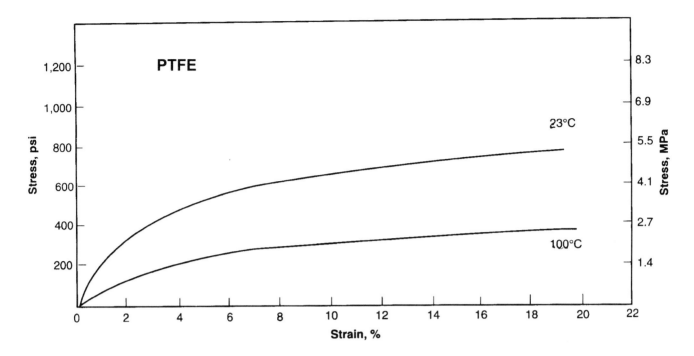

Figure 14.7 Stress vs. strain in shear to 20%.[3]

Poisson's ratio is defined as the ratio of the change in the width per unit width of a material, to the change in its length per unit length, as a result of strain.[8] Poisson's ratio for PTFE is about 0.5 at above the room temperature. At 23°C, it has a value of 0.46.

14.3.1 Deformation Under Load (Creep) and Cold Flow

This property is an important consideration in the design of parts from polytetrafluoroethylene. It deforms substantially over time when it is subjected to load. Metals similarly deform at elevated temperatures. *Creep* is defined as the total deformation under stress after a period of time, beyond the instantaneous deformation upon load application. Significant variables that affect creep are load, time under load and temperature. Creep data under various conditions in tensile, compressive, and torsional modes can be found in Figs. 14.8 through 14.15.

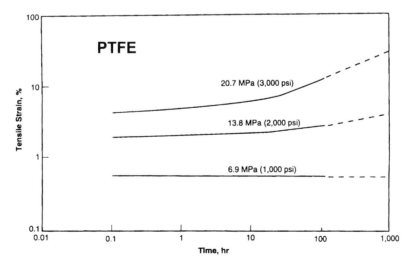

Figure 14.8 Total deformation vs. time under load at -54°C (-65°F).[3]

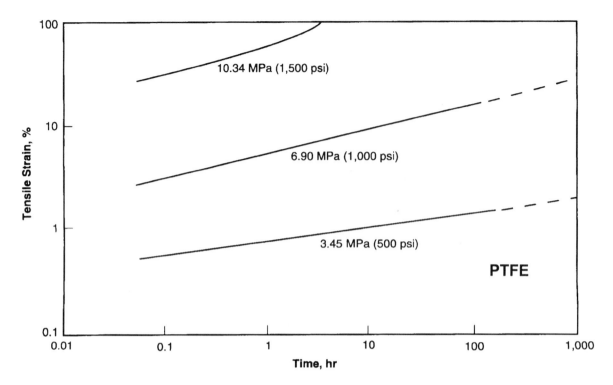

Figure 14.9 Total deformation vs. time under load at 23°C (73°F).[3]

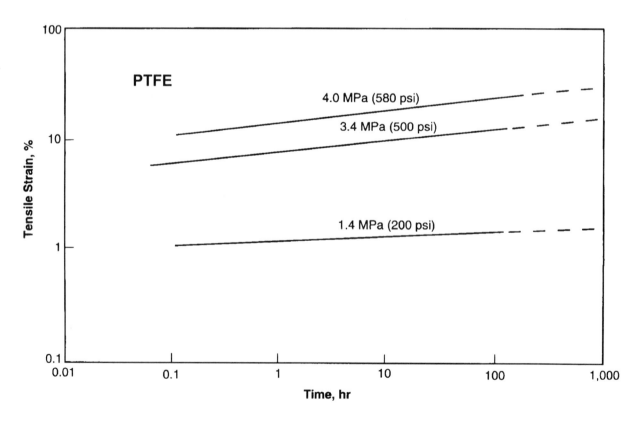

Figure 14.10 Total deformation vs. time under tensile load at 100°C.[3]

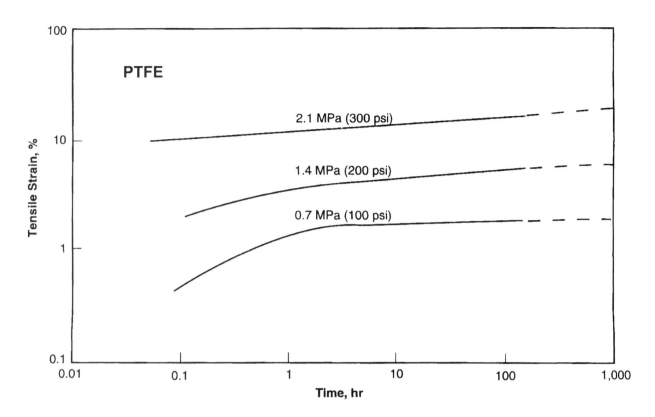

Figure 14.11 Total deformation vs. time under tensile load at 200°C.[3]

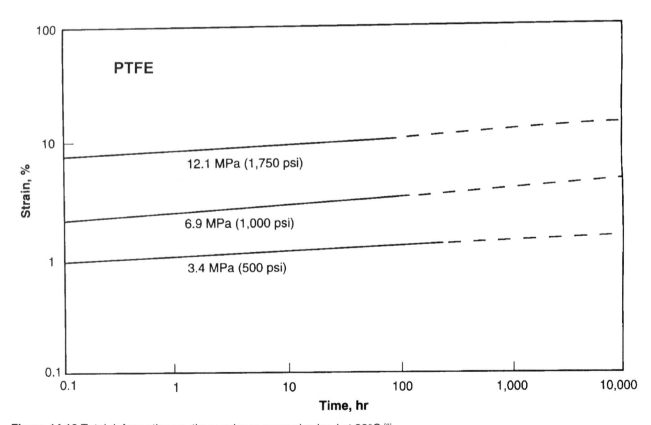

Figure 14.12 Total deformation vs. time under compressive load at 23°C.[3]

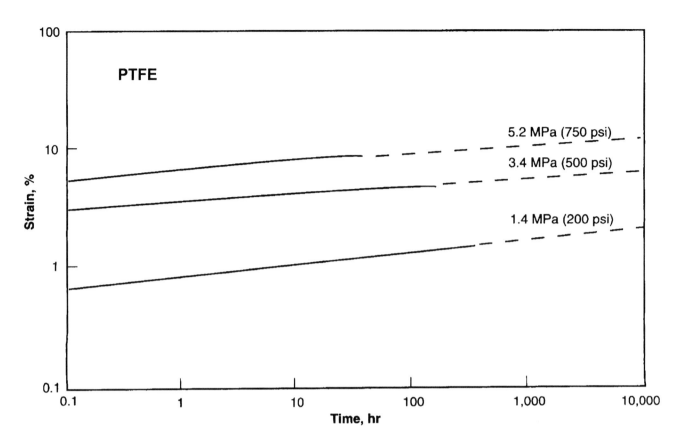

Figure 14.13 Total deformation vs. time under compressive load at 100°C.[3]

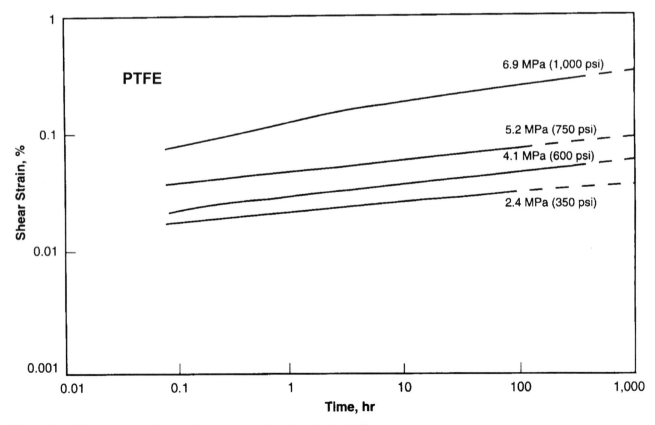

Figure 14.14 Total deformation vs. time under torsional load at 23°C.[3]

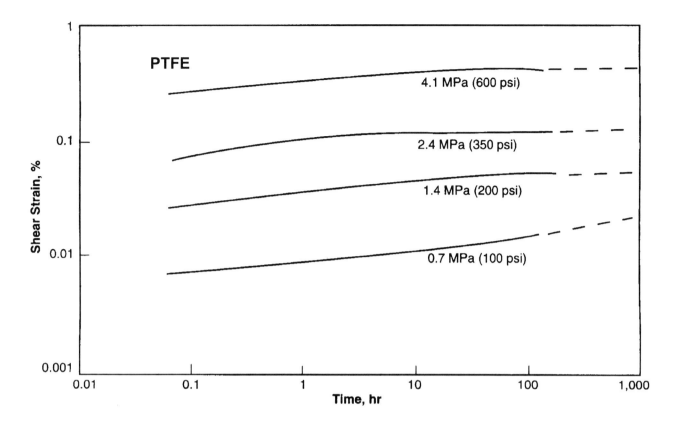

Figure 14.15 Total deformation vs. time under torsional load at 100°C.[3]

Resin manufacturers have long recognized the excessive deformation of polytetrafluoroethylene for applications where parts such as gaskets and seals experience high pressures. Copolymers of tetrafluoroethylene with small amounts of other fluorinated monomer are known as *Modified PTFE* resins and have been reported to exhibit reduced deformation under load. Examples of the properties of some of the commercial products can be seen in Tables 14.4–6 and Figs. 14.16 and 14.17. Significant reduction in deformation under load can be achieved, particularly at elevated temperatures and pressures.

Table 14.4. Deformation-Under-Load Data for Commercial Standard and Modified PTFE Resins[9]

Resin Type	Test	Test Conditions Temperature,°C/Pressure, MPa after 24 hours		
		23°C/3.4 MPa	23°C/6.9 MPa	23°C/14 MPa
Teflon® NXT 70	ASTM D621	0.2	0.4	3.2
Teflon® 7A	ASTM D621	0.7	1.0	8.2

Table 14.5. Deformation under Load Data for Commercial Standard and Modified PTFE Resins[9]

Resin Type	Test	Test Conditions Temperature,°C/Pressure, Mpa after 24 hours		
		25°C/6.9 MPa	100°C/3.4 MPa	200°C/1.4 MPa
Teflon® NXT 70	Dynamic Mechanical Analyzer (DMA)	5.3	5.4	3.6
Teflon® 7A	Dynamic Mechanical Analyzer (DMA)	6.7	8.5	6.4

Table 14.6. Deformation under Load Data for Commercial Standard and Modified PTFE Resins[10]

Resin Type	Test	Test Conditions Temperature,°C/Pressure, MPa after 100 hours		
		23°C/6 MPa	23°C/8 MPa	23°C/14 MPa
Hostaflon®1700	ASTM D621	2	2.5	8
Hostaflon®1750	ASTM D621	3	4	14.5

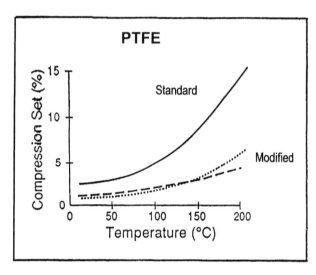

Figure 14.16 Conditions: based on ASTM D621 (load 6.9 MPa).[11]

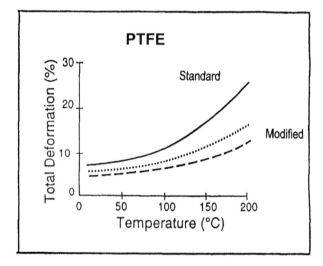

Figure 14.17 Conditions: based on ASTM D621 (load 6.9 MPa).[11]

Stress relaxation is important in applications where PTFE is subjected to a compressive load. For example, a gasket under bolt pressure will creep, resulting in a reduced bolt pressure. This reduction in load may lead to leakage at the joint. Tightening the flange after 24 hours will prevent leakage. Thereafter, stress relaxation will be negligible. Plots of tensile stress relaxation can be used (Figs. 14.18 and 14.19) to illustrate the decay rate at constant strain.

Alternative compressive loading and load removal indicates nearly complete recovery from strain, as long as the original strain is below the yield strain elastic range. PTFE does not experience work hardening.[3]

14.3.2 Fatigue Properties

Flexibility characteristics are of paramount importance in many applications involving motion. Valve diaphragms are a good example of a part where a polymer membrane experiences repeated movement. *Flex life* is defined as the number of cycles that a part can endure before catastrophic fatigue occurs; the higher the molecular weight the higher the flex life. Crystallinity has a detrimental effect on flex life; the higher the crystallinity the lower the flex life. Figure 14.20 illustrates flex life of PTFE as a function of molecular weight and crystallinity. Voids should be eliminated from any part expected to perform in demanding flex applications. Each void becomes the nuclei of failure after the part has been subjected to repeated movement.

14.3.3 Impact Strength

Impact strength of a part depends on its ability to develop an internal force multiplied by the deformation of the part as a result of impact. The shape of a part can enhance its ability to absorb impact such as a metal spring, as opposed to a flat metal plate. PTFE resins have excellent impact strength in a broad temperature range. Table 14.7 summarizes the results of tensile and Izod impact strength for polytetrafluoro-

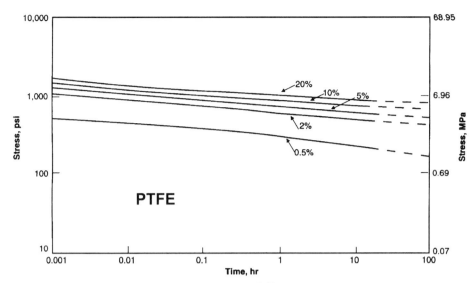

Figure 14.18 Tensile strength relaxation at 23°C.[3]

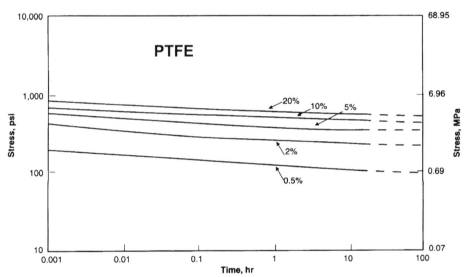

Figure 14.19 Tensile strength relaxation at 100°C.[3]

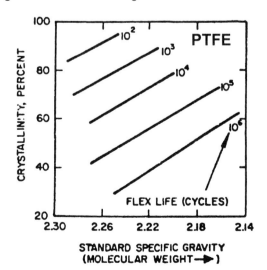

Figure 14.20 How crystallinity and molecular weight affect flex life (nonstandard test, 45 cycles per minute, 180-degree bend).[1]

ethylene (Izod, according to ASTM method D256). Even at extreme cold, PTFE retains a great deal of its impact toughness.

14.3.4 Hardness

Hardness of PTFE is determined by a number of methods, such as ASTM D758 or D2240 (Rockwell R Scale), or by Durameter scales. The numbers reported are Rockwell R Scale of 58, Durameter A Scale of 98[3] and Durameter D Scale of 50–65.[12] Figure 14.21 illustrates the variation of harness of PTFE and nylon as a function of temperatures. Both plastics become softer with increasing temperature but in all temperatures nylon is harder than PTFE.

Fillers improve the hardness of polytetrafluoroethylene by 10–15%, which is preserved over a wide range of temperature. Increasing the filler content, in general, elevates the hardness of the compound. The topic of fillers has been covered in detail in Ch. 12.

14.3.5 Friction

Polytetrafluoroethylene is a slippery material with a smooth surface due to its low coefficient of friction. Numerous mechanical applications have been developed with PTFE with slight or without lubrication, particularly at low velocities and pressures above 35 kPa. Table 14.8 contains values for coefficient of friction as a function of velocity. Dynamic coefficient of friction of PTFE is larger than its static coefficient of friction and grows with increasing speed until the motion is destabilized. Static coefficient of friction remains unchanged in the temperature range of 27–327°C which is important in applications where a polytetrafluoroethylene part may experience heat build up and temperature increase.

Polytetrafluoroethylene's coefficient of friction rises quickly with sliding speed (below 30 m/min, Fig. 14.22), which prevents "slipstick" behavior. No noise takes place even at slow speeds. Above 45 m/min sliding velocity has little effect at combinations of pressure and velocity before the PV limit of PTFE is reached.[3] Figure 14.23 indicates that static coefficient of friction decreases with increase in pressure.

Table 14.7. Tensile and Izod Impact Strength of Polytetrafluoroethylene[3]

Temp., °C	Tensile Impact Strength, ft.lb/in^2	Izod Impact Strength, ft.lb/in^2
23	320	2.9
-54	105	2.3

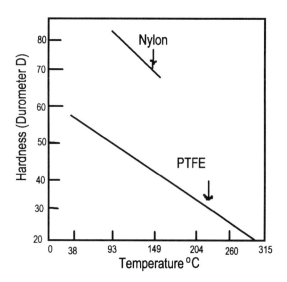

Figure 14.21 Hardness vs. temperature.[1]

Table 14.8. Coefficient of Friction[3]

Type of Coefficient	Condition	Value of Coefficient
Static	3.4 MPa Static Load	0.05-0.08
Dynamic (PV = 285-357 kg/cm^2.cm/sec.)	Velocity (m/min) 3	0.10
	30	0.13
	300	Unstable Operation

14.3.6 PV Limit

The PV convention is utilized to define the maximum combinations of pressure and velocity at which a given material will operate continuously without lubrication. The values are usually given for operation in air at temperatures of 21–27°C. PV limits do not always define the actual combinations of pressure and velocity where the material can be practically used, because wear is not considered in the determination of PV values. In other words, the application must not exceed PV limit and wear limits of a material. Such a limit can be determined by finding the pressure and velocity combination at which wear rate accelerates or exceeds the expected life of a part.

PV limits of polytetrafluoroethylene are given in Table 14.9. All PTFE reaches a PV value of zero at above 288°C, no matter whether the temperature is reached thermally or mechanically. Reducing the temperature below 21°C increases the limiting PV.

14.3.7 Abrasion and Wear

Polytetrafluoroethylene parts have good wear properties, as seen from the data in Table 14.9. The resistance of unfilled PTFE to wear is less than that of filled compositions, which has been discussed in Ch. 12. Data from tests measuring wear rate are presented in Tables 14.10–14.12. They should be viewed with an understanding that none of the techniques represent an actual wear situation. In all three methods a new surface is exposed to abrasion during the repeated motion of the abrading surface.

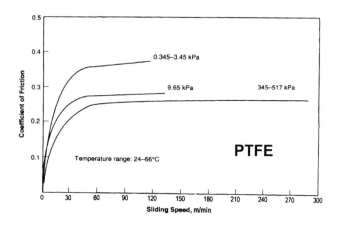

Figure 14.22 Coefficient of friction vs. sliding speed.[3]

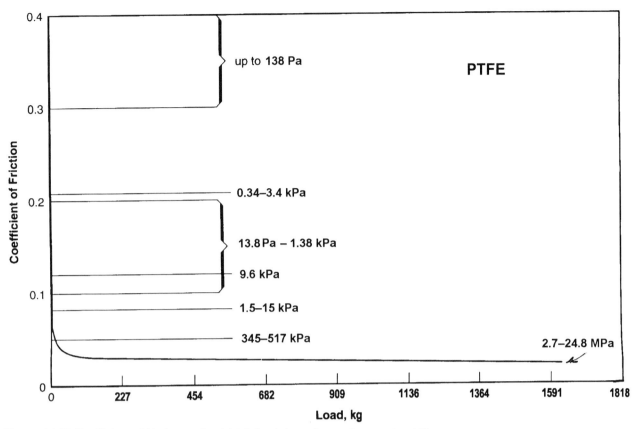

Figure 14.23 Coefficient of friction vs. load (at 0.6 m/min and room temperature).[3]

Table 14.9. PV and Wear Performance[3]

Velocity, m/min	PV Limit at 21-27°C
3	42.9
30	64.3
300	89.3

Table 14.10. Weight Loss Caused by Sliding Tape*[3]

Resin	Average Weight Loss, g/mm^2
PTFE	5.31

*Armstrong Abrasion Test (ASTM D1242): This test measures abrasion resistance of flat surfaces by drawing abrasive tape, under load, over test specimens at a slip rate of 15.75 g/cm^2. With No. 320 abrasive under a 6.8 kg load, weight loss was measured after 200 revolutions (1 hr, 40 min.).

Table 14.11. Weight Lost from Revolving Disk* (Cumulative Weight Loss in Milligrams)[3]

Resin	Test Cycles					
	10	50	100	500	1000	2000
PTFE	0.35	1.65	2.2	5.7	8.9	13.4

*Armstrong Abrasion Test: This test measures abrasion resistance of a flat surface by rotating a 10 cm diameter specimen disk beneath an abrasive under load. A 1,000 g load was used on a Calibrase wheel No. CS-17F.

Table 14.12 Tape Length Required to Abrade Through Wire Coating* (Average Tape Length in Centimeters)[3]

Resin	Heat Aging			
	None	96 hr at 150°C (302°F)	500 hr at 150°C (302°F)	96 hr at 200°C (392°F)
PTFE	191.5	196.6	247	211.7

*Armstrong Abrasion Test (MIL-T-5438): This test measures abrasion resistance of wire coating by drawing, under load, a clean abrasive cloth tape of continuous length across the test wire until the coating is worn through. A 0.45 load on No. 400 grit tape was used on a coating thickness of 0.038 cm.

14.4 Electrical Properties

Electrical stability of polytetrafluoroethylene is outstanding over a wide range of frequency and environmental conditions. This plastic makes an excellent electrical insulator at normal operating temperatures. Dissipation factor and dielectric constant values are virtually constant up to 10 MHz. Dielectric strength of PTFE drops off with increasing frequency slower than most other material.

PTFE dielectric constant and dissipation factors remain constant over a broad temperature range (-40 to 240°C) as seen in Fig. 14.24.[13] They are not affected by exposure to high frequency (>1 MHz) and high temperature. The value of dielectric constant is 2.1 essentially over the entire spectrum of frequency. The dissipation factor of PTFE resins remains <0.0004 up to 100 MHz. It reaches a peak value at 1 GHz. The peak value occurs at higher frequencies with increasing temperature.

The dielectric strength of polytetrafluoroethylene is quite high and remains fairly constant with temperature and heat aging. The short-term dielectric strength decreases slightly, up to 300°C (Fig. 14.25). Short-term dielectric strength is 24 kV/mm (for 1.5 mm thick film) according to ASTM method D149. Like all plastic material, dielectric strength decreases as thickness increases. Durability of insulation at high voltage is dependent on corona discharge.[3] In special wire constructions, absence of corona allows high voltages without any harm to the PTFE insulation.

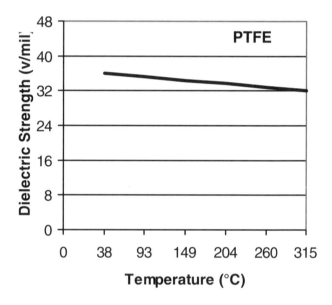

Figure 14.25 Dielectric strength (short-time) vs. temperature (1.1 mm thick molded sheet).[13]

The stability of PTFE is evidenced by a lack of change in its dielectric constant after aging at 300°C for a period of nine months (Table 14.13). There are no other plastics known to exhibit these properties. Exposure to weather has no effect on the dielectric constant and dissipation factor. Figure 14.26 summarizes the results of a ten-year study in south Florida where representative samples were exposed to the climatic elements. No change in dielectric constant or dissipation is detected as a result of weathering.

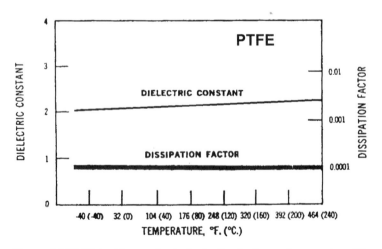

Figure 14.24 Dielectric constant, dissipation factor vs. temperature.[13]

Table 14.13. Effects of Oven-Aging at 300°C on the Electrical Properties of PTFE Resins[13]

Sample	Exposure Time at 300°C	Dissipation Factor	Dielectric Constant	Dielectric Strength, kV/mm (ASTM-D-149)
125-µm extruded PTFE film	As received	0.0001	2.1	117.1
	1 month	0.0001	2.1	
	3 months	0.0001	2.1	115.6
	6 months	0.0001	2.1	
	9 months			118

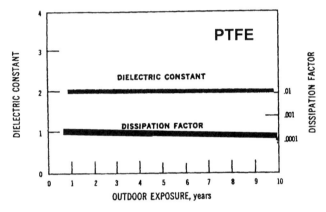

Figure 14.26 Dielectric constant, dissipation factor vs. exposure.[13]

14.5 Thermal Behavior

In this section heat and temperature related or dependent properties of polytetrafluoroethylene resins are discussed. These include thermal stability, thermal expansion, thermal conductivity and specific heat (heat capacity). These characteristics are important to both design and operation of PTFE parts.

14.5.1 Thermal Stability

Polytetrafluoroethylene resins are very stable at their normal use temperature range (<260°C). They exhibit a small degree of degradation at higher temperatures. The rate of decomposition is a function of the specific polymer, temperature, time at temperature, and, to some extent, on the pressure and nature of decomposition environment. In actual processing, degradation is tracked by indirect measurement of molecular weight. Thermal exposure leads to a reduction in the molecular weight, which can be quantified by an increase in the specific gravity and heat of fusion of PTFE, in controlled measurements.

Degradation is usually measured and characterized by weight loss using thermogravimetric analysis (TGA) technique while degradation products are identified by gas chromatography, infrared spectroscopy and mass spectroscopy. Initial rates of decomposition have been summarized in Table 14.14. This data is particularly helpful for the estimation of outgassing in applications where PTFE part is under vacuum exposure. It can clearly be seen that decomposition rates of PTFE are quite low at fairly high temperatures. The small amount of degradation requires TGA experiments to be conducted for several hours to allow accurate detection of weight loss.

Figure 14.27 and Table 14.15 provides a summary of degradation rates studies by Baker and Kasprzak.[14] In this figure, weight loss rate in air is plotted against temperature. It can be seen that polytetrafluoroethylene, disregarding of its type, is the most thermally stable of the perfluorinated fluoropolymers. As a matter of fact, there are very few organic matters approaching the thermal stability of PTFE. Degradation is accelerated in air compared to vacuum decomposition.

In vacuum, polytetrafluoroethylene degrades into nearly pure monomer. Products of PTFE degradation in air include carbonyl fluoride (CO_2F), tetrafluoroethylene and small amounts of perfluoroisobutylene (PFIB).[15][16][17] PFIB and CO_2F are highly toxic if they are inhaled (see Ch. 18).

14.5.2 Thermal Expansion

A polytetrafluoroethylene part contracts 2% when it is cooled from 23°C to -196°C and expands 4% upon heating from 23°C to 249°C. These dimensional changes are significant to the design, fabrication, and use, of PTFE parts. Tables 14.16 and 14.17 and Fig. 14.28 provide data for linear thermal expansion and volumetric coefficients of expansion. An abruptly large expansion occurs at the transition temperature (19°C) of polytetrafluoroethylene. It is important to take this dimensional change in mind, in the design, measurement of size, and machining, of PTFE parts. The best approach is to work at a consistent temperature (23–25°C) safely above the transition temperature.

Table 14.14. Decomposition Rates of PTFE at Elevated Temperatures[3]

	Rate of Decomposition, %/hr		
	Fine Powder	Granular Resin	
Temperature, °C	Initial	Initial	Steady State
232	0.0001–0.0002	0.00001–0.00005	1×10^{-11}
260	0.0006	0.0001–0.0002	100×10^{-11}
316	0.005	0.0005	0.000002
371	0.03	0.004	0.0009

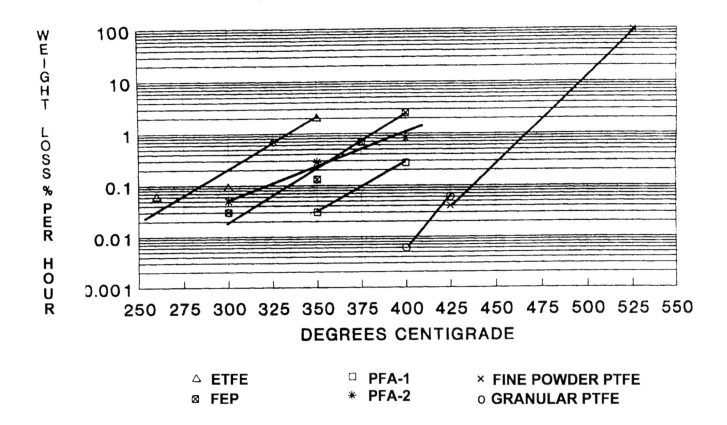

Figure 14.27 Percent weight loss on heating in air.[14]

Table 14.15. Fluoropolymer Weight Loss Data[14]

Resins	Temp °C	% Weight Loss/Hr		
		TE to 15 min.	15 to 65 min.	TE + 60 min.
ETFE	150			<0.05
	260	0.31	0.06	0.11
	300	0.42	0.09	0.14
	325			0.67
	350	~2		6.8
FEP	205			<0.05
	300		~0.03	<0.05
	350	0.45	0.13	0.18
	375			0.67
	400			3.2
PFA-1			<0.05	
	300	0.18	0.05	0.07
	350			0.22
	400			0.58
PFA-2			<0.05	
	300			<0.05
	350	0.12	~0.03	0.05
	400			0.26
FINE POWDER	400			~0.06
	425			0.15
	425			0.04*
	525	255**		95.0
GRANULAR	350			0.02
	350			0.005#
	400			0.03
	400			0.006#
	425			0.06
	425			0.06*

* Hourly rate from 8 to 11.8 hours after beginning run.
\# Hourly rate from 3.3 to 6.6 hours after beginning run.
** Gross decomposition in one hour. Initial rate 255% per hour.

Table 14.16. PTFE Resins Linear Coefficients of Expansion[3]

Temperature Range, °C.	Linear Coefficient of Expansion, 10^{-5} mm/mm·°C
25 to −190	8.6
25 to −150	9.6
25 to −100	11.2
25 to −50	13.5
25 to 0	20
10 to 20	16
20 to 25	79
25 to 30	16
25 to 50	12.4
25 to 100	12.4
25 to 150	13.5
25 to 200	15.1
25 to 250	17.5
25 to 300	22

Table 14.17. PTFE Resins Cubical Coefficients of Expansion[3]

Temperature Range, °C	PTFE Resins Cubical Coefficients of Expansion cm^3/cm^3·°C
−40 to 15	2.6×10^{-4}
15 to 35*	1.7%
35 to 140	3.1×10^{-4}
140 to 200	6.3×10^{-4}
200 to 250	8.0×10^{-4}
250 to 300	1.0×10^{-3}

*Quinn et al., J. Applied Phys. 22, 1085 (1951)

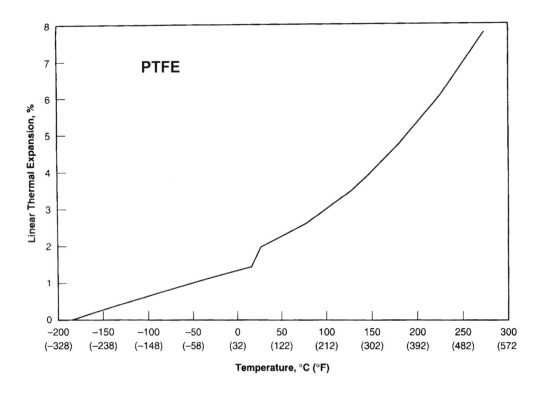

Figure 14.28 Linear thermal expansion vs. temperature.[3]

Polytetrafluoroethylene has a somewhat higher coefficient of expansion than other plastics. This can cause problems when it is combined with other materials such as leaking of joints as PTFE warms. Addition of fillers such as glass, fiber, graphite, bronze and molybdenum disulfide, alters the coefficient of expansion of polytetrafluoroethylene compounds (Table 14.18). A compound containing 25% has a coefficient of expansion about half that of the unmodified resin.

14.5.3 Thermal Conductivity and Heat Capacity

Polytetrafluoroethylene resins have very low thermal conductivity and are considered good insulators. The values of thermal conductivity are given in Table 14.19.

Fillers modify, usually by increasing the thermal conductivity of polytetrafluoroethylene as with other plastics (Table 14.20). Specific heat of PTFE at various temperatures is given in Table 14.21. Enthalpy of molded PTFE is given in Fig. 14.29.

14.5.4 Heat Deflection Temperature

Heat deflection temperature of polytetrafluoroethylene is given in Table 14.22.

Table 14.18. The Effect of Fillers on the Linear Thermal Expansion of TFE Resins[1]

Property	Test	Unfilled TFE	15% Glass Fiber	25% Glass Fiber	15% Graphite	60% Bronze	20% Glass 5% Graphite	15% Glass 5% MoS_2
Filler loading vol.%			13.3	22.2	14.6	27.1	17.6, 5.0	13.3, 2.4
Bulk density, lb./cu. in.		0.0129	0.0103	0.0107	0.0087	0.0103	0.0104	0.0107
Spec. gravity Theoretical Measured	ASTM D1457-66T	2.18	2.22 2.20	2.26 2.22	2.19 2.12	3.97 3.85	2.25 2.18	2.28 2.27
Coeff. of linear therm. exp. (x 10^{-5}), in./in./°F. 78°-200°F MD CD 78°-300°F MD CD 78°-400°F MD CD 78°-500°F MD CD	ASTM D696-44	6.81 7.04 7.60 9.12	8.02 2.94 8.40 2.96 9.05 3.42 10.31 4.15	6.97 4.19 7.33 4.19 8.00 4.72 9.35 5.55	6.97 4.38 7.46 4.67 8.14 5.12 9.50 5.97	5.40 4.38 5.72 4.37 6.35 4.99 7.82 5.75	7.73 2.61 7.73 3.01 8.12 3.22 9.24 3.88	8.33 3.51 8.77 3.55 9.61 3.85 11.11 4.45

MD = Parallel to molding direction
CD = Cross section perpendicular to molding direction

Table 14.19. Thermal Conductivity of PTFE[1]

Temperature, °C	Thermal Conductivity, W/mK
-253	0.13
-128 to 182	0.25

Table 14.20. Thermal Conductivity of PTFE Compounds[1]

Filler Type	Filler Concentration, weight %	Thermal Conductivity, W/mK
Unfilled	0	0.25
Glass Fiber	15	0.37
Glass Fiber	25	0.46
Graphite	15	0.46
Bronze	60	0.48
Glass Graphite	20 5	0.37
Glass Molybdenum Disulfide	15 5	0.33

Table 14.21. Specific Heat of PTFE[3]

Temperature, °C	Specific Heat, kJ/kg.K
20	1.4
40	1.2
150	1.3
260	1.5

Table 14.22. Heat Deflection Temperature of PTFE[1] (ASTM Method D648)

Pressure, kPa	Temperature, °C
450	73
1800	45

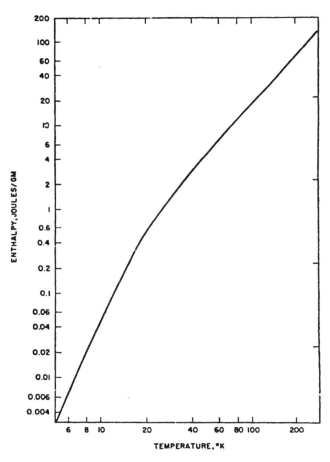

Figure 14.29 Enthalpy of molded PTFE.[1]

14.6 Irradiation Behavior

Polytetrafluoroethylene and other perfluorinated fluoropolymer are quite susceptible to radiation. Exposure to high energy radiation such as x-ray, gamma-ray, and electron beam, degrades PTFE by breaking down the molecules and reducing its molecular weight. As in thermal degradation, radiation stability of PTFE is much better under vacuum compared to air. For example, exposure of PTFE film to a dose of one megarad of gamma rays from a Co^{60} in air resulted in 87% loss of initial elongation and 54% loss of tensile strength. Significantly less loss of tensile properties occurs when PTFE is exposed to the same radiation dose in vacuum. (See Table 14.23.)

Chapter 11 of this book, presents a review of the chemistry and data of the radiation behavior of polytetrafluoroethylene. Other sources are available for performance in radiation related applications.[18][19]

14.7 Standard Measurement Methods

Table 14.24 summarizes the standard test methods for inspection of polytetrafluoroethylene parts. These tests are often specified by the customer and suppliers of parts are expected to meet them consistently.

Table 14.23. Heat Deflection Temperature of PTFE[1]

Environment	Elongation Loss, %	Tensile Strength Loss, %
Air	87	54
Vacuum	44	17

Table 14.24. ASTM Tests Applicable to Fabricated PTFE Parts[3]

Property	Extruded Rod	Molded Sheet	Molded Parts	Films and Tapes	Extruded Tubing
Tensile Strength	D1710	D3293	D3294	D3308 D3369	D3295
Ultimate Elongation	D1710	D3293	D3294	D3308 D3369	D3295
Measured Specific Gravity	D1710	D3293	D3294	D3308	D3295
Dielectric Strength	D1710	D3293	D3294	D3308 D3369	D3295
X-Ray Inspection	D1710	—	D3294	—	—
Melting Point	D4894	D3293 D4895	D3294	D3308 D3369	D3295
Dye Penetrant	—	D3293	D3294	—	—
Dimensional Stability	D1710	D3293	D3294	—	D3295
Pinhole Count	—	—	—	D3308 D3369	—

*Central section of test specimens machined to 60% of nominal diameter. Tested at 2 in/min crosshead speed

Reference

1. Milek, J. T., *A Survey of Materials Report for TFE Fluoroplastics,* by Electronic Properties in Formation Center, EPIC Report No. S-3, 1964.
2. The *J. of Teflon,*® Reprint No. 30, 1967.
3. "Teflon® PTFE Fluoropolymer Resin," *Properties Handbook,* DuPont Co., July, 1996.
4. Moynihan, R. E., *J. Am. Chem. Soc.*, 81:1045, 1959.
5. McGrum, N. G., *J. Poly. Science,* 34:355, 1959.
6. Pierce, R. H. H., Bryant, W. N. D., and Whitney, J. F., Chicago Meeting of Amer. Chem. Soc., Sep., 1953.
7. Simeral, W. G., Fluoropolymers Div. of Society of Plastics Industry Conference, Sep., 1957.
8. *Properties of Polymers - Their Estimation and Correlation With Chemical Structure,* Van Krevelen, 2nd edition, Elsevier, 1976.
9. DuPont Teflon® NXT Resin Profile, Modified Teflon® NXT Lets You Do More, No. H-77937, Mar., 1998.
10. "DYNEON Hostaflon® Polytetrafluoroethylene," Product Comparison Guide, Feb., 1998.
11. "Superior Performance PTFE- Daikin Polyflon® TFE M-111, M-112," Daikin America, Inc., presentation at the Meeting of Society of Plastics Industry, Fall 1995 Meeting.
12. Ausimont Fluoropolymer Resins Comparison Data, AUSIMONT Montedison Group, June, 1998.
13. The *J. of Teflon*®, Reprint No. 25, Apr. 10, 1965.
14. Baker, B. B., and Kasprzak, D. J., "Thermal Degradation of Commercial Fluoropolymer in Air," *Polymer Degradation and Stability,* 42:181–188, 1994.
15. U. S. Department of Health, Education, and Welfare, Public Health Service, Center for Disease Control, National Institute for Occupational Safety and Health, *Criteria for a Recommended Standard Occupational Exposure to Decomposition Products of Fluorocarbon Polymers,* DHEW (NIOSH) Publication # PB274727, Sep., 1977.
16. Kaplan, H. L., Grand, A. F., Switzer, W. C. and Gad, S. C., "Acute Inhalation Toxicity of the Smoke Produced by Five Halogenated Polymers," *Journal of Fire Science,* 2:153–172, 1984.
17. Williamd, S. J., Baker, B. B. and Lee, K. P., "Formation of Acute Pulmonary Toxicants Following Thermal Degradation of Perfluorinated Polymers: Evidence for a Critical Atmospheric Reaction," *Food Chem. Toxicology,* pp. 177–185, 1987.
18. Loy, W. E., Jr., "Effects of Gamma Radiation on Some Electrical Properties of TFE-Fluorocarbon Plastics," paper in MATERIALS IN NUCLEAR APPLICATIONS, ASTM-STP-276, pp. 68–78, 1959.
19. "The Effect of High Energy Radiation on Plastics and Rubber, Part 1: Polytetrafluoroethylene," Report No. ERDE 21/R.60, AD 252333, Nov., 1960.

15 Properties of Chlorotrifluoroethylene Homopolymers

15.1 Introduction

Polychlorotrifluoroethylene (PCTFE) properties depend on processing conditions in addition to its molecular structure and composition. PCTFE invariably degrades to some extent during fabrication. Degradation of this polymer is due to molecular weight reduction when it is exposed to processing temperatures. Mechanical properties are degraded. It is important to plan for and take into account this loss of performance. Zero strength time, crystallinity and stress levels are the factors which affect mechanical properties and performance characteristics of PCTFE parts. Higher molecular weight PCTFE resins are more suitable for applications requiring stress crack resistance.

15.2 Crystallinity

PCTFE is a *semicrystalline* polymer, which means that crystal type and the extent to which it recrystallizes depends on its thermal exposure and cooling rate during fabrication into a part. A semicrystalline material is comprised of crystalline and amorphous phases. A highly crystalline polychlorotrifluoroethylene part has high specific gravity along with high mechanical strength and low elongation. Mostly amorphous PCTFE has relatively low specific gravity and is transparent. It has lower mechanical strength and higher elongation than a mostly crystalline part. To increase amorphous phase, a part must be cooled rapidly. Only thin wall parts should be cooled quickly because of formation of voids and cracks in a thick part.

15.3 Mechanical Properties

Figure 15.1 shows a typical stress-strain curve for two commercial types of polychlorotrifluoroethylene at 23°C. Type M-300H has a lower molecular weight than M-400H. The specimens were made by compression molding. The elastic region extends to nearly 4–5% strain for these resins.

Tensile strength and elongation are summarized as a function of temperature in Fig. 15.2 and 15.3. Tensile strength decreases as temperature is raised. Elongation increases up to 100–125°C before decreasing. Tensile modulus of elasticity is given in Fig. 15.4.

Temperature softens plastics, including polychlorotrifluoroethylene, which is demonstrated in Fig. 15.5 where the effect of temperature on hardness is demonstrated (Shore D). Creep (flow under load) curves for the two types of resin are shown in Fig. 15.6 at a load of 7 MPa. Creep grows rapidly beyond 80°C. At the room temperature (25°C), both resins have very low creep.

15.4 Electrical Properties

Polychlorotrifluoroethylene has excellent electrical properties. It is quite different from polytetrafluoroethylene in that it can get polarized when it is subjected to an electrical field due to the asymmetry created by substitution of a fluorine atom by chlorine. Table 15.1 contains a number of electrical properties of PCTFE including dielectric breakdown strength, dielectric constant, dissipation factor and others. Figures 15.7 through 15.9 present dielectric constants and dissipation factors as a function of frequency and temperature. Figures 15.10–15.11 illustrate dielectric breakdown strength and arc resistance as a function of film thickness. Volume resistivity versus temperature is plotted in Fig. 15.12.

15.5 Thermal Properties

Polychlorotrifluoroethylene is believed to be stable up to 204°C just below its melting point at 215.6°C. Slow degradation of PCTFE begins at 260°C and accelerates at 299°C. Hydrochloric and hydrofluoric acids are produced by reaction of the decomposition fragments and any moisture. Table 15.2 and Fig. 15.13 provide data for the rate of thermal degradation of PCTFE.

Some of the thermal properties of PCTFE are listed in Table 15.3.

Figure 15.14 depicts the coefficient of linear thermal expansion of polychlorotrifluoroethylene as a function of temperature. As the figure shows, this coefficient increases steadily beyond -100°C.

15.6 Irradiation Behavior

This plastic is more stable than polytetrafluoroethylene and most other fluoropolymers. After exposure to 24 Mrads of gamma radiation, a sample of low crystallinity PCTFE retained 60% of its ultimate tensile strength, 80% of tensile yield strength and elongation, and 100% of its tensile modulus.[3] More crystalline PCTFE embrittles more easily after exposure to radiation. The outgassing characteristics of PCTFE are superior to many other materials. Increased dose of gamma ray irradiation alters the crystallinity of PTFE and PCTFE. Initially, PTFE shows an increase in crystallinity as a result of degradation into smaller molecules. Beyond, 200–300 megarads, crystallinity decreases as shown in Fig. 15.15.

15.7 Properties of PCTFE Films

Table 15.4 summarizes the mechanical, thermal, optical and performance properties of PCTFE films with a thickness range of 0.019–0.038 mm.

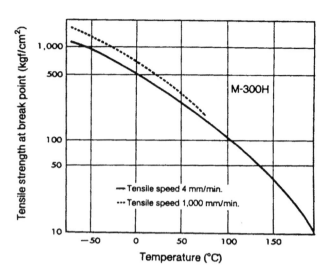

Figure 15.2 Tensile strength (at break point) vs temperature for two commercial grades of PCTFE.[1]

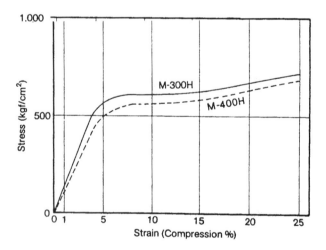

Figure 15.1 Compression stress-strain curves for two commercial grades of PCTFE[1] Test conditions: compression speed = 1 mm/min; size of specimen = 12.7 mm dia. × 25.4 mm height; temperature = 23°C.

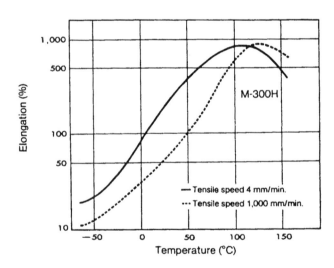

Figure 15.3 Elongation vs temperature for two commercial grades of PCTFE.[1]

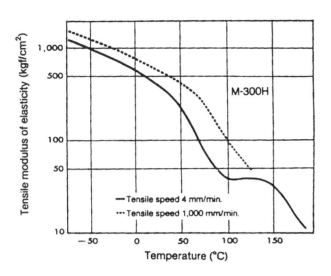

Figure 15.4 Tensile modulus of elasticity for two commercial grades of PCTFE.[1]

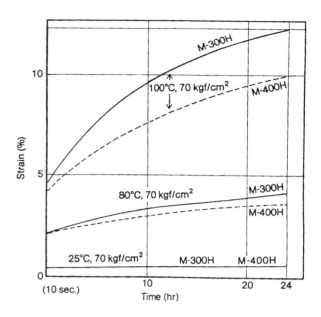

Figure 15.6 Flow (creep) of PCTFE under load.[1] Size of the specimen = 11.3 mm dia. × 10 mm height.

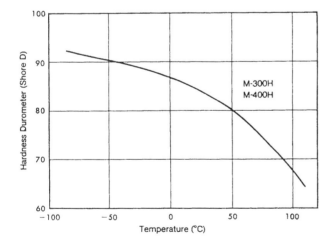

Figure 15.5 Effect of temperature on the hardness of PCTFE.[1]

Table 15.1. Electrical Properties of Polychlorotrifluoroethylene[1]

Property	Test Method, ASTM	Value
Dielectric Break-down Strength (short time), kV/mm 0.1 mil thickness 1.7 mil thickness	D 149	120 20
Dielectric Constant at 10^3 Hz	D 150	2.6
Dissipation Factor at 10^3 Hz, %	D 150	0.02
Arc Resistance, Seconds	D 495	360
Volume Resistivity at 50% relative humidity, ohm.cm	D 257	2×10^{17}
Surface Resistivity at 100% relative humidity, ohm.cm	D 257	1×10^{15}

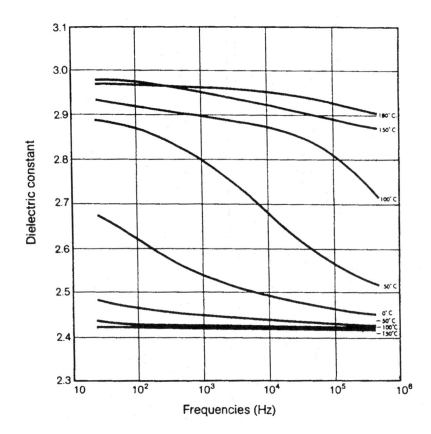

Figure 15.7 Dielectric constant versus frequency.[1]

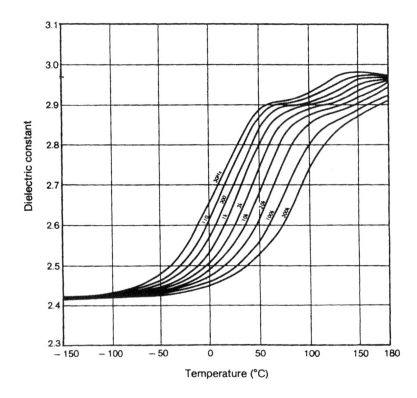

Figure 15.8 Dielectric constant versus temperature for PCTFE.[1]

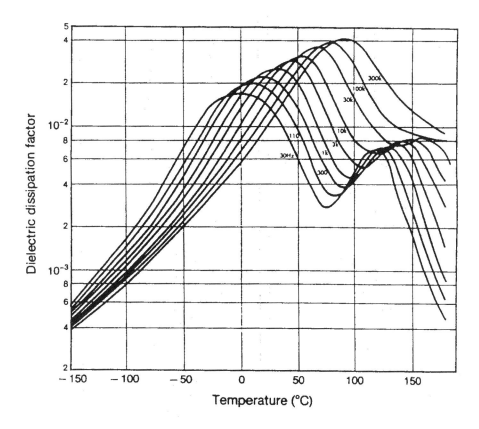

Figure 15.9 Dissipation factor versus temperature for PCTFE.[1]

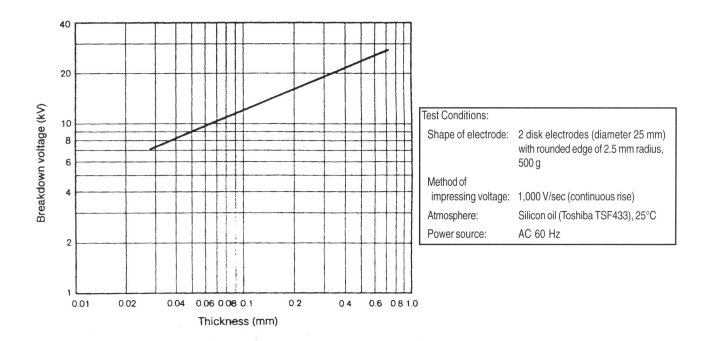

Figure 15.10 Dielectric strength as a function of PCTFE thickness.[1]

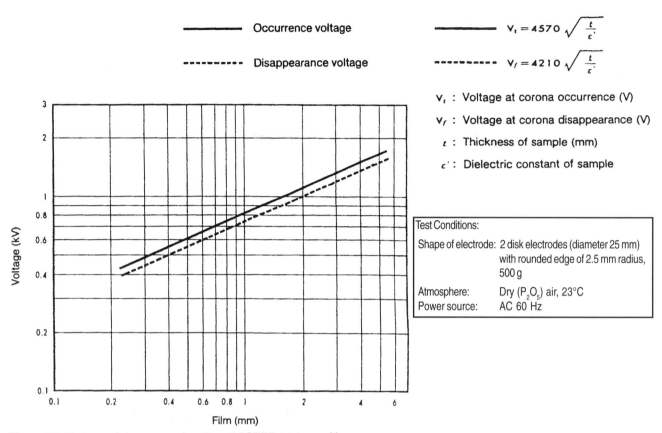

Figure 15.11 Arc resistance as a function of PCTFE thickness.[1]

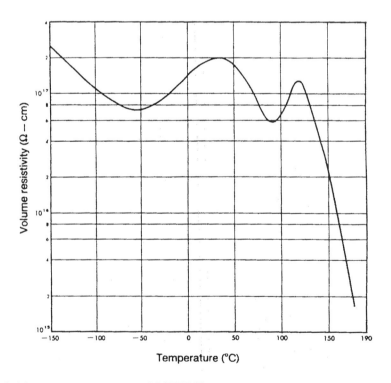

Figure 15.12 Volume resistivity versus temperature of PCTFE.[1]

Table 15.2. Rates of Degradation of Polychlorotrifluoroethylene[2]

Temp, °C	Test Duration, min	Total Volatilized, %	Initial Volatilization Rate[1], %/min
365	400	78.1	0.20
370	300	82.9	0.28
375	200	75.0	0.42
380	160	82.3	0.58
385	130	83.2	0.84

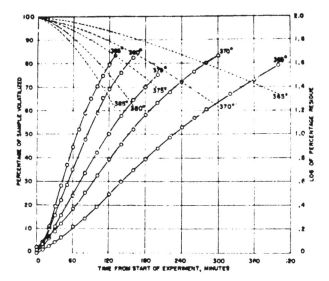

Figure 15.13 Thermal degradation of Polychlorotrifluoroethylene. Percentage of sample volatilized vs time, —; percentage residue vs time, ······.[2]

Table 15.3. Thermal Properties of Polychlorotrifluoroethylene[3]

Property	Test Method, ASTM	Quenched	Slowly Cooled
Heat Deflection Temperature,°C 0.45 MPa 1.82 MPa	D 648	129.4 75	- -
Thermal Conductivity, W/m.K		0.26	0.27
Specific Heat, J/g.K		-	0.85

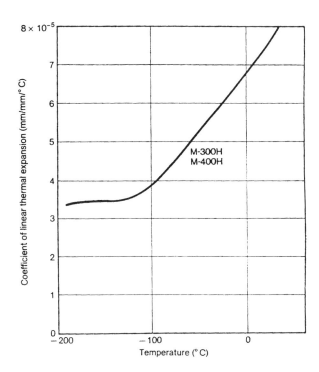

Figure 15.14 Coefficient of linear thermal expansion vs. temperature for compression molded PCTFE. Size of specimen: 7 mm dia. × 10 mm length.[1]

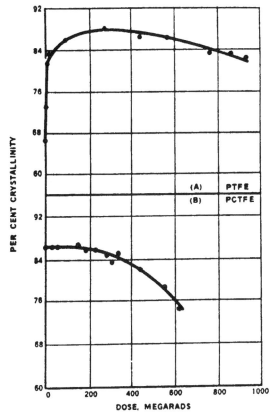

Figure 15.15 Crystallinity change as a function of gamma ray dose.[4]

Table 15.4. Properties of PCTFE Films[5]

Material Family	POLYCHLOROTRIFLUOROETHYLENE			
Material Supplier/ Trade Name	ALLIED SIGNAL ACLAR			
Grade	88A	22A	22C	33C
Product Form	FILM			
Features	transparent	transparent	transparent	transparent
MATERIAL CHARACTERISTICS				
Sample Thickness	0.019 mm	0.038 mm	0.19 mm	0.019 mm
Specific Gravity	2.10	2.10	2.11	2.12
Yield (in^2/lb)	17,760	8880	1750	17,360
MECHANICAL PROPERTIES				
Modulus Of Elasticity - MD (MPa)	970-1100	970-1100	900-1100	1310-1380
Modulus Of Elasticity - TD (MPa)	837-970	1040-1100	900-1100	1310-1380
Tensile Strength - MD (MPa)	48.3-69	51.7-75.9	27.6-41.4	65.5-79.3
Tensile Strength - TD (MPa)	27.6-41.4	37.9-55.2	27.6-41.4	37.9-55.2
Ultimate Elongation - MD (%)	150-250	115-225	200-300	50-150
Ultimate Elongation - TD (%)	200-300	200-300	200-300	50-150
Drop Dart Impact Strength (g)	93	347	1200	<57
Tear Strength, Propagated, Elmendorf - MD (g)	24	40	>1600	12
Tear Strength, Propagated, Elmendorf - transverse (g)	24	130	>1600	33
Tear Strength, Graves - MD (g)	240	380	465	400
Tear Strength, Graves - TD (g)	300	360	415	380
Abrasion Resistance - weight loss (mg)	14	16	7	15
THERMAL PROPERTIES				
Melting Point (°C)	183-186	183-186	183-186	202-204
Thermal Conductivity (cal-cm/cm^2-sec-°C)	5.3×10^{-4}	5.3×10^{-4}	5.3×10^{-4}	4.7×10^{-4}
OPTICAL PROPERTIES				
Haze (%)	<1	<1	<1	<1
PERFORMANCE PROPERTIES				
Unrestrained Shrink - 149°C Air, MD (%)	12-15	12-15	≤ 2	≤ 2.5
Unrestrained Shrink - 149°C Air, TD (%)	-12 to -15	-12 to -15	≤ 2	≤ 2.5

References

1. Technical Information, Fluoroplastics of Daikin Industries, Neoflon® CTFE Molding Powders, Daikin, Oct., 1987.
2. Madorsky, S. L., and Straus, S., "Thermal Degradation of Polychlorotrifluoroethylene, Poly-alpha, beta, beta-Trifluorostyrene, and Poly-*p*-Xylylene in a Vacuum," *J. Res. National Bureau of Standards*, 55:4, Oct., 1955.
3. *EPST*, Vol. 7, 1967.
4. Peffley, W. M., Honnold, V. R., and Binder, D., "X-ray and NMR Measurements on Irradiated Polytetrafluoroethylene and Polychlorotrifluoroethylene," *J. Polymer Science*, Part A-1, 4:977–983, 1966.
5. *Permeability and Other Film Properties of Plastics and Elastomers,* 1st ed., PDL Handbook Series, *Plastics Design Library*, Norwich, NY, 1995.

16 Fabrication Techniques for Fluoropolymers

16.1 Introduction

Polytetrafluoroethylene cannot be processed by melt processing techniques such as injection molding due to its extremely high viscosity. Intricate shapes required by various applications can be achieved by secondary machining of molded stock shapes or rough-molded parts. It is sometimes necessary to bond polytetrafluoroethylene to itself or other materials. Of course, this polymer is known for its nonstick properties and must be rendered adherable. This chapter describes a number of techniques commonly applied to finish parts made from fluoropolymers.

16.2 Machining

All customary high speed machining operations can be carried out on PTFE as long as the cutting tools are very sharp. The closest match for the free turning of fluoropolymers among metals is brass. Tool wear is similar to stainless machining. The low thermal conductivity of fluoropolymers such as PTFE causes heating of the tool and the charge material during turning. This can cause deformation of the polytetrafluoroethylene and excessive wear of the tools. PTFE parts can be machined to a depth of less than 1.5 mm without the use of a coolant. It is necessary to use a coolant to achieve critical tolerances or to machine by automatic lathes.

Dimensions of PTFE parts should be measured at a specified temperature due to the large dimensional change (1.3%) that takes place between 0 and 100°C.[1]

All standard machining operations including turning, tapping, facing, boring, drilling, threading, reaming, grinding, etc., can be performed on polytetrafluoroethylene and other fluoropolymers. Special machinery is not required for any of these operations. Speed selection, and tool shape and use, are important considerations in successful machining.

Other than poor thermal conductivity, polytetrafluoroethylene has a ten times higher coefficient of linear thermal expansion than metals. This means that any type of heat buildup will cause significant expansion of the part at that point resulting in overcuts or undercuts, thus deviating from the desired part design. Coolants should be used if surface speed of the tool exceeds 150 m/min. At higher speed, low feeds are helpful to reducing heat generation. Surface speeds between 60–150 m/min are satisfactory for fine-finish turning. At these speeds, feed should be run between 0.05–0.25 mm per revolution. At higher speeds than 150 m/min, feed must be dropped to a lower value.[2]

Choice of tools is important to the control of heat buildup. Standard tools can be used but best results are obtained with specially shaped tools. A single point tool can be designed according to the following information. Top rake should have a positive angle of 0–15° with a 0–15° side rake and side angle. Front or end rake should have an angle of 0.5–10°. Boring tools require angles on the higher end of these ranges.

A dull tool affects tolerances achieved during machining. The tool that is not sharp pulls the stock out of alignment, thereby causing overcutting and excessive resin removal. An improperly edged tool tends to compress which results in shallow cuts. A very sharp tool is desirable, particularly for turning filled parts. Carbide and Stellite (a cobol alloy tool steel material) tipped tools reduce tool sharpening frequency.

Other considerations include material support, especially when turning long thin stocks due to flexibility of PTFE. Another issue is the characteristics of machined resin which tends to be continuous and curly. It should be removed to prevent blocking coolant flow or pushing the work away from the tool.

Polytetrafluoroethylene parts can be finished to tolerances in the range of ±12 μm to ±25 μm by machining. Finishing to low tolerances is often not required for PTFE parts because they can be press-fitted at a lower cost. Resilience of this plastic allows its conformation to the working dimensions. It is usually essential to relieve stress in the stock.

The annealing procedure entails heating the stock shape 10°C above its service temperature (always below 327°C, melting point of PTFE). It should be held at temperature at the rate of 25 min/cm of thickness. Stresses are relieved during this operation. At the end of the hold time, the part should be cooled slowly down to room temperature. After rough cutting the part to about 300–500 μm, re-annealing before the final cut is made will help remove stresses induced by the tool.

Lapping and grinding compounds could be used for finishing the surface of PTFE. These powders can become embedded in the surface of the part and may not be easily removed. This is true of any contaminants from machinery that is not dedicated to polytetrafluoroethylene finishing.

16.2.1 Sawing and Shearing

Polytetrafluoroethylene parts of any size can be sawed. Coarse saw blades are preferable to fine toothed blades, which can become blocked with resin. Longer saw blades perform better than a short blade such as a hack saw. A band saw operated at moderate speeds is ideal because the long blade can remove the heat.

Shearing of rods and sheets of PTFE can be done as long as the work and the blade are firmly supported to prevent angular cuts. The limit for shearing sheets is a thickness of 10 mm and for cutting rods is a diameter of 20 mm.

16.2.2 Drilling, Tapping and Threading

Drilling can be performed on PTFE parts using normal high-speed drills. A speed of 1,000 rpm for up to 6 mm diameter and 600 rpm for up to 13 mm diameter holes are recommended.[1] To improve the accuracy of drilling, heat should be removed. This can be accomplished by reducing friction by taking a sharply angled "back-off" to the cutting edge and to the polished flutes. Another consideration is the relaxation of the polymer. To increase accuracy of the holes, a coarse hole should be drilled following a finishing cut, after allowing the part to relax for twenty-four hours at the room temperature.

16.2.3 Skiving

A popular method for producing polytetrafluoroethylene tapes is by skiving a cylindrical molding of the resin. *Skiving* is similar to the peeling of an apple where a sharp blade is used at a low angle to the surface. A comparable industrial operation is production of wood veneer. This operation removes layers of PTFE by applying a sharp blade to the surface of the cylinder. A grooved mandrel is pressed into the center hole of the billet and the assembly mounted on a lathe. A cutting tool mounted on a rigid cross-slide is advanced towards the work at a constant speed to peel off the continuous tape with a constant thickness. The cutting tool must have an extremely sharp blade that has been finished on a fine stone and honed or stropped to prevent formation of "tram lines" on the skived film.

A typical skiving tool can be seen in Fig. 16.1. For relatively thin tapes (50–250 µm) should be set on the center line of the stock, but for higher thickness tape, the blade should be set above the center line of the molding. Sheets up to 400 µm thickness and 30 cm width can be skived on a 20 cm wide lathe.

The blade should be fabricated from high-quality tool steel or tungsten, and regularly sharpened. Skiving speed should be low and the chuck speed should be in the range of 20–30 rpm.

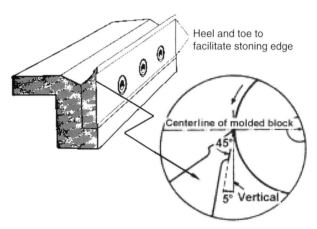

Figure 16.1 Cutting tool for machining tape from a molded block.[1]

16.3 Adhesive Bonding Methods

One of the attributes of fluoropolymers is the non-stick property of their surfaces, which is useful for innumerable applications. This characteristic prevents bonding of these plastics to themselves and other materials. Many applications require bonding of fluoropolymer parts to themselves or other substrates. There are two types of solutions to the bonding problem of fluoropolymers: with and without an adhesive. Adhesive bonding is described in this section.

There are two types of adhesives by which fluoropolymers are bonded: *contact adhesives* and *bonding adhesives*. The distinction between these two adhesives is that contact adhesives can be applied without modification of the surface of the plastic, while bonding adhesives require surface modification.

16.3.1 Contact Adhesives

Contact or pressure sensitive adhesives are suitable for applications where large surface areas are to be adhered together such as lining process equipment. Examples include hoppers, chutes, and conveyor belts lined with sheet PTFE. Tapes of polymer can be reinforced with glass cloth by using pressure sensitive adhesives. These adhesives can be applied to the PTFE surface after thorough cleaning and removal of all contaminants without surface modification.

Bond strength is relatively low due to the surface energy of polytetrafluoroethylene. The range of bond strength is 0.1–1.8 kg/cm depending on the type of adhesive. The adhesive can be applied in the form of a fluid by working it on the surface to achieve an even thickness. A more convenient alternative is two-sided tapes to which the pressure sensitive adhesive has already been applied. All bubbles should be removed to lessen the chance of delamination and peeling.

16.3.2 Bonding Adhesives

It is necessary to modify the surface of fluoropolymers to obtain stronger adhesive bonds. Modification or surface treatment alters the structure of the polymer at the surface, enabling formation of true adhesive bonds. Mechanical abrasion imparts little improvement and chemical etching is required. Chemical resistance of fluoropolymers such as PTFE, PCTFE, and FEP mandates the use of highly potent agents. Alkali metals, particularly sodium, can carry out the surface modification. Another method is plasma treatment in which the plastic surface is struck by energized atomic fragments. Both technologies are described.

Sodium is the most economical choice, but it must be handled with utmost care. There are two ways to prepare solutions of sodium. It can be dissolved in anhydrous liquid ammonia or made into a complex with naphthalene followed by dissolution in an ether like tetrahydrofuran or dimethyl glycol ether. Special precautions must be taken while working with sodium etching solutions. Fluoropolymers surface treated by sodium etching should be stored in a cold dark atmosphere free from oxygen and moisture. The useful shelf life of etched polymer stored under these conditions at <5°C is three to four months.

Once the fluoropolymer sheet, film etc., has been etched, it is an adherable surface. The choice of adhesive depends on the required service conditions and, to some extent, the substrate. Some of the considerations in the selection of adhesive include chemical resistance, flexibility, and temperature resistance.

16.3.3 Sodium Etching

The original method for surface treatment of PTFE for adhesive bonding is etching in a sodium solution in anhydrous liquid ammonia.[3][4] The reagent is prepared by simply dissolving metallic sodium in liquid ammonia to obtain a 0.5–1% by weight concentration. The solution has a dark blue color and should be stirred thoroughly before use. The surface of fluoropolymer should be cleaned carefully with an organic solvent such as acetone to remove oils or greases and other contamination, which can cause poor treatment and weak bonding. Moisture must be kept from the solution by storing it under positive pressure in protective packaging.

The fluoropolymer needs to be in the solution for a brief duration ranging from 2–10 seconds.[1] Ammonia rapidly volatilizes after the article is removed from the bath. Sodium can be removed by dipping the treated article into ethyl alcohol. An immersion time which is too long actually weakens the adhesive bond. The optimum time depends on the freshness of the etching solution. Treated fluoropolymer has a shiny dark brown color which grows into a dull brown after exposure to air. Analysis of the bath shows the presence of fluoride ions suggesting defluorination of the surface during treatment. Electron Spectroscopy for Chemical Analysis (ESCA) shows (Table 16.1) very low fluorine to carbon ratio and high oxygen to carbon ratio, indicating defluorination and oxidation of PTFE.

The second solution consists of a bath consisting of mixture of 1:1 molar ratio of sodium in naphthalene dissolved in tetrahydrofuran.[6] For example, 23 g of sodium is mixed with 128 g of naphthalene flakes (sometimes called sodium naphthalenide) then dissolved in one liter of tetrahydrofuran. The reaction begins slowly and accelerates with time accompanied by a color change of the solution from inky blue to dark brown and finally black. It takes about two hours to prepare the solution.

Liquid ammonia and tetrahydrofuran present serious safety and toxicity issues. Over the years, replacement solvents have been developed to overcome these problems. Commercial examples of these solvents include ethylene glycol dimethyl ether *(monoglyme)*, diethylene glycol dimethyl ether *(diglyme)* and tetraethylene glycol dimethyl ether *(tetraglyme)*. Monoglyme has the greatest solubility for sodium naphthalenide and produces a very active etching solution. Its disadvantage is low flash point (-1°C) and relative lower stability; it produces ten

Table 16.1. Effect of Sodium Etching on the Surface Composition of Fluoropolymers[5]

Polymer	Treatment	Surface Chemical Analysis by ESCA[1]				Bond Strength[2], MPa
		F/C Ratio	C	F	O	
PTFE	None	1.60	38.4	61.6	-	2.1
PTFE	Tetra-Etch®[3] (1 minute)	0.011	82.2	0.9	16.9	21.3
PVF[4]	None	0.42	70.4	29.6	-	1.8
PVF	Tetra-Etch®[3] (30 minute)	0.21	80.7	17.2	2.1	20.8

[1]Electron Spectroscopy for Chemical Analysis [2]Composite lap shear test [3]Supplied by W. L. Gore Corporation [4]Polyvinylfluoride

times more methyl vinyl ether than diglyme, due to degradation. Monoglyme is most suitable for low temperature etching.

Elevation of temperature (e.g., to 50°C) increases the etching effect of the solutions of polytetrafluoroethylene. This allows the use of diglyme, which has a flash point of 56.7°C, but lower solubility of sodium naphthalenide than monoglyme. Viscosity of etching preparations decreases with temperature to nearly water-like consistency at 50–60°C. Lower viscosity permits penetration of the solution into confined areas such as small diameter tubing and woven PTFE fabrics. Lower viscosity etching solutions produce more uniform etched surfaces.

Using a commercial dispersion of sodium in naphthalene can reduce preparation time. The shelf life of the bath is sixty days provided that it is stored in a closed container and isolated from air. The polymer is immersed in the solution for one to five minutes followed by rinsing in alcohol or acetone. Bond strength using epoxy adhesives ranges from 7–14 MPa in tensile mode. This broad range of adhesive bond strength is obtained in butt-tensile, disk tensile and lap shear test configurations seen in Fig. 16.2.

Sodium-etched surfaces are highly active and could lose activity if they are not stored properly and adhered to within a reasonable time after etching. The treated part has to be stored, within hours after etching, in an opaque container in a cool and humidity- and temperature-controlled environment. Temperature, humidity and ultraviolet light have a detrimental effect on the adhesion bond strength of the etched surface. Proper storage of a treated part can extend its shelf life to weeks and possibly months.

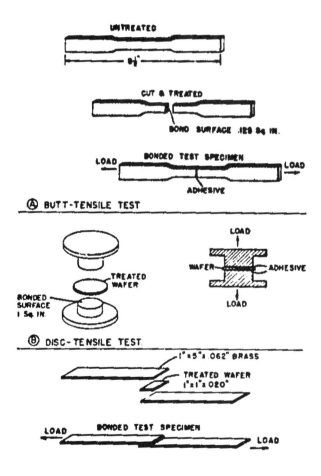

Figure 16.2 Test specimens for determination of bondability of treated Teflon®.[6]

Research has shown that the defluorination depth of the polymer surface extends to about 3000 Å as a result of sodium naphthalenide treatment.[7] They reported that exposure to x-rays decreased the defluorination depth to 30–150 Å. The structure of the defluorinated layer is highly porous. The suggested adhesion mechanism is mechanical interlocking of adhesive with this porous structure. Bond failure nearly always occurs by stripping the etched layer as opposed to cohesive failure of the adhesive.

Etching solutions can be purchased from a number of sources. They include Fluorotech® by Acton Corp. and Tetra-etch® by W. L. Gore & Associates. Some companies such as Acton Corporation provide surface treatment service. Operational safety and waste disposal are two issues concerning parties who deal with etching solutions.

16.3.4 Plasma Treatment

Plasma (glow discharge) is sometimes referred to as the fourth state of the matter. It is produced by exciting a gas with electrical energy (Fig. 16.3). It is a collection of charged particles containing positive and negative ions. Other types of fragments such as free radicals, atoms, and molecules may also be present. Plasma is electrically conductive and is influenced by a magnetic field. Plasma is intensely reactive which is precisely the reason that it can modify surfaces of plastics.[8] Plasma can be used to treat parts to make their surfaces harder, rougher, more wettable, less wettable, and easier to adhere to. Plasma treatment can be carried out on a variety of plastic parts and even powder additives like pigments and fillers.

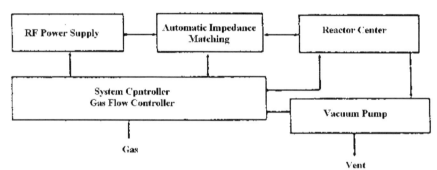

Figure 16.3 Schematic diagram of a plasma system.[9]

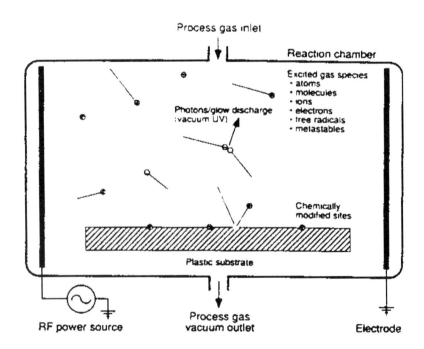

Figure 16.4 Schematic of the surface modification of plastic in a gas plasma reactor.[10]

The plasma used for treating material surfaces is called *cold plasma,* which means its temperature is about the room temperature. Cold plasma is created by introducing the desired gas into a vacuum chamber (Fig. 16.4) followed by radio frequency (13.56 MHz) or microwave (2450 MHz) excitation of the gas. The energy dissociates the gas into electrons, ions, free radicals and metastable products. Practically any gas may be used for plasma treatment, but oxygen is most common. The electrons and free radicals created in the plasma collide with the polymer surface and rupture covalent bonds thus creating free radicals on the surface of the polymer. The free radicals in the plasma may then recombine to generate a more stable product. After a predetermined time or temperature is reached, the radio frequency is turned off. The gas particles recombine rapidly and the plasma is extinguished.

Table 16.2 contains the data comparing the results of plasma treatment and sodium etching for four fluoropolymers. Peel strength of untreated and treated samples were measured by bonding them into T-peel specimen using the flexibilized epoxy adhesive Scotch-Weld® 3553 (available from 3M Corporation). The laminates were cured for several hours at 70°C and peel tested at 12.5 cm/min pull rate. Polytetrafluoroethylene does not accept plasma treatment as well as PFA and FEP, as indicated by its relatively low peel strength. Sodium etching is the only effective method of modifying the surface of PTFE.

Table 16.2. Peel Strength of Adhesive Bonded Fluoropolymers[11]

Treatment	Material			
	PTFE	FEP	ETFE	PFA
Untreated	*	0.1	*	0.04
Sodium Etched[1]	5	8.2	-	6.4
Plasma Treated	2.2	10.4	15.8	8.3

[1]Tetra-Etch® by W. L. Gore & Associates, Inc.
* Too low to measure.

Plasma treatment is a dry process, that is, it does not utilize solvents and generates little waste compared to sodium etching. It is a more expensive process due to equipment requirement and vacuum operation. It also does not impart sufficiently strong adhesive bond in polytetrafluoroethylene, the most common perfluoropolymer.

16.4 Welding and Joining

The bonding techniques involving adhesives are normally suitable for applications where the fluoropolymer does not carry large loads such as those experienced by chemical processing equipment. Welding or adhesiveless joining is a method by which parts for load-bearing applications are manufactured. The load could consist of temperature, chemical corrosion, and force. This method also known as *welding* or *joining* allows economical fabrication of complex parts by joining individual components.

It is possible to obtain a good bond between fluoropolymers themselves, without the use of adhesives, by application of heat and pressure. Pressure can help drive the molten polymer into the pores of the substrate. Bond strength is dependent on the mechanical interlocking that is achieved by the adhesion mechanism, improving with increased surface roughness of the substrate. Examples of parts made by this technology include glass cloth-backed polytetrafluoroethylene sheet, or multi-ply circuit board and coated aluminum or copper sheet. Achieving this type of bonding is more complex with polytetrafluoroethylene than melt processable polymers. PTFE does not flow after melting due to its extremely high viscosity.

It is possible to achieve adhesiveless bonding using standard PTFE in special applications where the polymer can be heated to a temperature well above its melting point. It can then be forced under pressure into the substrate surface. These polymers are not suitable for applications where the geometry of the joining objects must be preserved, contact surfaces are smooth, or the objects being bonded are too large. In such cases, a different type of polytetrafluoroethylene is required.

Polytetrafluoroethylene for these applications is known as "modified" which refers to the presence of a small amount of a second perfluorinated monomer, known as a *modifier*, in its structure. The modifier molecule always contains a pendent group. The preparation method of this type of PTFE has been described in Ch. 5. Its commercial grades have been described in Ch. 6.

How does it work? A simple explanation is offered here, based on the author's own experience. The modification reduces the molecular weight of the polymer, which in turn reduces its melt viscosity. Lower melt viscosity increases the mobility of the polytetrafluoroethylene chains. This facilitates diffusion and entanglement of polymer chains at the bonding interface. The pendent group of the modifier disrupts the crystals of PTFE, thus preventing excessive crystallization. Crystallinity which is too high results in poor mechanical properties such as poor tensile and flex properties. An optimally modified PTFE has good mechanical properties in addition to weldability.

Welding can be achieved using PTFE made by dispersion or suspension polymerization. Most applications involve welding of parts made from granular resins (suspension polymer). Dispersion polymerized PTFE is also used for application such as wire coating. A thin (50–100 µm) tape of the "modified" polytetrafluoroethylene is wrapped around the conductor followed by sintering. The layers of the tape adhere to each other and form a solid insulation, due to its good interlayer adhesion, around the conductor at the completion of sintering cycle.

16.4.1 Welding Technique

Quality of a welded area is defined by the strength of the bond. One of the ways to measure bond strength is to cut a microtensile bar specimen in such a way that the weld line would fall near its center (Fig. 16.5). Tensile strength and elongation can be determined by extensiometry. *Weld factor* is defined by Eq. 16.1 as the ratio of tensile strength of the welded specimen (T_w) to the tensile strength of the material (T_p). The weld factor is defined for the weakest polymer, if two different polymers are welded together.

Eq. (16.1) $$WF = \frac{T_w}{T_p}$$

The closer the weld factor (*WF*) is to unity the better the weld quality.

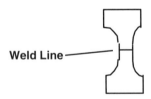

Figure 16.5 Schematic diagram of a microtensile bar for weld strength measurement.

Three variables are significant in welding a given modified PTFE part: welding temperature, pressure and time. Optimal combinations of these three parameters must be found for successful welding of parts. Temperature should be well above the melting point (320–330°C), typically in the range of 360–380°C. Little pressure is required to weld the parts after reaching gel state. Less than 350 kPa, and often less than 35 kPa, pressure is required for welding. It is normally not possible to trade higher welding pressure for lower temperature and vice versa. Time, the third variable of the process, is dependent on the size and shape of the part. The actual weld time, defined as time at the final temperature, is of the order of 1–2 minutes. It often takes a great deal longer to heat up the part to the welding temperature. High heating rates do not accelerate the process due to the low thermal conductivity of PTFE. Heat rates similar to those of sintering cycles of preforms can be expected.

The mating surfaces should be smooth and uniform and free from any contamination. Unsintered preforms and sintered parts of modified polytetrafluoroethylene can be welded. Sintering and welding can be combined. Parts can often be stacked up in the sintering oven without additional pressure. A weld factor of one can be routinely obtained in the combined process. A higher pressure is required for welding sintered parts to counteract the residual stresses, which tend to move the parts upon release. It is important to cool the welded parts slowly to minimize stresses stored in the part. Figure 16.6 illustrates a device for hot-tool welding films and sheets.

Figure 16.7 shows a comparison of the stress-strain curves of a conventional and a modified PTFE for the original and welded material. The weld line in conventional PTFE when welded to itself, at best, fails at very low strains. In the case of modified resin welded to itself, the weld factor attains value of 0.80–0.85. Weld factors for welding of conventional and modified PTFE have been reported in the range of 0.66–0.87.[13]

Another method is welding with the help of a PFA (melt processable) rod. In this case, the conventional or modified PTFE is heated by hot air near the seam until it is in gel state. The PFA rod is molten and used to fill the seam.

16.5 Thermoforming

Modified polytetrafluoroethylene can be thermoformed by vacuum forming, pressure forming, and matched-die forming (Fig. 16.8). In both methods, a sheet of PTFE is heated until it reaches gel point. It requires more "soak" (heat up) time than conventionally thermoformed plastics due to its low thermal conductivity. Vacuum forming uses the pulling force generated by vacuum to force conformation of the molten PTFE sheet to the contours of a mold. In pressure forming, hot pressurized air is used to generate the conformation force. In the matched die technique, mechanical force is generated by the male half of the mold, which drives the polytetrafluoroethylene gel into conformation with the contours of the female half of the mold.

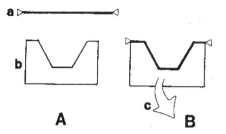

Vacuum Forming. A: Preheated Sheet Prior to Forming. B: Formed Sheet into Female Mold. a: Preheated, Clamped Sheet. b: Female Mold with Vacuum Holes. c: Vacuum.

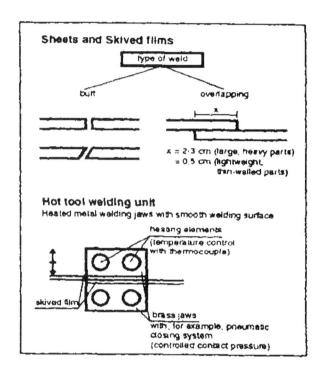

Figure 16.6 Welding sheets and skived films.[12]

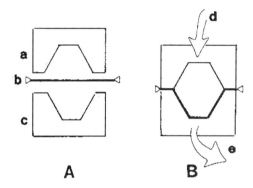

Pressure Forming. A: Preheated Sheet Prior to Forming. B: Formed Sheet into Female Mold. a: Pressure Box. b: Preheated, Clamped Sheet. c: Female Mold with Vacuum/Vent Holes. d: Applied Air Pressure. e: Venting or Vacuum.

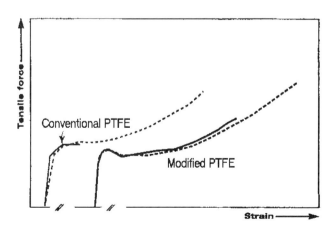

Figure 16.7 Tensile stress-strain graph of welded specimens and original material.[14]

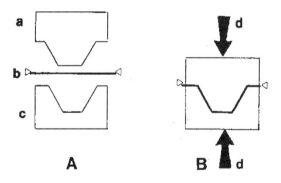

Matched Die Forming. A: Preheated Sheet Prior to Forming. B: Sheet Formed by Simultaneous Motion of Two Mold Halves. a: Male Mold Half. b: Preheated, Clamped Sheet. c: Female Mold Half. d: Applied Force.

Figure 16.8 Schematic configuration of different thermoforming methods.[15]

16.6 Other Processes

Polytetrafluoroethylene parts can be processed by other common techniques such as encapsulation, hot stamping, and ink printing.

Encapsulation of metal parts can be easily achieved with modified polytetrafluoroethylene due to the excellent bondability of this grade of PTFE to itself. These parts are used in applications where extreme mechanical integrity and rigidity combined with chemical resistance are required such as butterfly valves.

Hot stamping is conducted using a stamp which is heated well above (> 360°C), the melting temperature of polytetrafluoroethylene. Application of pressure to the hot stamp for a period of time will imprint the desired pattern. Temperature, pressure and time should be determined by trial and error.

Inks having a fluorocarbon base can be used to print stripe patterns on wire for identification. In practice a wheel coated with the desired color runs along the wire prior to the sintering step of the manufacturing process. For more than one stripe additional wheels are needed. The ink is sintered at the same time as the insulation. PTFE dispersion can be used as the base to produce ink. Inorganic pigments stable under the sintering conditions of polytetrafluoroethylene must be used.

Reference

1. Fluon® Technical Service Notes F9, "Finishing Processes," Imperial Chemical Industries, Ltd.
2. *The Journal of Teflon®*, 1(11), Nov., 1960.
3. Purvis, R. J., and Beck, W. R., US Patent 2,789,063, assigned to Minnesota Mining and Manufacturing Co., Apr. 16, 1957.
4. Brit. Patent 765,284, Jan., 1957.
5. Brewis, D. M., Loughborough University, UK, "Surface Modification of Fluoropolymers for Adhesion," paper presented at the Fluoropolymers Conference, 1992.
6. Benderly, A. A., "Treatment of Teflon to Promote Bondability," *J. of Applied Polymer Science, VI* (20):221–225, 1962.
7. Rye, R. R., and Arnold, G. W., "Depth Dependence of Alkali Etching of Polytetrafluoroethylene: Effect of X-ray Radiation," *Langmuir*, 5:1331–1334, 1989.
8. Schut, J. H., "Plasma Treatment," *Plastics Technology*, pp. 64–69, Oct., 1992.
9. Liston, E. M., "Plasma Treatment for Improved Bonding: A Review," *J. Adhesion*, 30:199–218, 1989.
10. Kaplan, S. L., and Rose, P. W., "Plasma Treatment of Plastics to Enhance Adhesion: An Overview," Technical Paper, Plasma Science, Inc.
11. GaSonics/IPC Applications Notes, Gasonics International, San Jose, Ca.
12. Hostaflon® TFM Series, Dyneon High Chem Polymer Materials, Dyneon Corporation 1998.
13. Teflon® NXT, Modified Teflon® NXT Lets You Do More, DuPont Company, Bulletin No. H-77937, March, 1998.
14. Dyneon® TFM®PTFE – the Second Generation, Dyneon Corporation, 1998.
15. Throne, J. L., *Thermoforming*, Hanser Publishers, New York, 1987.

17 Typical Applications of Fluoropolymers

17.1 Chemical Processing

In the chemical process industry (CPI), much manufacturing is carried out with chemically aggressive fluids. Fluoropolymers have major applications in chemical processing equipment to prevent corrosion and product contamination.

Increasing the resistance of equipment to corrosion is highly beneficial because it extends service life and cuts unscheduled downtime. Even though fluoropolymers may increase the initial equipment cost, lifetime costs can be reduced, allowing manufacturers to be more competitive; and in especially severe service such as exposure to hydrogen fluoride, carbon steel components lined with PTFE or PFA have replaced exotic metal alloys at a lower initial cost.

Contamination of process streams by corrosion byproducts or ions from metallic equipment can be an issue. Problems range from flaws due to rust particles in finished products to reduction in yields caused by unexpected reactions promoted by contaminants.

Other polymers, such as polyesters, polypropylene, and PVC, are used to resist corrosion and contamination. However, fluoropolymers are compatible with a much wider range of chemicals and can serve at higher temperatures than other polymers, so they are suitable for demanding environments.

Linings and coatings for piping, vessels and components are the principal applications for fluoropolymers in the CPI. Support is provided by economical carbon steel or fiberglass reinforced plastic (FRP) structures. Fluoropolymers are used in this manner for two reasons:

1. They lack the mechanical properties most of these structures need.
2. Even when polymer properties are adequate, linings are usually more cost effective.

Fluoropolymers are also used in accessories including seals and gaskets, and in a few relatively small self-supporting components that require chemical resistance on all surfaces.

17.2 Piping

Flanged steel piping is lined with PTFE tubes formed from granular or fine powder resins (Fig. 17.1).

Figure 17.1 PTFE lined pipe. *(Courtesy Crane Resistoflex and DuPont Companies.)*

The ends of the tubes are flared over flanges by thermoforming so that fluids contact only PTFE surfaces when piping sections are joined, usually by bolting. The lining thickness is typically less than 0.25 in. (6.4 mm).

Tubing extruded from PFA, FEP, PVDF, ETFE and ECTFE is used in similar fashion to lined steel piping. The choice of polymers depends on which offers the best combination of chemical and temperature resistance and cost for the expected service conditions.

FRP piping lined with fluoropolymer is manufactured by forming FRP over a fluoropolymer tube. The resulting structure is a called a *dual laminate*. Typically, this tube is a thermoplastic fluoropolymer such as PFA or FEP so that glass fabric can be readily embedded in its outer wall. This fabric bonds with the FRP structure so that the lining is held firmly in place.

Lined FRP piping can be installed by thermoplastic welding of linings and then covering the joints with additional FRP. This procedure is also used to form complex manifolds and transition pieces that are difficult or impossible to produce using lined steel.

This installation method reduces the number of flanges in a piping system, which is desirable because flanges must be monitored for emissions and maintained by periodically checking bolts for tightness.

Forming lined piping in the desired configuration can reduce the number of flanges required for lined steel piping. This is accomplished by a proprietary process with pipe manufactured to withstand forming operations.

17.3 Vessels

Steel vessels such as scrubbers and tanks are lined with fluoropolymer sheeting joined by welding. PTFE linings can be installed in this way, including those that have a glass fabric backing applied with proprietary technology. This backing allows adhesive bonding to the steel structure for better support of the lining, an important consideration when pressures below atmospheric are expected.

Thermoplastic fluoropolymers are also used in this way to line vessels. Compared with PTFE, they are easier to weld, and unlike PTFE, they can be readily thermoformed to fit vessel heads and provide connections for nozzles, or inlets and outlets.

Like piping, lined vessels can be constructed as dual laminates with FRP. In some uses, these are preferred to lined steel vessels because the FRP vessels often have sufficient chemical resistance to prevent damage by spills, and they do not require periodic painting to prevent rusting. Also, because they weigh less, they need less support so installations can cost less.

The thermoplastic fluoropolymers also can be applied as coatings or linings by powder coating and rotational lining. Both processes can provide relatively thick, void-free fluoropolymer layers compared with coatings that can be achieved with PTFE dispersions.

Fluoropolymers are used in powder coating in much the same way as other polymers. The principal differences are that higher temperatures are required to melt the fluoropolymers, and surface preparation requirements may be more stringent. In powder coating, powdered resin is applied to a surface using an electrostatic process and then heated in an oven so that the resin melts to form a continuous layer. Additional layers are applied to achieve the desired thickness.

Rotational lining, or rotolining, is similar to rotational molding or casting, a process used to form shapes from polymers. The difference is that in rotolining, the product is a lining, while in rotational casting, the product is a formed part that is removed from the mold. Because it produces a seamless lining that can include multiple openings and complex shapes, rotolining is well suited for CPI applications.

To line a vessel or component, powdered resin is placed in the component's cavity and all openings are covered. Then it is rotated on two axes in an oven so that the resin melts and flows to cover all surfaces. The amount of resin used depends on the desired lining thickness.

ETFE has been used with good success in rotolining (Fig. 17.3 and 17.4). Manufacturers report good adhesion between the resin and the steel shell.

Figure 17.2 Powder coated PFA-lined vessel. (*Courtesy General Plastics and DuPont Companies.*)

Figure 17.3 Seamless ETFE lined vessel. (*Courtesy RMB and DuPont Companies.*)

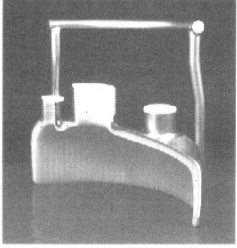

Figure 17.4 Seamless ETFE lined vessels. (*Courtesy RMB and DuPont Companies.*)

17.4 CPI Components

Valves, pumps and other components can be lined with PTFE through isostatic molding in which the resin is formed within the component's cavity by pressure from inflatable elastomeric tooling. After tooling is deflated and removed, the lining is sintered in place (Fig. 17.5). These components are also lined with the thermoplastic fluoropolymers (Fig. 17.6) by conventional molding using metal tooling that is removed after molding of the lining is completed.

Both lining methods may require secondary finishing to achieve appropriate dimensions for mating surfaces such as flanges.

Unlined valves made of stainless steel and other metals often have components of PTFE such as seats, packings and diaphragms. Along with chemical resistance, PTFE in seats and packings provides conformability, that is, the parts conform to mating surfaces for good sealing, and low friction for ease of operation. Special designs are used, particularly for higher pressures, to prevent PTFE parts from creeping away from contact areas. For diaphragms, PTFE provides good flexural life.

17.5 Seals and Gaskets

Seals for rotating or sliding shafts use elements of PTFE compounded with fillers such as graphite. Fillers must have the chemical resistance for the service conditions that will be encountered. Labyrinth seals that do not contact the shaft depend on a tortuous interface, fluid surface tension, and a pump-like design to prevent leakage.

Most PTFE seals are supported by metal springs or elastomeric rings to assure good shaft contact. Used by itself, PTFE, even when compounded with fillers, may creep under load and reduce sealing force. PTFE seals may have surface features that, with shaft rotation, produce a pumping action in the direction of the component being sealed.

PTFE is a standard material for gasketing (Fig. 17.7 and 17.8) due to its nearly universal chemical compatibility, good performance at temperatures from cryogenic to over 260°C, and conformability. PTFE is used alone, or with fillers or metal supports, or as a dispersion for impregnating fibers such as aramids for gasketing. In addition to excellent sealing performance, gaskets made with PTFE can be used in a broad range of applications so inventories can be reduced and there is less risk of misapplication.

Expanded PTFE that is fibrous in structure and contains microvoids is especially useful for gasketing against uneven or delicate surfaces. With increasing pressure, fillers are added to PTFE to counteract its tendency to creep and cause sealing to deteriorate. For higher pressures, gaskets incorporate metal rings or flat spirals to support sections of PTFE.

Figure 17.5 Examples of PTFE lined components. (*Courtesy Crane Resistoflex and DuPont Companies.*)

Figure 17.6 A PFA lined valve. (*Courtesy Xomox and DuPont Companies.*)

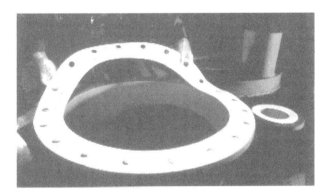

Figure 17.7 A Gore-Tex GR® sheet gasketing. (*Courtesy W. L. Gore and DuPont Companies.*)

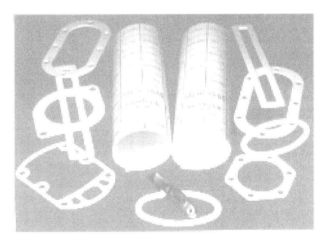

Figure 17.8 A Gylon® sheet gasketing material. (*Courtesy Garlock and DuPont Companies.*)

17.6 Self-Supporting Components

Some relatively small components for the CPI are made entirely of fluoropolymer. Typically, they do not bear significant loads and require uniform chemical resistance for all surfaces. Examples include dip tubes and distribution manifolds that blend fluids in vessels, and steam spargers that employ steam under pressure to pump aggressive fluids from tanks.

17.7 Trends in Using Fluoropolymers in Chemical Service

Improvements in processibility and technology are making it easier to use fluoropolymers to protect chemical equipment. Relatively new modified PTFE resins, for example, allow welding and thermoforming to reduce manufacturing costs.

For the thermoplastic fluoropolymers, rotolining is proving to be a reliable and versatile method for lining complex parts. The need to eliminated flanged connections wherever possible is increasing interest in dual laminate constructions that permit welding of joints.

In general, the CPI is moving towards extended maintenance intervals to reduce cost of operation, a move that supports the use of long-lasting equipment made with fluoropolymers. Productivity gain programs that depend on higher pressures and temperatures along with reduced contamination to increase yields also favor increasing use of fluoropolymers.

17.8 Semiconductor Processing

From its onset, the semiconductor industry has relied on fluoropolymers for wet processing equipment, fluid transport systems, and wafer handling tools (Fig. 17.9–14). Semiconductor manufacturing processes are extremely intolerant of particulate and chemical contamination which, even in trace amounts, can cause severe decrease in yields. Therefore, fluoropolymers' purity and resistance to chemical attack are important assets.

Wafer carriers, the devices that carry silicon wafers through various chemical processes, were first made from PTFE. This material is still used for some carriers, but because machining is required to create carrier designs in PTFE, PFA is often preferred, especially high-purity grades of PFA. PFA can be efficiently fabricated in the required configurations by injection molding method without costly secondary machining operations.

Aggressive ultrapure acids, solvents and water are required for processing semiconductors. To maintain their purity, these fluids are transported from fluoropolymer-lined storage containers to use points through fluoropolymer tubing, often high-purity PFA, and their flow is managed with valves, pumps, and pressure regulators made from PFA, PTFE, and sometimes other fluoropolymers.

These fluid handling components are nearly always solid fluoropolymer constructions. These designs are possible because sizes are relatively small compared with functionally similar but far larger components used in the CPI. Therefore, mechanical properties and material cost are not particularly sensitive issues. In addition, semiconductor processes are easily compromised by the presence of tiny amounts of metallic ions, so manufacturers avoid having metal components in the vicinity of process fluids.

The designers of these all-fluoropolymer components employ such unusual features as machined PTFE bellows for valves actuated by air pressure and rods incorporating carbon fiber that provide stiffness for wafer carriers.

Wet benches, the facilities in semiconductor plants where wet processing is carried out, typically use fluoropolymers for all surfaces that can affect fluid purity. Storage containers and sinks, along with all other plumbing, are constructed from fluoropolymers.

17.8.1 Trends for the Use of Fluoropolymers in the Semiconductor Industry

Circuit density is increasing, so tiny defects that did not interfere with the operation of less-dense circuits can cause failures. Such defects result from chemical and particulate contamination of fluids so there is growing demand for fluoropolymers with even higher purity and resistance to particulate formation. In addition, manufacturers of fluid handling equipment are working to assure that their processing methods do not contaminate fluoropolymers.

In addition, wafer size is increasing, and more processing steps are required to build more powerful circuits. These factors greatly increase the value of wafers in work, so upsets such as contamination of process fluids have a severe financial impact. To help guard against contamination, there is a growing use of double containment designs for fluid transport systems that feature two fluoropolymer barriers between fluids and the external environment.

17.9 Electrical and Mechanical

17.9.1 Electrical Applications

Fluoropolymers are widely used for wire and cable insulation (Fig. 17.15 and 17.16), but not always for the same reasons. It is true that all applications make use of a polymer's dielectric properties, but not to the same degree. Some uses exploit the ability of fluoropolymers to serve over a wide temperature range, and particularly at high temperatures. Others rely on their resistance to chemicals or their resistance to changes in properties over time.

Figure 17.9 High purity fluid handling system. (*Courtesy Atlantic Tubing and DuPont Companies.*)

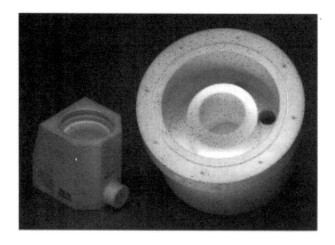

Figure 17.10 High purity flow controller and buret. (*Courtesy Mace Products and DuPont Companies.*)

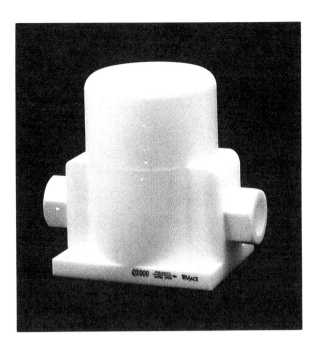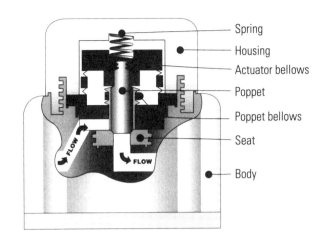

Figure 17.11 Poppet valves of PTFE for high purity applications. (*Courtesy Mace Products and DuPont Companies.*)

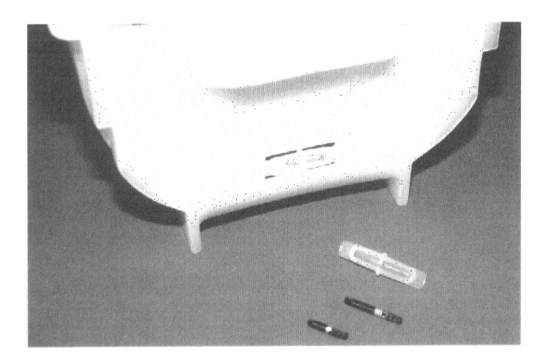

Figure 17.12 PFA semiconductor tracking system. (*Courtesy Entegris and DuPont Companies.*)

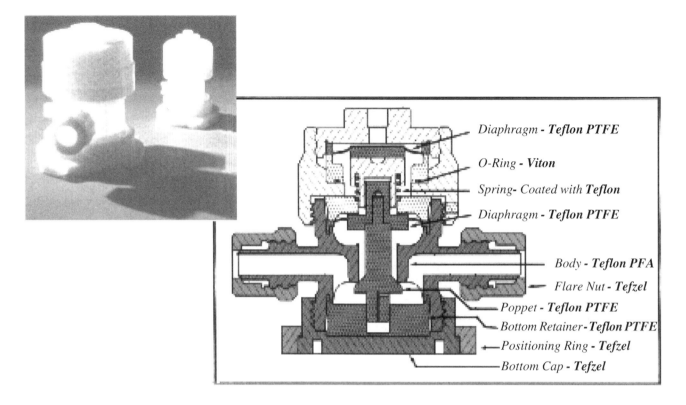

Figure 17.13 High purity fluid flow regulators. (*Courtesy Furon Fluid Handling and DuPont Companies.*)

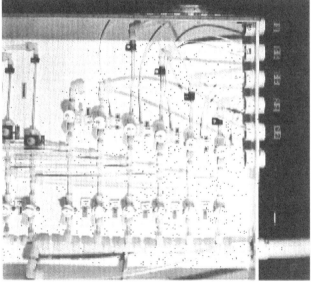

Figure 17.14 High purity fluid handling system. (*Courtesy Advance Micro Devices and DuPont Companies.*)

Figure 17.15 High temperature wire insulated with ETFE. (*Courtesy DuPont Company.*)

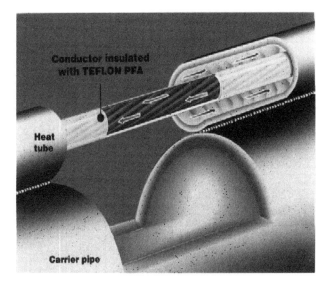

Figure 17.16 High temperature PFA insulation. (*Courtesy DuPont Company.*)

In electronic connectors for use at high frequencies, PTFE is a standard material due to its low dielectric constant. This dielectric property help minimize the loss of strength of signals being transmitted through the connectors. In thermocouple connectors, ETFE provides resistance to elevated temperatures.

A major use of FEP is in insulation for data cables that serve computer networks in office buildings. They often are called "plenum cables" because they are installed in air-handling plenum spaces. FEP is used for two reasons, its excellent high-frequency dielectric properties and fire performance. Its low dielectric constant and dissipation factor at high frequencies helps assure good data signal transmission at the high frequencies required for computer networks. Its fire performance helps these data cables meet building code safety requirements for low flame spread and low smoke generation.

FEP is also used to insulate coaxial cables that must meet the stringent building code requirements. Such cables are used for video transmission in broadcasting studios and other demanding high-frequency applications.

In some code jurisdictions, fire alarm cables are insulated with PVDF because of its good fire performance and mechanical toughness. In this use, high-frequency dielectric properties are unimportant.

For several decades, virtually all commercial and military aircraft have used signal, control and power wire and cables insulated entirely or partly with fluoropolymers. PTFE, FEP, ETFE, ECTFE and PVDF have been used for this insulation. Critical performance characteristics include service at extreme temperatures, good fire performance and resistance to chemicals such as hydraulic fluids, fuels, and cleaning solutions.

The resistance of fluoropolymers to property changes over time is also important for aircraft wiring insulation. PVC and many other common polymers used for insulation must be compounded with plasticizers, fire retardants and other materials to achieve required levels of performance. Some additives may migrate out of the insulation or undergo changes that will reduce performance to unacceptable levels. This deterioration can occur more rapidly at the high temperatures at which some aircraft wiring operates.

Most fluoropolymers do not require additives to enhance performance, and they are chemically stable at relatively high service temperatures, so their performance does not change with aging.

17.9.2 Mechanical Applications

PTFE is used in bearings due to its extremely low friction and inherent lubricity (Fig. 17.17). Compounded with graphite, bronze powder or other fillers to reduce creep and improve wear resistance, PTFE is molded or machined into bearings for unlubricated service.

PTFE offers unusual performance in bearings because its static coefficient of friction is lower than its dynamic coefficient. Therefore, PTFE bearings do not exhibit "stick-slip;" the jerking action that occurs in overcoming a higher static coefficient before movement begins.

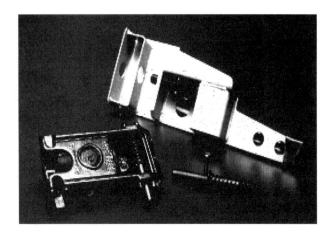

Figure 17.17 PTFE coated locking pins. (*Courtesy Youngblood and Associated and DuPont Companies.*)

In motion against a PTFE surface, a metal component causes fairly rapid initial wear. After a time, the rate of wear diminishes. That is because PTFE transfers to the metal surface so that the contact is PTFE against PTFE, a combination that produces little wear.

Bearings of PTFE usually require the added support of metal structures because PTFE lacks mechanical strength and stiffness. Compared with metal bearings, PTFE bearings are suitable only for relative low loads and velocities.

PTFE bearings are used in instruments, aircraft and aerospace vehicles control systems, office machines, and other applications where lubrication is difficult or undesirable.

Slide bearings called *bearing pads* are used in support systems for bridges and some buildings to accommodate thermal and seismic movement without damage to the structures they support. Compared with similar elastomeric supports, they allow greater range of movement, and compared with lubricated metal bearings, PTFE bearing pads require no lubrication and are highly resistant to moisture and chemical attack. Designs often have stainless steel plates riding against a PTFE sheet surface. As for other bearings, the PTFE is usually compounded to improve creep performance.

Bearings made with PTFE fiber are used in packaging machinery, pulp and paper processing equipment, and other applications. The PTFE fiber is woven into a fabric with some other fiber that will bond well with adhesives. This fabric is incorporated in a RTP (reinforced thermoplastic) structure to form a spherical bearing that will accommodate misalignment.

Low-friction linings for push-pull control cables that require no lubrication can be made with PTFE. Such cables are used in automotive and aerospace applications.

17.10 Automotive

For applications in automobiles and other vehicles, fluoropolymers are selected for their resistance to high temperatures and chemicals. Rising underhood temperatures and the need to prevent the release into the atmosphere of aggressive fuel components are causing these applications to grow (Figs. 17.18, 17.19).

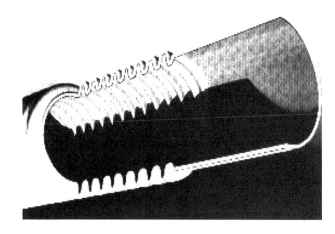

Figure 17.18 Convoluted ETFE tubing for automotive filling and vapor management. (*Courtesy Pilot Industries and DuPont Companies.*)

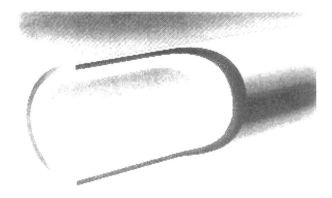

Figure 17.19 ETFE lined composite tubing for hydrocarbon transfer. (*Courtesy Pilot Industries and DuPont Companies.*)

Wire insulated with PTFE is used for connections to oxygen sensors mounted in vehicle exhaust manifolds. In this hot environment, PTFE provides reliable dielectric protection for wiring that is vital to control of exhaust emissions. ETFE is used to insulate wiring for other high-temperature locations near engines, and wiring exposed to hot hydraulic fluid within automatic transmissions.

The resistance of ETFE to permeation of fuel components is critical in applications in fuel tank filler necks and fuel and vapor management hoses. These components help automobile manufacturers meet strict emissions regulations. In one successful design, these components are elastomeric structures lined with ETFE.

PTFE tubing jacketed with braided stainless steel wire is used for brake lines and coolant circulation in heavy-duty vehicles. The wire jacket helps prevent damage due to high fluid pressures and abrasion, and the PTFE tubing resists elevated temperatures, aging and exposure to chemicals in fluids.

18 Safety, Disposal, and Recycling of Fluoropolymers

18.1 Introduction

Fluoropolymers are among the most versatile plastics thanks to their properties. Thermal stability is a major feature of these polymers spurring their applications where high temperature exposures are encountered. Fluoropolymers can produce toxic products if they are overheated. Precautions should be taken to remove any degradation fragments produced during the processing and fabrication of parts from fluoropolymers.

This chapter contains information about safe handling and processing of fluoropolymers. The material in this chapter is in no way intended as a replacement for the specific information and data supplied by the manufacturers of fluoropolymers. A source of information to which frequent reference is made in this chapter is *The Guide to Safe Handling of Fluoropolymers Resins,* published by The Society of Plastics Industry, Inc. Disposal and recycling issues are also reviewed.

18.2 Toxicology of Fluoropolymers

Fluoropolymers are chemically stable and inert or relatively unreactive. Reactivity decreases as fluorine content of the polymer increases. Fluorine induces more stability than chlorine. This family of plastics has low toxicity and almost no toxicological activity. No fluoropolymers have been known to cause skin sensitivity and irritation in humans. Polyvinyl fluoride contains one fluorine atom and three hydrogen atoms per monomer unit and has been shown to cause no skin reaction in human beings.[1] Excessive human exposure to fluoropolymer resin dust resulted in no toxic effects, although urinary fluoride content increased.[2]

Filled or compounded resins contain pigments, surfactants, and other additives to modify the plastic properties. These additives are likely to present risks and hazards in the fluoropolymer compound. For example, aqueous dispersions of fluoropolymers contain surfactants that may produce adverse physiological symptoms. The hazards of using these additives should be considered by themselves and in conjunction with fluoropolymers. Safety information provided by manufacturers of the additives and the compounds should be consulted.

18.3 Thermal Properties of Fluoropolymers

Thermal decomposition of fluoropolymers has been discussed in Ch. 14 and 15. The reader should refer to these sections for a review of this topic. Fluoropolymers are heated to high temperatures during processing and degrade to some extent. It is important to remember that the type of degradation products and the extent of decomposition depend on several factors. One must consider the following variables during processing and fabrication:

Temperature

Presence of oxygen

Physical form of the article

Residence time at temperature

Presence of additives

The products of decomposition of fluoropolymers fall in three categories: fluoroalkenes, oxidation products, and particulates of low molecular weight fluoropolymers. These products must be removed by adequate ventilation from the work environment to prevent human exposure. A major oxidation product of PTFE is carbonyl fluoride, which is highly toxic and hydrolyzes to yield hydrofluoric acid and carbon dioxide. At 450°C in the presence of oxygen, PTFE degrades into carbonyl fluoride and hydrofluoric acid. At 800°C, tetrafluoromethane is formed. It has been suggested that tetrafluoroethylene (TFE) is the only product that is produced when PTFE is heated to melt stage.[3] Some studies have proposed that TFE, trapped in the polymer matrix, is released upon heating.

It is important for service protocols of fluoropolymer parts to follow the recommendations and specifications of the resin and part suppliers. From a thermal exposure standpoint, the maximum continuous use temperature should comply with the values specified by *The Guide to Safe Handling of Fluoropolymers Resins,* published by The Society of Plastics Industry, Inc. (Table 18.1).

Table 18.1. Maximum Continuous Use and Processing Temperatures[3]

Polymer	Maximum Continuous Use Temperature, °C	Typical Processing Temperature, °C
PTFE	260	380
PFA	260	380
FEP	205	360
ETFE	150	310
ECTFE	150	280
PCTFE	120	265

18.4 Emission During Processing

Fluoropolymers degrade during processing and generate effluents with an increasing rate with temperature. Operation of process equipment at high temperatures may result in generation of toxic gases and particulate fume. The most common adverse effect associated with human exposure to degradation products of fluoropolymers is *polymer fume fever* (PFF). This exposure presents itself by a temporary (about 24 hours) flu-like condition similar to metal fume fever.[1] Fever, chills, and occasionally coughs are among the observed symptoms

Other than inhalation of degradation products, fume fever may also be caused by fluoropolymer contaminated smoking material. It is prudent to ban tobacco products from fluoropolymer work areas. Local exhaust ventilation should be installed to remove the process effluents from the work areas. It has been suggested that no health hazards exist unless the fluoropolymer is heated above 300°C.[4]

Johnston and his coworkers[5] have proposed that heating PTFE gives rise to fumes which contain very fine particulates. The exposure of lung tissues to these particulates can result in a toxic reaction causing pulmonary edema or excessive fluid build up in the lung cells. Severe irritation of the tissues along with the release of blood from small vessels is another reaction to exposure. In controlled experiments, animals were exposed to filtered air from which fumes had been removed and the unfiltered air. Unfiltered air produced the expected fume fever response. Animals exposed to the filtered air did not develop any of the symptoms of polymer fume fever.

The products of fluoropolymer decomposition produce certain health effects upon exposure, summarized in Table 18.2. The risks associated with exposure to these effluents have prompted the establishment of a number of exposure limits by various regulatory agencies (Table 18.3). Resin manufacturers can supply available exposure information.

18.5 Safety Measures

A number of measures can be taken to reduce and control exposure to monomers and decomposition products during the processing of fluoropolymers. It is important to monitor processing plants and take measures where necessary according to legal requirements established under the Occupational Safety and Health Act (OSHA). The customary precautionary actions for safe fluoropolymer processing are described in this section. They include ventilation, processing measures, spillage clean up and equipment cleaning, and maintenance procedures. A number of general measures should be taken while handling fluoropolymers including protective clothing, personal hygiene, fire hazard, and material incompatibility.

18.5.1 Ventilation

Removal of the decomposition products from the work environment is one of the most important actions taken to reduce and control human exposure. Even at room temperature, small amounts of trapped monomers or other gases can diffuse out of the resin particles. It is a good practice to open the fluoropolymer container in a well-ventilated area. All processing equipment should be ventilated by local exhaust ventilation schemes.

The most effective method of controlling emissions is to capture them close to the source before they are dispersed in the workspace. A fairly small volume of air has to be removed by local exhaust compared to the substantially larger volume of air that must be removed from the entire building. Correct design and operation of local exhaust systems can minimize human exposure. Examples of hood devices available for ventilation are shown in Figs. 18.1–5.

Exhaust air must enter the hood to carry the contaminants with it and convey them to the exhaust point. The required air velocity at the point the contaminants are given off to force these contaminants into an exhaust hood is called *capture velocity* and should be at least 0.5 m/sec. An airflow meter can be used to measure the air velocity. A static pressure gauge can be installed to continuously monitor the air velocity in the hood by pressure drop.

Table 18.2. Health Hazards of Fluoropolymers Decomposition Products[3]

Decomposition Product	Associated Health Effects
Hydrogen fluoride (HF)	Symptoms: choking, coughing, and sever eye, nose and throat irritation, fever, chills, breathing difficulty, cyanosis and pulmonary edema. HF is corrosive to the eyes, skin and respiratory tract. May be absorbed through skin in toxic quantities. Sometimes-invisible delayed burns. Overexposure to HF may cause kidney and liver injury.
Carbonyl fluoride (COF_2)	Overexposure causes skin irritation with rash, eye corrosion or conjunctival ulcer, irritation of upper respiratory passage, and temporary lung irritation with coughs, discomfort, breathing difficulty or shortness of breath.
Tetrafluoro-ethylene (TFE) ($CF_2=CF_2$)	Causes acute effects when inhaled, including irritation of upper respiratory tract and eyes, mild depression of central nervous system, nausea and vomiting and dry cough. Massive inhalation produces cardiac arrhythmia, cardiac arrest and death. A study by National Toxicology Program has reported kidney and liver tumors in rats and mice, which had exposed to lifetime inhalation of TFE. Relationship to human response has not been established. An exposure limit of 5ppm has been established by fluoropolymer producers.
Perfluoro-isobutylene $CF_3C=CF_2$ CF_3 (PFIB)	Animal studies of PFIB inhalation indicate occurrence of severe adverse including pulmonary edema as a result of exposure to high concentrations and death. Wheezing, sneezing, difficulty breathing and deep or rapid breathing are among the symptoms. Animals that survived 24 hours after exposure recovered with no after-effects.

Table 18.3. Exposure Limit Types

Limit	Type	Source
Permissible Exposure Limit (PEL)	Legal occupation exposure limit (OEL)	US Code of Federal Regulations, Title 29, Part 1910 (29 CFR 1910)
Threshold Limit Values (TLV)	Recommended exposure limit (REL)	American Conference of Governmental Industrial Hygienists or National Institute for Occupational Safety and Health (NIOSH)

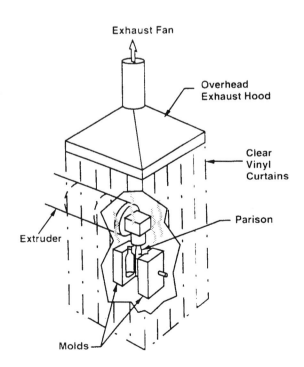

Figure 18.1 Exhaust scheme for blow molding.[6]

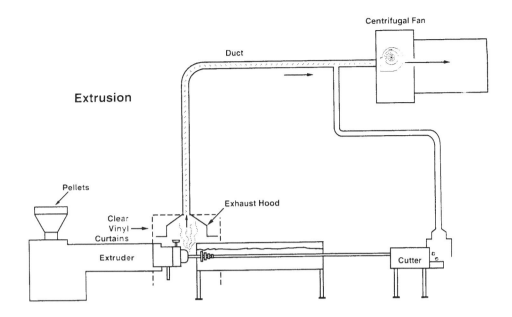

Figure 18.2 Typical setup for exhaust.[6]

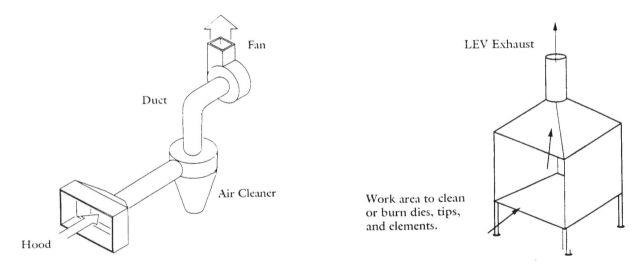

Figure 18.3 Exhaust system.[6]

Figure 18.4 Burnout hood.[6]

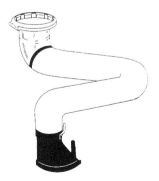

Figure 18.5 Adjustable ducting hood.[6]

Three publications by The Society of Plastics Industry,[3] American Conference of Governmental Industrial Hygienists[7] and Canadian Center for Occupational Health and Safety[8] provide detailed information on various aspects of industrial ventilation.

18.5.2 Processing and Fabrication

Processing and fabrication of different fluoropolymers may involve one or more of a number of processes. These include sintering, paste extrusion, dispersion coating, melt processing, machining, soldering and melt stripping, welding, and welding and flame cutting fluoropolymer-lined metals. This section covers each of these processing steps.

18.5.2.1 Sintering

This operation requires heating the polymer in ovens at high temperatures where decomposition products are formed to different extents. Ovens must be equipped with sufficiently strong ventilation to remove the gaseous products and prevent them from entering the work area. It is important that ventilation prevent entrance of the contaminants into the plant area during the operation of the oven and when the door is open.

Ovens operate at high temperatures approaching 400°C. Overheating should be avoided by installing limit switches to avoid oven runoff that can result in high temperatures at which accelerated decomposition may occur. It is a good practice to operate the sintering ovens at the lowest possible temperature that is adequate for the completion of the part fabrication. An overheated oven must be cooled before opening the doors. Proper personal protective equipment, including a self-contained breathing apparatus, must be worn prior to opening the oven doors.

Compounds containing fillers are usually more sensitive to thermal decomposition due to the acceleration of thermo-oxidative reactions by a number of additives at elevated temperatures. It may also be possible to sinter compounds at lower temperatures due to changes in conductivity of the part. For example, a metal filled PTFE compound (steel, lead, or bronze) has a significantly higher thermal conductivity than PTFE, which leads to rapid heating of the part.

18.5.2.2 Paste Extrusion

Dispersion polymerized polytetrafluoroethylene is processed by *paste extrusion* in which the polymer is mixed with a hydrocarbon lubricant (see Ch. 8). These petroleum-based lubricants have low flash points posing potential fire hazard. Lubricants should be stored in electrically conductive containers and the process equipment should be grounded to eliminate static electricity as a source of ignition. Inhalation of vapors and skin contact are the human exposure hazards. Removal of the lubricant from the extrudate takes place in a batch oven or continuously. Action must be taken to minimize the risk of forming explosive mixtures of lubricant and air. In continuous operation, sintering zones immediately follow the drying oven. Incorrect operation may carry flammable vapors into the very high temperature oven zones, resulting in a most certain ignition and fire. Ventilation of the ovens by adequate air flow would minimize fire hazards. Chapter 8 includes a more detailed discussion of safety issues.

18.5.2.3 Dispersion Coating

Aqueous dispersions of fluoropolymers are applied to substrates by coating techniques, as described in Ch. 9. Water and surfactant are normally removed in a heating step prior to sintering during which surfactants may decompose. The degradation fragments and the surfactant may be flammable. They may also have adverse health effects. Forced ventilation of the drying oven is necessary to remove the surfactant vapors and minimize build up of degradation products. Some coating formulations contain organic solvents. Combustion hazards and health effects of these substances should be considered during the handling and processing of the coating.

18.5.2.4 Melt Processing

Melt processing of fluoropolymers exposes them to very high temperatures mainly to lower viscosity to improve flow. Typical melt processes such as extrusion and injection molding confine the melt in a closed environment. Any decomposition as a result of extended high temperature exposure may produce gases and generate pressure causing blow-back into the feed section or equipment rupture in the absence of a vent. Discoloration of the polymer is an indication of thermal degradation. Rupture disks are installed on melt process equipment to act as emergency pressure relief. Special metal alloys are specified for the construction of contact surfaces due to the corrosive properties of the molten fluoropolymer.

18.5.2.5 Machining

Fabricators of fluoropolymer articles machine, saw, and grind the plastic shapes into their final geometry. Sharp tools can be used in high speed machining of these plastics. Coolants are recommended to increase production rates without overheating. Dust is generated, particularly during the grinding and sawing of shapes. It is recommended that dust particulates be removed from the workspace. Occupational exposure limit of 10 mg/m^3 and respirable limit of 5 mg/m^3 are recommended.[2] These limits may have to be lowered when machining parts containing fillers, which may form hazardous dust. Material safety data sheets (MSDS) provide further information about the additives.

18.5.2.6 Soldering and Melt Stripping

In electronic applications, wires are commonly stripped by heat and soldered for hook up of circuits. Fluoropolymer insulation is subjected to heat and decomposition occurs. It is important to remove the fumes by local ventilation to avoid exposure to gases.

18.5.2.7 Welding Fluoropolymer

Welding fluoropolymer parts to each other should be done carefully due to the generation of large quantities of hydrofluoric acid. Appropriate protective clothing including self-contained breathing apparatus must be worn during welding.

18.5.2.8 Welding and Flame-Cutting Fluoropolymer-lined Metals

Welding arcs and torches are capable of massive destruction of fluoropolymers due to the excessive heat that they generate. The plastic part should be removed before metal is cut or welded. Local exhaust ventilation should be provided whenever it is not possible to remove the polymer from the area to be welded or cut.

18.5.3 Spillage Cleanup

Fluoropolymers can create a slippery surface when they are rubbed against a hard surface because they are soft and easily abrade away and coat the surface. Any spills during handling should be cleaned up immediately. It is helpful to cover the floors of the processing area with anti-slip coatings.

18.5.4 Equipment Cleaning and Maintenance

Dies, screens, molds, screws, mandrels, screen packs and other components of processing equipment should be periodically cleaned up and polymer residues removed. The cleaning methods includes a pyrolysis step which should be conducted under adequate ventilation.

18.5.5 Protective Clothing

Appropriate protective clothing should be worn to avoid burns at the processing temperatures of fluoropolymers. They include safety glasses, gloves and gauntlets (arm protection). Dust masks or respirators should be worn to prevent inhalation of dust and particulates of fluoropolymers during grinding and machining. Additional protection may be required when working with filled compounds. Skin contact with fluoropolymer dispersions should be avoided by wearing gloves, overalls, and safety glasses due to their surfactant or solvent content. Fluoropolymer coatings must be sprayed in a properly equipped spray booth. Overspray should be captured in a water bath. The spray operator should wear a disposable (Tyvek®)* suit, goggles, gloves and a respirator or self-contained breathing apparatus.

18.5.6 Personal Hygiene

Tobacco products should be banned from the work areas to prevent polymer fume fever. Street clothing should be stored separately from work clothing. Thorough washing after removal of work clothing will remove powder residues from the body.

18.5.7 Fire Hazard

Fluoropolymers do not ignite easily and do not sustain flame. They can decompose in a flame and evolve toxic gases. For example, PTFE will sustain flame in an ambient of >95% oxygen (Limiting Oxygen Index by ASTM D 2863). In less oxygen rich environments, burning stops when the flame is removed. Underwriters Laboratory rating of fluoropolymers is 94–V0. Self ignition temperature of PTFE is 500–560°C according to ASTM D1929, far above most other organic materials.[2] PTFE does not form flammable dust clouds under normal conditions as determined in the Godwert Greenwald test at 1000°C. Polytetrafluoroethylene falls in the explosion class ST1.[2]

18.5.8 Material Incompatibility

Small particles of fluoropolymers become highly combustible in the presence of metal fines. Aluminum and magnesium produce fire and/or explosion with PTFE powder when exposed above 420°C. Practical situations where a similar reaction is possible include pumps with fluoropolymer packing used to pump aluminum flake dispersions, or grinding or sanding metal parts coated with fluoropolymers.[3]

---

*Tyvek® is a trademark of DuPont Co.

18.6 Food Contact and Medical Applications

Fluoropolymer resins are covered by Federal Food, Drug and Cosmetic Act, 21 CFR & 177.1380 & 177.1550 in the United States and EC Directive 90/128 in the European Union. The U.S. Food and Drug Administration has approved many fluoropolymers (e.g., PTFE, PFA and FEP) for food contact. Additives such as pigments, stabilizers, antioxidants, and others must be approved under a food additive regulation, if they do not have prior clearance.

Medical applications of fluoropolymers have been a sensitive area for the resin manufacturers and medical device fabricators. Some fluoropolymers have been used in the construction of FDA regulated medical devices. FDA only grants approval for a complete device, not components such as resin. Resin suppliers usually have specific policies regarding the use of their products in medical devices. Thorough review of these policies and regulatory counsel advice would be prudent before initiation of any activity in this area.

18.7 Fluoropolymer Scrap and Recycling

Fluoroplastics described in this book are thermoplastics and can be reused under the right circumstances. There are a few sources of waste fluoropolymer. Various processing steps of fluoropolymers such as preforming, molding, machining, grinding, cutting create debris and scrap. Some of the scrap material is generated prior to sintering, but the majority is produced after sintering. A third category of scrap is polymer that does not meet specifications and cannot be used in its intended applications. Efforts have been made to recycle PTFE soon after its discovery. The incentive in the early days to recycle scrap was economic due to the high cost of polytetrafluoroethylene. Today, a small industry has evolved around recycling fluoropolymers.

Scrap PTFE has to be processed for conversion to usable feedstock. The extent of processing depends on the amount of contamination in the debris. The less contaminant in the scrap, the higher the value of the material will be. Machine cuttings and debris usually contain organic solvents, metals, moisture and other contaminants. Conversion of this material to useful feedstock requires chemical and thermal treatment. The clean PTFE feedstock can be converted to a number of powders.

A large quantity of scrap PTFE is converted into micropowder (fluoroadditive) by methods discussed in Ch. 11. Fluoroadditives are added to plastics, inks, oils, lubricants, and coatings to impart fluoropolymer-like properties such as reduced wear rate and friction. Part of the PTFE is converted back into molding powders, which are referred to as "Repro," short for *reprocessed*. Unmelted new polymer is, by contrast, called "virgin." These powders are molded under pressure by ram extrusion or sheet molding as described in Ch. 7.

Removal of organic contaminants is accomplished by heating the scrap material to above its melting point and oxidizing them to gases and volatile products. Inorganic compounds are usually removed by heating the material in a mixture of nitric and perchloric acids. The highly oxidizing nature of the acid blend converts the inorganic material into water-soluble salts, which are removed by subsequent washing with water. Repeated heating reduces the molecular weight of repro PTFE molding powders with ramifications for the properties of parts made from them

Studies of repro PTFE properties have long indicated that tensile properties are lower due to molecular weight reductions.[9][10] Figures 18.6 and 18.7 and table 18.4 show properties measured in two comparative studies of repro and virgin PTFE. Repro resins have lower tensile strength and break elongation than virgin resin at the room temperature. The property differences grow as temperature increases. Dielectric constant and dielectric breakdown strength of repro are higher than virgin. Impact strength of repro is lower than virgin's as seen in Fig. 18.8. Molded parts of repro and virgin can, therefore, be expected to exhibit differences in service performance. Less demanding or mechanical applications are suitable for repro parts where full properties of polytetrafluoroethylene are not required for satisfactory performance.

In powder form, repro and virgin PTFE can be differentiated by the difference in their melting points (327°C for repro versus 342°C for virgin). In part form, higher specific gravity and lower tensile properties of repro products are clues to the source of PTFE. Sometimes uneven appearance or off-white discoloration of the part, due to in-mold sintering, are signs that the part is made from repro PTFE.

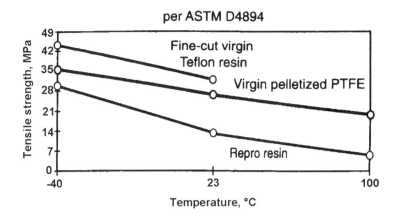

Figure 18.6 Variation in tensile strength with temperature.[10]

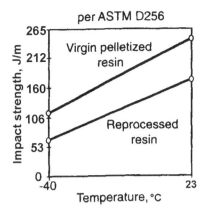

Figure 18.8 Impact strength of virgin vs repro PTFE.[10]

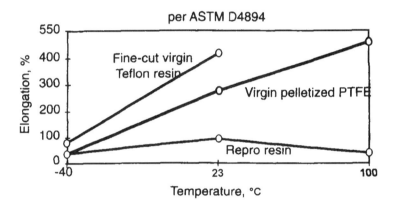

Figure 18.7 Temperature changes vs elongation.[10]

Table 18.4. Comparison of Virgin and Reprocessed PTFE Properties[9]

Property	Virgin	Repro
Specific Gravity	2.14-2.20	2.15-2.20
Tensile Strength, MPa	24-42	11-16.5
Break Elongation, %	150-450	100-200
Dielectric Breakdown Strength, kV/mm	60	48
Dielectric Constant	2.10	2.26

18.8 Environmental Protection and Disposal Methods

None of fluoropolymers or their decomposition products poses any threats to the ozone layer. None are subject to any restrictive regulations under the Montreal Protocol and the US Clean Air Act. Reacting HF with chloroform (see Ch. 4 for details) produces the main fluorinated ingredient of tetrafluoroethylene synthesis $CHClF_2$. It has a small ozone depleting potential but is excluded from the Montreal Protocol regulation due to its intermediate role and destruction from the environment.

The preferred method of disposing fluoropolymers are recycling and landfilling according to the various regulations. In the case of suspensions and dispersions, solids should be removed from the liquid and disposed. Liquid discharge to waste water systems should be according to the permits. None of the polymers should be incinerated unless the incinerator is equipped to scrub out hydrogen fluoride, hydrogen chloride and other acidic products of combustion.

In the disposal of fluoropolymer scrap containing pigments, additives, or solvents, additional consideration must be given to the regulation for the disposal of the non-fluoropolymer ingredients. Some of the compounds and mixtures may require compliance with Hazardous Material Acts.

Reference

1. Harris, L. R. and Savadi, D. G., "Synthetic Polymers," *Patty's Industrial Hygiene and Toxicology*, 4th ed., Vol. 2, Part E, George D. Clayton and Florence E. Clayton, Eds., John Wiley & Sons, New York, 1994.
2. "Guide for the Safe Handling of Fluoropolymer Resins," Association of Plastics Manufacturers in Europe, Brussels, Belgium, 1995.
3. *The Guide to Safe Handling of Fluoropolymers Resins*, published by The Society of Plastics Industry, Inc., 1998.
4. Rose, C. A., *Inhalation Fevers*, in *Environmental and Occupational Medicine*, 2nd ed., Rom, W. N., ed., pp. 373–380, Little, Brown and Company, Boston, 1992.
5. Johnston, C. J., Finkelstein, J. N., Gelein, R., Baggs, R., and Obrduster, G., "Characterization of Early Pulmonary Inflammatory Response Associated with PTFE Fume Exposure," *Toxicology and Applied Pharmacology*, Article No. 0208, Academic Press, May, 1996.
6. "Proper Use of Local Exhaust Ventilation during Hot Processing of Plastics," DuPont Publication, Oct., 1992.
7. *Industrial Ventilation: A Manual of Recommended Practice*, published by American Conference of Governmental Industrial Hygienists.
8. "A Basic Guide to Industrial Ventilation," published by Canadian Center for Occupational Health and Safety, Hamilton, Ontario, Canada L8N 1H6, Pub. No. 88-7E.
9. Arkles, B., "Recycling Polytetrafluoroethylene," *Polymer Science and Technology*, Vol. 3, J. Guillet, Ed, Plenum Press, New York, 1973.
10. Ebnesajjad, S. and Lishinsky, V., "Can Reprocessed Resins Do the Job?" *Machine Design,* Feb. 11, 1999.

Appendix I: Polytetrafluoroethylene

Table A.I.1. Hydrogen Permeability vs. Temperature and Pressure Through Carbon-Filled DuPont Teflon® Polytetrafluoroethylene

Material Family	POLYTETRAFLUOROETHYLENE								
Material Supplier/ Grade	DUPONT TEFLON								
MATERIAL CHARACTERISTICS									
Sample Thickness	0.05 mm	0.05 mm	0.05 mm	0.05 mm	0.05 mm	0.05 mm	0.05 mm	0.05 mm	0.05 mm
MATERIAL COMPOSITION									
Note	carbon filled	carbon filled	carbon filled	carbon filled	carbon filled	carbon filled	carbon filled	carbon filled	carbon filled
TEST CONDITIONS									
Penetrant	hydrogen								
Temperature (°C)	-15	25	68	-11	25	67	-14	25	65
Pressure Gradient (kPa)	1724	1724	1724	3447	3447	3447	6895	6895	6895
Test Method	mass spectrometry and calibrated standard gas leaks; developed by McDonnell Douglas Space Systems Company Chemistry Laboratory								
PERMEABILITY (source document units)									
Gas Permeability ($cm^3 \cdot mm/cm^2 \cdot kPa \cdot sec$)	3.95×10^{-9}	1.34×10^{-8}	3.53×10^{-8}	4.51×10^{-9}	1.27×10^{-8}	3.42×10^{-8}	4.17×10^{-9}	1.23×10^{-8}	3.32×10^{-8}
PERMEABILITY (normalized units)									
Permeability Coefficient ($cm^3 \cdot mm/m^2 \cdot day \cdot atm$)	346	1173	3090	395	1112	2994	365	1077	2906

Table A.I.2. Hydrogen Permeability vs. Temperature and Pressure Through DuPont Teflon® Polytetrafluoroethylene

Material Family	POLYTETRAFLUOROETHYLENE								
Material Supplier/ Grade	DUPONT TEFLON								
MATERIAL CHARACTERISTICS									
Sample Thickness	0.03 mm	0.03 mm	0.03 mm	0.03 mm	0.03 mm	0.03 mm	0.03 mm	0.03 mm	0.03 mm
TEST CONDITIONS									
Penetrant	hydrogen								
Temperature (°C)	-16	25	68	-17	25	67	-18	25	63
Pressure Gradient (kPa)	1724	1724	1724	3447	3447	3447	6895	6895	6895
Test Method	mass spectrometry and calibrated standard gas leaks; developed by McDonnell Douglas Space Systems Company Chemistry Laboratory								
PERMEABILITY (source document units)									
Gas Permeability ($cm^3 \cdot mm/cm^2 \cdot kPa \cdot sec$)	1.7×10^{-9}	6.34×10^{-9}	1.88×10^{-8}	1.63×10^{-9}	5.9×10^{-9}	1.86×10^{-8}	1.59×10^{-9}	5.94×10^{-9}	1.64×10^{-8}
PERMEABILITY (normalized units)									
Permeability Coefficient ($cm^3 \cdot mm/m^2 \cdot day \cdot atm$)	149	555	1646	143	516	1628	139	520	1436

Table A.I.3. Nitrogen Permeability vs. Temperature and Pressure Through DuPont Teflon® Polytetrafluoroethylene

Material Family	POLYTETRAFLUOROETHYLENE								
Material Supplier/ Grade	DUPONT TEFLON								
MATERIAL CHARACTERISTICS									
Sample Thickness	0.03 mm	0.03 mm	0.03 mm	0.03 mm	0.03 mm	0.03 mm	0.03 mm	0.03 mm	0.03 mm
TEST CONDITIONS									
Penetrant	nitrogen								
Temperature (°C)	-23	25	71	-25	25	70	-23	25	68
Pressure Gradient (kPa)	1724	1724	1724	3447	3447	3447	6895	6895	6895
Test Method	mass spectrometry and calibrated standard gas leaks; developed by McDonnell Douglas Space Systems Company Chemistry Laboratory								
PERMEABILITY (source document units)									
Gas Permeability ($cm^3 \cdot mm/cm^2 \cdot kPa \cdot sec$)	9.46×10^{-11}	7.87×10^{-10}	2.9×10^{-9}	8.89×10^{-11}	7.88×10^{-10}	2.89×10^{-9}	9.47×10^{-11}	7.84×10^{-10}	2.87×10^{-9}
PERMEABILITY (normalized units)									
Permeability Coefficient ($cm^3 \cdot mm/m^2 \cdot day \cdot atm$)	8.3	68.9	254	7.8	69	253	8.3	68.6	251

Table A.I.4. Nitrogen Permeability vs. Temperature and Pressure Through Carbon-Filled DuPont Teflon® Polytetrafluoroethylene

Material Family	POLYTETRAFLUOROETHYLENE								
Material Supplier/ Grade	DUPONT TEFLON								
MATERIAL CHARACTERISTICS									
Sample Thickness	0.05 mm	0.05 mm	0.05 mm	0.05 mm	0.05 mm	0.05 mm	0.05 mm	0.05 mm	0.05 mm
MATERIAL COMPOSITION									
Note	carbon filled	carbon filled	carbon filled	carbon filled	carbon filled	carbon filled	carbon filled	carbon filled	carbon filled
TEST CONDITIONS									
Penetrant	nitrogen								
Temperature (°C)	-14	25	68	-17	25	71	-17	25	67
Pressure Gradient (kPa)	1724	1724	1724	3447	3447	3447	6895	6895	6895
Test Method	mass spectrometry and calibrated standard gas leaks; developed by McDonnell Douglas Space Systems Company Chemistry Laboratory								
PERMEABILITY (source document units)									
Gas Permeability (cm^3 · mm/ cm^2 · kPa · sec)	2.5×10^{-10}	1.46×10^{-9}	5.28×10^{-9}	2.34×10^{-10}	1.52×10^{-9}	5.32×10^{-9}	2.34×10^{-10}	1.42×10^{-9}	4.78×10^{-9}
PERMEABILITY (normalized units)									
Permeability Coefficient (cm^3 · mm/m^2 · day · atm)	21.9	128	462	20.5	133	466	20.5	124	418

Table A.I.5. Oxygen and Ammonia Permeability vs. Temperature and Pressure Through Carbon-Filled DuPont Teflon® Polytetrafluoroethylene

Material Family	POLYTETRAFLUOROETHYLENE								
Material Supplier/ Grade	DUPONT TEFLON								

MATERIAL CHARACTERISTICS

Sample Thickness	0.03 mm	0.03 mm	0.03 mm	0.03 mm	0.03 mm	0.03 mm	0.03 mm	0.03 mm	0.03 mm

TEST CONDITIONS

Penetrant	ammonia			oxygen					
Temperature (°C)	-3	25	63	-17	25	51	-17	25	51
Pressure Gradient (kPa)	965	965	965	1724	1724	1724	3447	3447	3447
Test Method	mass spectrometry and calibrated standard gas leaks								
Test Note	developed by McDonnell Douglas Space Systems Company Chemistry Laboratory								

PERMEABILITY (source document units)

Gas Permeability ($cm^3 \cdot mm / cm^2 \cdot kPa \cdot sec$)	4.71×10^{-10}	1.73×10^{-9}	8.62×10^{-9}	5.27×10^{-10}	2.55×10^{-9}	5.38×10^{-9}	4.55×10^{-10}	2.54×10^{-9}	5.46×10^{-9}

PERMEABILITY (normalized units)

Permeability Coefficient ($cm^3 \cdot mm/m^2 \cdot day \cdot atm$)	41.2	151	755	46.1	223	471	39.8	222	478

Table A.I.6. Oxygen and Ammonia Permeability vs. Temperature and Pressure Through Carbon-Filled DuPont Teflon® Polytetrafluoroethylene

Material Family	POLYTETRAFLUOROETHYLENE								
Material Supplier/ Grade	DUPONT TEFLON								

MATERIAL CHARACTERISTICS

Sample Thickness	0.05 mm	0.05 mm	0.05 mm	0.05 mm	0.05 mm	0.05 mm	0.05 mm	0.05 mm	0.05 mm

MATERIAL COMPOSITION

Note	carbon filled	carbon filled	carbon filled	carbon filled	carbon filled	carbon filled	carbon filled	carbon filled	carbon filled

TEST CONDITIONS

Penetrant	ammonia			oxygen					
Temperature (°C)	-2	25	62	-16	25	55	-15	25	53
Pressure Gradient (kPa)	965	965	965	1724	1724	1724	3447	3447	3447
Test Method	mass spectrometry and calibrated standard gas leaks								
Test Note	developed by McDonnell Douglas Space Systems Company Chemistry Laboratory								

PERMEABILITY (source document units)

Gas Permeability ($cm^3 \cdot mm/cm^2 \cdot kPa \cdot sec$)	7.77×10^{-10}	2.75×10^{-9}	1.21×10^{-8}	9.28×10^{-10}	5.05×10^{-9}	1.16×10^{-8}	9.56×10^{-10}	5.15×10^{-9}	1.01×10^{-8}

PERMEABILITY (normalized units)

Permeability Coefficient ($cm^3 \cdot mm/m^2 \cdot day \cdot atm$)	68.0	241	1059	81.2	442	1015	83.7	451	884

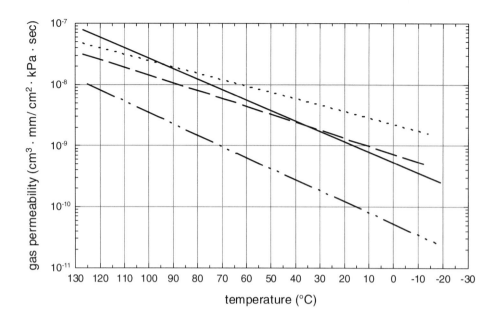

Figure A.I.1 Gas permeability vs. temperature through polytetrafluoroethylene.

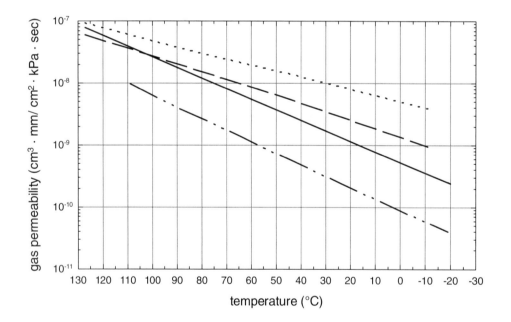

Figure A.I.2 Gas permeability vs. temperature through carbon-filled polytetrafluoroethylene.

Appendix II: Polychlorotrifluoroethylene

Permeability

Allied Signal: ACLAR® (mfeatures: transparent; product form: film)

ACLAR has an outstanding ability to prevent the passage of water vapor and liquids. This means that ACLAR provides product protection and, because of its transparency, permits inspection viewing of the product at the same time. These combined properties have lead to new product designs for moisture-sensitive items.

Reference: *Aclar Performance Films,* supplier technical report (SFI-14 Rev. 9-89), Allied-Signal Engineered Plastics, 1989.

Film Properties and Applications

Allied Signal: ACLAR® (features: transparent; product form: film)

ACLAR is a flexible thermoplastic film made from fluorinated-chlorinated resins. The film is produced in three basic types: ACLAR 22, ACLAR 88, and ACLAR 33. These products are copolymers, and a terpolymer, respectively, and consist primarily of CTFE. The presence of fluorine and chlorine in the structure leads to a clear, heat-sealable film with excellent thermal and chemical stability.

ACLAR can be heat-sealed, laminated, printed, thermoformed, metallized, and sterilized. The unsupported and laminated varieties can be handled and processed on most common converting and packaging machinery.

Reference: *ACLAR Performance Films,* supplier technical report (SFI-14 Rev. 9-89), Allied-Signal Engineered Plastics, 1989.

Allied Signal: ACLAR® 22A (features: transparent; product form: film)

ACLAR 22A is a copolymer film consisting primarily of CTFE. It is used primarily for pharmaceutical packaging applications. ACLAR 22A film thermoforms at a lower temperature than ACLAR 33C and may be formed on a vacuum forming machine. ACLAR 22A provides excellent moisture barrier properties. Standard product thicknesses include 0.0381 mm and 0.127 mm.

Reference: *ACLAR Performance Films,* supplier technical report (SFI-14 Rev. 9-89), Allied-Signal Engineered Plastics, 1989.

Allied Signal: ACLAR® 22C (features: transparent; product form: film)

ACLAR 22C is a copolymer film consisting primarily of CTFE. It is used primarily as an encapsulating film for electroluminescent lamps and for cleanroom packaging. Standard product thicknesses include 0.051 mm.

Reference: *ACLAR Performance Films,* supplier technical report (SFI-14 Rev. 9-89), Allied-Signal Engineered Plastics, 1989.

Allied Signal: ACLAR® 33C (features: transparent; product form: film)

ACLAR 33C is a terpolymer film consisting primarily of CTFE. It is used for military and pharmaceutical packaging applications. It thermoforms satisfactorily on equipment having a preheat station and a plug assist system. ACLAR 33C provides superior moisture barrier properties. It is available in standard thicknesses of 0.019 mm, 0.025 mm, 0.051 mm, 0.076 mm, 0.127 mm, and 0.19 mm.

Reference: *ACLAR Performance Films,* supplier technical report (SFI-14 Rev. 9-89), Allied-Signal Engineered Plastics, 1989.

Allied Signal: ACLAR® 88A (features: transparent; product form: film)

ACLAR 88A is a copolymer film consisting primarily of CTFE. It is used for pharmaceutical packaging applications. This product thermoforms at the same temperature as ACLAR 22A on a vacuum forming machine. It is available in a thickness of 0.019 mm.

Reference: *ACLAR Performance Films,* supplier technical report (SFI-14 Rev. 9-89), Allied-Signal Engineered Plastics, 1989.

Table A.II.1. Carbon Dioxide, Hydrogen and Hydrogen Sulfide Permeability Through 3M Kel-F 81 Polychlorotrifluoroethylene Film

Material Family	POLYCHLOROTRIFLUOROETHYLENE
Material Supplier/ Grade	3M KEL-F 81
Product Form	FILM

MATERIAL COMPOSITION

Note	amorphous form of polymer

TEST CONDITIONS

Penetrant	carbon dioxide				hydrogen			hydrogen sulfide	
Temperature (°C)	0	25	50	75	0	25	50	50	75

PERMEABILITY (source document units)

Gas Permeability (1×10^{-10} cm^3 · mm / cm^2 · sec · cm Hg)	0.35	1.4	2.4	15	3.2	9.8	24	0.35	2.0

PERMEABILITY (normalized units)

Permeability Coefficient (cm^3 · mm/m^2 · day · atm)	2.3	9.2	15.8	98.5	21.0	64.3	158	2.3	13.1

Table A.II.2. Hydrogen Permeability vs. Temperature and Pressure Through 3M Kel-F 81 Polychlorotrifluoroethylene

Material Family	POLYCHLOROTRIFLUOROETHYLENE								
Material Supplier/ Grade	3M KEL-F								

MATERIAL CHARACTERISTICS

Sample Thickness	0.01 mm	0.01 mm	0.01 mm	0.01 mm	0.01 mm	0.01 mm	0.01 mm	0.01 mm	0.01 mm

TEST CONDITIONS

Penetrant	hydrogen								
Temperature (°C)	-15	25	68	-12	25	67	-16	25	70
Pressure Gradient (kPa)	1724	1724	1724	3447	3447	3447	6895	6895	6895
Test Method	mass spectrometry and calibrated standard gas leaks; developed by McDonnell Douglas Space Systems Company Chemistry Laboratory								

PERMEABILITY (source document units)

Gas Permeability ($cm^3 \cdot mm/cm^2 \cdot kPa \cdot sec$)	6.39×10^{-11}	4.07×10^{-10}	2.33×10^{-9}	6.69×10^{-11}	4.13×10^{-10}	2.25×10^{-9}	5.77×10^{-11}	4.14×10^{-10}	2.49×10^{-9}

PERMEABILITY (normalized units)

Permeability Coefficient ($cm^3 \cdot mm/m^2 \cdot day \cdot atm$)	5.6	35.6	204	5.9	36.2	197	5.1	36.2	218

Table A.II.3. Nitrogen Permeability vs. Temperature and Pressure Through 3M Kel-F 81 Polychlorotrifluoroethylene

Material Family	POLYCHLOROTRIFLUOROETHYLENE					
Material Supplier/ Grade	3M KEL-F					

MATERIAL CHARACTERISTICS

Sample Thickness	0.01 mm	0.01 mm	0.01 mm	0.01 mm	0.01 mm	0.01 mm

TEST CONDITIONS

Penetrant	nitrogen					
Temperature (°C)	25	68	25	69	25	70
Pressure Gradient (kPa)	1724	1724	3447	3447	6895	6895
Test Method	mass spectrometry and calibrated standard gas leaks; developed by McDonnell Douglas Space Systems Company Chemistry Laboratory					

PERMEABILITY (source document units)

Gas Permeability ($cm^3 \cdot mm/cm^2 \cdot kPa \cdot sec$)	1.77×10^{-13}	4.15×10^{-11}	1.77×10^{-13}	4.36×10^{-11}	1.77×10^{-13}	4.45×10^{-11}

PERMEABILITY (normalized units)

Permeability Coefficient ($cm^3 \cdot mm/m^2 \cdot day \cdot atm$)	0.02	3.63	0.02	3.82	0.02	3.9

Table A.II.4. Oxygen and Ammonia Permeability vs. Temperature and Pressure Through 3M Kel-F 81 Polychlorotrifluoroethylene

Material Family	POLYCHLOROTRIFLUOROETHYLENE					
Material Supplier/ Grade	3M KEL-F					
MATERIAL CHARACTERISTICS						
Sample Thickness	0.01 mm	0.01 mm	0.01 mm	0.01 mm	0.01 mm	0.01 mm
TEST CONDITIONS						
Penetrant	ammonia	ammonia	oxygen	oxygen	oxygen	oxygen
Temperature (°C)	25	59	25	52	25	52
Pressure Gradient (kPa)	965	965	1724	1724	3447	3447
Test Method	mass spectrometry and calibrated standard gas leaks; developed by McDonnell Douglas Space Systems Company Chemistry Laboratory					
PERMEABILITY (source document units)						
Gas Permeability ($cm^3 \cdot mm/cm^2 \cdot kPa \cdot sec$)	1.2×10^{-11}	2.76×10^{-10}	2.95×10^{-12}	9.42×10^{-11}	2.84×10^{-12}	9.42×10^{-11}
PERMEABILITY (normalized units)						
Permeability Coefficient ($cm^3 \cdot mm/m^2 \cdot day \cdot atm$)	1.05	24.2	0.26	8.2	0.25	8.25

Table A.II.5. Oxygen, Nitrogen, and Carbon Dioxide Permeability Through Allied Signal ACLAR Polychlorotrifluoroethylene

Material Family	POLYCHLOROTRIFLUOROETHYLENE							
Material Supplier/ Grade	ALLIED SIGNAL ACLAR							
Material Supplier/ Grade	33C		22C			22A		
Product Form	FILM							
Features	transparent	transparent	transparent	transparent	transparent	transparent	transparent	transparent
TEST CONDITIONS								
Penetrant	oxygen	carbon dioxide	oxygen	nitrogen	carbon dioxide	oxygen	nitrogen	carbon dioxide
Temperature (°C)	25	25	25	25	25	25	25	25
Test Note	STP conditions							
PERMEABILITY (source document units)								
Gas Permeability ($cm^3 \cdot mil/100\ in^2 \cdot day$)	7	16	15	2.5	40	12	2.5	30
PERMEABILITY (normalized units)								
Permeability Coefficient ($cm^3 \cdot mm/m^2 \cdot day \cdot atm$)	2.8	6.3	5.9	1.0	15.7	4.7	1.0	11.8

Table A.II.6. Oxygen, Nitrogen, and Helium Permeability Through 3M Kel-F 81 Polychlorotrifluoroethylene

Material Family	POLYCHLOROTRIFLUOROETHYLENE							
Material Supplier/Grade	3M KEL-F 81							
Product Form	FILM							

MATERIAL COMPOSITION

Note	amorphous form of polymer							

TEST CONDITIONS

Penetrant	nitrogen			helium	oxygen			
Temperature (°C)	25	50	75	25	0	25	50	75

PERMEABILITY (source document units)

Gas Permeability (1×10^{-10} cm^3 · mm / cm^2 · sec · cm Hg)	0.05	0.30	0.91	21.7	0.07	0.40	1.40	5.70

PERMEABILITY (normalized units)

Permeability Coefficient (cm^3 · mm/m^2 · day · atm)	0.33	1.97	5.98	142.5	0.46	2.63	9.19	37.43

Table A.II.7. Water Vapor Permeability Through Allied Signal ACLAR Polychlorotrifluoroethylene

Material Family	POLYCHLOROTRIFLUOROETHYLENE						
Material Supplier/Trade Name	ALLIED SIGNAL ACLAR						
Grade	33C	22C		22A	88A		
Product Form	TRANSPARENT FILM						

MATERIAL CHARACTERISTICS

Sample Thickness	0.019 mm	0.051 mm	0.0254 mm	0.051 mm	0.19 mm	0.038 mm	0.019 mm

TEST CONDITIONS

Penetrant	water vapor						
Temperature (°C)	37.8	37.8	37.8	37.8	37.8	37.8	37.8
Relative Humidity (%)	90	90	90	90	90	90	90
Test Method	ASTM E96, method E; measured on sealed pouches						

PERMEABILITY (source document units)

Vapor Transmission Rate (g/m^2 · day)	0.43 - 0.59	0.15 - 0.31	0.47 - 0.93	0.24 - 0.62	0.09 - 0.13	0.32 - 0.62	0.70 - 0.86
Vapor Transmission Rate (g/day · 100 in^2)	0.028 - 0.038	0.010 - 0.020	0.030 - 0.060	0.016 - 0.040	0.006 - 0.007	0.020 - 0.040	0.045 - 0.055

PERMEABILITY (normalized units)

Vapor Transmission Rate (g · mm/m^2 · day)	0.008 - 0.011	0.0077 - 0.0158	0.0119 - 0.0236	0.0122 - 0.0316	0.0171 - 0.0247	0.0122 - 0.0236	0.0133 - 0.0163

Table A.II.8. Water Vapor Transmission Through 3M Kel-F 81 Polychlorotrifluoroethylene

Material Family	POLYCHLOROTRIFLUOROETHYLENE			
Material Supplier/ Grade	3M KEL-F 81			
Product Form	FILM			

MATERIAL COMPOSITION

Note	amorphous form of polymer			

TEST CONDITIONS

Penetrant	water vapor			
Temperature (°C)	25	50	75	100

PERMEABILITY (source document units)

Gas Permeability (1×10^{-10} cm$^3 \cdot$ mm / cm$^2 \cdot$ sec $\cdot$ cm Hg)	1	10	28	100
Vapor Transmission Rate (g $\cdot$ mil/ m$^2 \cdot$ atm $\cdot$ day)	0.19	1.76	4.56	15.20

PERMEABILITY (normalized units)

Permeability Coefficient (cm$^3 \cdot$ mm/m$^2 \cdot$ day $\cdot$ atm)	6.57	65.7	184	657
Vapor Transmission Rate (g $\cdot$ mm/ m$^2 \cdot$ day)	0.005	0.043	0.116	0.386

Table A.II.9. Film Properties of Allied Signal ACLAR Polychlorotrifluoroethylene

Material Family	POLYCHLOROTRIFLUOROETHYLENE			
Material Supplier/ Trade Name	ALLIED SIGNAL ACLAR			
Grade	88A	22A	22C	33C
Product Form	FILM			
Features	transparent	transparent	transparent	transparent
MATERIAL CHARACTERISTICS				
Sample Thickness	0.019 mm	0.038 mm	0.19 mm	0.019 mm
Specific Gravity	2.10	2.10	2.11	2.12
Yield (in^2/lb)	17,760	8880	1750	17,360
MECHANICAL PROPERTIES				
Modulus Of Elasticity - MD (MPa)	970-1100	970-1100	900-1100	1310-1380
Modulus Of Elasticity - TD (MPa)	837-970	1040-1100	900-1100	1310-1380
Tensile Strength - MD (MPa)	48.3-69	51.7-75.9	27.6-41.4	65.5-79.3
Tensile Strength - TD (MPa)	27.6-41.4	37.9-55.2	27.6-41.4	37.9-55.2
Ultimate Elongation - MD (%)	150-250	115-225	200-300	50-150
Ultimate Elongation - TD (%)	200-300	200-300	200-300	50-150
Drop Dart Impact Strength (g)	93	347	1200	<57
Tear Strength, Propagated, Elmendorf - MD (g)	24	40	>1600	12
Tear Strength, Propagated, Elmendorf - transverse (g)	24	130	>1600	33
Tear Strength, Graves - MD (g)	240	380	465	400
Tear Strength, Graves - TD (g)	300	360	415	380
Abrasion Resistance - weight loss (mg)	14	16	7	15
THERMAL PROPERTIES				
Melting Point (°C)	183-186	183-186	183-186	202-204
Thermal Conductivity (cal-cm/cm^2-sec-°C)	5.3×10^{-4}	5.3×10^{-4}	5.3×10^{-4}	4.7×10^{-4}
OPTICAL PROPERTIES				
Haze (%)	<1	<1	<1	<1
PERFORMANCE PROPERTIES				
Unrestrained Shrink - 149°C Air, MD (%)	12-15	12-15	≤ 2	≤ 2.5
Unrestrained Shrink - 149°C Air, TD (%)	-12 to -15	-12 to -15	≤ 2	≤ 2.5

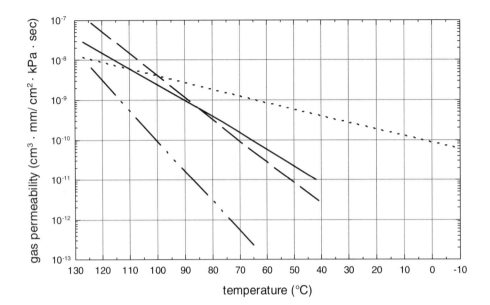

Figure A.II.1 Gas permeability vs. temperature through polychlorotrifluoroethylene.

Appendix III: Trademarks

Trademark	Property of	Trademark	Property of
Aclar®	Allied Signal Corp.	Neoflon®	Daikin Corp.
Acrysol®	Rohm & Haas Corp.	Polyflon®	Daikin Corp.
Algoflon®	Ausimont Corp.	Polymist®	Ausimont Corp.
Delrin®	DuPont Corp.	Ryton®	Phillips 66
Ekonol®	Norton Chemplast	Scotch-Weld®	3M Corp.
Fluon®	ICI Asahi Corp.	Tedlar®	DuPont Co.
Fluoroplast®	KCCE Corp.	Teflon®	DuPont Co.
Fluorotech®	Acton Corp.	Teflon® AF	DuPont Co.
Genetron®	Allied Signal Corp.	Tefzel®	DuPont Corp.
Gore-Tex®	W.L. Gore & Assoc.	Tetra-etch®	W.L. Gore & Assoc.
Gore-Tex GR®	W.L. Gore & Assoc.	Torlon®	Amoco Performance Prod.
Gylon®	B.F. Goodrich Corp.	Tornado®	Solus Industries
Hostaflon®	Dyneon Corp.	Triton®	Union Carbide Corp.
Hiflon®	Hindustan Fluoropolymer Co.	Turbula®	Bachofen Maschinenfabrik
Isopar®	Exxon-Mobile Corp.	Tyvek®	DuPont Corp.
Kapton®	DuPont Corp.	Vespel®	DuPont Corp.
Kel F®	3M Corp	Viton®	DuPont Corp.
Kevlar®	DuPont Corp.	Zonyl®	DuPont Corp.
Natrosol®	Hercules Corp.		

Glossary

A

Abrasion Resistance - Wear rate or abrasion rate is an important property of materials during motion in contact with other materials. Abrasion or wear resistance is measured by a number of methods such as ASTM Method D3389, also known as *Taber Test*.

Abused Particles - Particles of fine powder (dispersion or emulsion polymerized) polytetrafluoroethylene subjected to shear resulting in fibrillation during handling and transportation prior to paste extrusion.

Acyl Fluoride - An end group of polytetrafluoroethylene polymers commonly present in the polymers after they have been subjected to radiation in the presence of oxygen.

$$\left(-\overset{\overset{\displaystyle O}{\|}}{C}-F\right)$$

Adhesion Promoter - A coating applied to a substrate prior to adhesive application, in order to improve adhesion of the material, such as plastic. Also called a *primer*.

Adhesive - A material, usually polymeric, capable of forming permanent or temporary surface bonds with another material as-is or after processing such as curing. Used for bonding and joining. Some of the classes of adhesives include hot-melt, pressure-sensitive, contact, UV cured, emulsion, etc.

Adhesive Bonding - A method of joining two plastics or other material in which an adhesive is applied to the part surfaces. Bonding occurs through mechanical or chemical interfacial forces between the adhesive and adherend and/or by molecular interlocking. Surface preparation of the adherends and curing of the adhesive may be required.

Adhesive Bond Strength - Also see *Adhesive Bonding*. The strength of a bond formed by joining of two materials using an adhesive. Bond strength can be measured by a technique such as extensiometry.

Adhesive Failure - Failure of an adhesive bond at the adhesive-adherend interface. An example is an adhesive failure that leaves adhesive all on one adherend, with none on the other adherend. Adhesive failure is less desirable than cohesive failure because it is indicative of a joint with lower adhesive strength. See also *Cohesive Failure*.

Adhesive Joining - See *Adhesive Bonding*.

Adhesiveless Joining - Also see *Welding by Distance* and *Welding by Pressure*. Joining or welding of two materials without the use of an adhesive.

Agglomerates - Polytetrafluoroethylene fine powder particles (dispersion polymerized) are small (<0.25 μm) and form loosely associated clusters that are called *agglomerates*.

Amorphous Phase - Also see *Amorphous Polymer* and *Semicrystalline Plastic*.

Amorphous Polymer - Amorphous polymers are polymers having noncrystalline or amorphous supramolecular structure or morphology. Amorphous polymers may have some molecular order but usually are substantially less ordered than crystalline polymers and subsequently have inferior mechanical properties.

Annealing - A process in which a material, such as plastic, metal, or glass, is heated then cooled slowly. In plastics and metals, it is used to reduce stresses formed during fabrication. The plastic is heated to a temperature at which the molecules have enough mobility to allow them to reorient to a configuration with less residual stress. Semicrystalline polymers are heated to a temperature at which retarded crystallization or recrystallization can occur.

Asbestos - Fillers made from fibrous mineral silicates, mostly chrysotile. Used in thermosetting resins and laminates in fibrous form as reinforcements and in thermoplastics such as polyethylene in finer form as a filler. Asbestos fillers resist heat and chemicals while providing reinforcement, but pose health hazards and therefore their use has been declining.

Attractive Intermolecular Forces - Also see *van der Waals Forces*.

Attrition Bar - Also see *V-Shaped Blender*. A bar parallel to the rotation axis of V-shaped blender which can be operated at different speeds. Various attrition bar designs are available which subject the powder being mixed to different shear rates.

Autopolymerization Inhibitor - A variety of terpenes, such as α-pinene, Terpene B and d-limonene are useful to inhibit the polymerization of tetrafluoroethylene or other monomers in an uncontrolled manner during storage and transportation. The inhibitor is removed prior to polymerization in the reactor.

B

Ball Mill - The function of a ball mill is reduction in size of solid ingredients of dispersion. This media mill is in the shape of a cylinder made of metal or ceramic. It contains a media usually in the form of pebbles made from glass, steel, or zirconium oxide. The ball mill is loaded with the liquid and solid ingredients

Bar - A metric unit of measurement of pressure equal to 1.0×10^6 dynes/cm^2 or 1.0×10^5 pascals. It has a dimension of unit of force per unit of area. Used to denote the pressure of gases, vapors and liquids.

Barrier Material - Materials such as plastic films, sheeting, wood laminates, particle board, paper, fabrics, etc., with low permeability to gases and vapors. Used in construction as water vapor insulation, food packaging, protective clothing, etc.

Beam Current - This is a characteristic of electron beam irradiation and is calculated from the surface dose and area flow rate of the material being irradiated.

Beam Power - In electron beam irradiation, beam power is calculated from the average dose of irradiation and mass flow rate of polytetrafluoroethylene using the following equation: Beam power = 1/3 (average dose, MRad) × (mass flow rate, kg/hr).

Bearing Pad - Slide bearings used in support systems for bridges and some buildings to accommodate thermal and seismic movement without damage to structures they support.

Biaxial Orientation - Orientation in which the material is drawn in two directions, usually perpendicular to one another. Commonly used in films and sheets. See also *Orientation*.

Billet - Refers to a solid or hollow cylindrical object usually made from polytetrafluoroethylene and occasionally made from other fluoropolymers.

Bisulfite Initiator - This is a redox initiator for dispersion polymerization of tetrafluoroethylene. For example, sodium bisulfite is a redox initiator from the bisulfite family and may be combined with ferrictriphosphate (activator) for polymerization.

Bleaching - A process to obtain white polytetrafluoroethylene yarn, accomplished by heating the yarn in an oven at 300°C for a period of five days.

Blender - This is the name of the family of equipment used for blending combinations of solid and liquid ingredients. A specific variety is V-cone blenders which are common in producing mixtures of polytetrafluoroethylene fine powder, lubricant and pigment. Some blenders have a simple cylindrical shape where blending is achieved by tumbling or rolling.

Braiding - The name of the process in which fluoropolymer tubes, wires and cables are reinforced. Strands of metal and plastic and other thin wire shape materials are formed into braids and wrapped around the fluoropolymer tube or wire to improve pressure rating and wear and puncture resistance.

Breaking Elongation See *Elongation*.

Buckling - This is a phenomenon that occurs in elastic bags, in isostatic process. Straining the bag excessively can result in formation of ridges or *buckling*. Buckling can occur when the bag is empty and unsupported. Buckling is most likely when the direction of pressure is from outside-to-inside in isostatic molding.

Burst Strength - The pressure at which a tube fails mechanically (i.e., breaks open) is called *burst strength*.

Bursting Strength - Bursting strength of a material, such as plastic film, is the minimum force per unit area or pressure required to produce rupture. The pressure is applied with a ram or a diaphragm at a controlled rate to a specified area of the material held rigidly and initially flat but free to bulge under the increasing pressure.

C

C8 - An alternative name for perfluoroammonium octanoate.

Calendar - This is the equipment by which a lower thickness is obtained from a thicker bead. The equipment consists of twin rolls with adjustable gap. The thicker sheet is fed into the calendar opening where it is "squeezed" by the force of the rolls into a thinner sheet or other forms.

Capture Velocity - The air velocity that generates sufficient air flow to remove contaminated air being given off from the source and force it to flow into an exhaust hood.

Carbon Black - A black colloidal carbon filler made by the partial combustion and/or thermal cracking of natural gas, oil, or another hydrocarbon. Depending upon the starting material and the method of manufacture, carbon black can be called acetylene black, channel black, furnace black, etc. For example, channel black is made by impinging gas flames against steel plates or channel irons, from which the deposit is scraped at intervals. The properties and the uses of each carbon black type can also vary. Thus, furnace black comes in high abrasion, fast extrusion, high modulus, general purpose, and semireinforcing grades among others. Carbon black is widely used as a filler and pigment in PVC, phenolic resins, and polyolefins. It increases the resistance to UV light and electrical conductivity, and sometimes acts as a cross-linking agent. Also called *colloidal carbon*.

Carbon Disulfide - $S=C=S$

Carbon Fiber - Carbon fibers are high-performance reinforcement consisting essentially of carbon. They are made by a variety of methods including pyrolysis of cellulosic (e.g., rayon) and acrylic fibers, burning-off binder from a pitch precursor, and growing single crystals (whiskers) via thermal cracking of hydrocarbon gas. The properties of carbon fibers depend on the morphology of carbon in them and are the highest for crystalline carbon (graphite). These properties include high modulus and tensile strength, high thermal stability, electrical conductivity, chemical resistance, wear resistance, and relatively low weight. Used as continuous or short fibers and in mats in autoclave and die molding, filament winding, injection molding, and pultrusion. Carbon fibers are used at a loading levels of 20–60 vol% or more in both thermosets and thermoplastics such as epoxy resins and ABS. Carbon fibers are often used in combination with other fibers such as glass fibers to make hybrid composites. The end products containing carbon fibers include wheel chairs, tennis racquets, auto parts, machine tools, and support structures in electronic equipment. Also called *graphite fiber*.

Carbon Filler - Carbon fillers are a family of fillers based on carbon in various forms, such as carbon black and graphite. Used as a black pigment, to improve lubricity, and to increase electrical conductivity of plastics. Also called *powdered carbon, carbon powder*.

Carbonyl Fluoride -

$$F-\underset{\underset{O}{\|}}{C}-F$$

Carbon Powder - See *Carbon Filler*.

Carboxylic Acid -

$$-\underset{\underset{O}{\|}}{C}-O-H$$

Cast Film - Film produced by pouring or spreading resin dispersion, resin solution or melt over a suitable temporary substrate, followed by curing via solvent evaporation or melt cooling and removing the cured film from the substrate.

Casting - Method to produce a cast film. See *Cast Film*

Charge-to-Charge Weld - This is the mechanism by which a continuous rod or tube is formed in ram extrusion of polytetrafluoroethylene. Under heat and pressure, the polymer melts and is welded to the previous charge.

Chemical Resistance - Degradation of a material caused by chemical reaction.

Chlorotrifluoroethylene (CTFE) - $CF_2=CFCl$

Cleanroom - Special room where the number of particles in air are reduced and maintained below a given level. Cleanrooms have class designations such as Class 100 or Class 10 which indicates the number of particles larger than a certain size in one cubic meter of air. Special filters are used to reduce the number of particles.

Cobalt-60 - One of the unstable isotopes of Co used widely as a source of gamma radiation.

Coagulated Dispersion Powder - Also see *Fine Powder*. An alternative name for fine powder polytetrafluoroethylene.

Coagulation - This is a process for separation of PTFE (polytetrafluoroethylene) solids from its dispersion. The emulsion or dispersion containing this polymer (dispersion polymerization) has to be broken (destabilized) in order to cause precipitation of PTFE particles. Dilution to reduce solids concentration below 20%, addition of water-soluble organic compounds and addition of soluble inorganic salts are the common techniques to break PTFE emulsions.

Coalescence - Refers to the mechanism for melting and consolidation of polytetrafluoroethylene parts. After the polymer melts, adjacent particles begin to combine (i.e., coalesce) under the driving force of surface tension.

Co-Coagulation - This is a process in which other species are added to polytetrafluoroethylene dispersion, followed by coagulation. See also *DU Bearing Process* and *Coagulation*.

Coefficient of Linear Thermal Expansion - The change in unit of length or volume that occurs due to a unit change in temperature. The expansion and contraction of a material with changes in temperature depend on its coefficient of linear thermal expansion, and movement of a part that is attached to another part with a lower CLTE value may be restricted.

Cohesive Failure - Failure of an adhesive bond that occurs within the adhesive leaving adhesive present on both adherends. Optimum failure is 100% cohesive failure when both shear areas are completely covered. See also *Adhesive Failure.*

Cold Flow - See *Creep.*

Cold Plasma - Plasma is used for treating material surfaces. It is made of a stream of ions, free radicals and other atomic particles produced by introducing a gas into a vacuum chamber followed by radiofrequency or microwave excitation of the gas. The energy dissociates the gas into ions and other particles. Plasma treatment modifies surfaces to make them harder, rougher, more or less wettable, and more adherable.

Compressive Strain - The relative length deformation exhibited by a specimen subjected to a compressive force. See also *Strain, Flexural Strain, Tensile Strain.*

Conditioning - Process of bringing the material or apparatus to a certain condition, e.g., moisture content or temperature, prior to further processing, treatment, etc. Also called *conditioning cycle.*

Conduction - In heat transfer, migration of energy due to a temperature gradient. Heat energy is transferred by the movement of molecules at hotter or colder temperatures, with different degrees of thermal motion, into colder or hotter regions, respectively.

Contact Adhesive - See also *Pressure Sensitive Adhesive.* An adhesive that will adhere to itself on contact. When applied to both adherends, it forms a bond after drying, without sustained pressure on the adherends. Composed of neoprene or, less commonly, nitrile elastomers.

Contact Angle - The angle that the droplet or edge of the liquid forms with the solid plane is called the *contact angle.*

Convection - See also *Radiation, Conduction.* The mass movement of particles arising from the movement of a streaming fluid due to difference in a physical property such as density, temperature, etc. Mass movement due to a temperature difference results in heat transfer, as in the upward movement of a warm air current.

Copolymer - See *Copolymerization.*

Copolymerization - A polymerization where more than one monomer takes part in the reaction and form the polymer chain.

Corona Discharge Treatment - See also *Plasma Arc Treatment.* In adhesive bonding, a surface preparation technique in which a high electric potential is discharged by ionizing the surrounding gas, usually air. The gas reacts with the plastic, roughening the surface to provide sites for mechanical interlocking and introducing reactive sites on the surface. Functional groups such as carbonyls, hydroxyls, hydroperoxides, aldehydes, ethers, esters, carboxylic acids, and unsaturated bonds have been proposed as reactive sites. Commonly used for polyolefins, corona discharge increases wettability and surface reactivity. In processing plastics, treating the surface of an inert plastic such as polyolefin with corona discharge to increase its affinity to inks, adhesives or coatings. Plastic films are passed over a grounded metal cylinder with a pointed high-voltage electrode above it to produce the discharge. The discharge oxidizes the surface, making it more receptive to finishing. Also called *corona treatment.*

Corona Treatment - See *Corona Discharge Treatment*.

Covalent Bond - A bond formed by the sharing of two or more electrons between two atoms. Covalent bonds can be single (two electrons shared), double (four shared electrons), or triple (six shared electrons).

Cracking - Appearance of external and/or internal cracks in the material as a result of stress that exceeds the strength of the material. The stress can be external and/or internal and can be caused by a variety of adverse conditions: structural defects, impact, aging, corrosion, etc., or a combination thereof. Also called *resistance to cracking, grazing, cracking resistance*.

Creep - Nonrecoverable deformation in a part subjected to a continuous load. Creep is dependent on temperature and the duration and amount of the load.

Creep Viscosity - See *Melt Creep Viscosity*.

Critical Cracking Thickness - The maximum thickness which can be coated in a single layer (pass) of polytetrafluoroethylene dispersion without crack formation. This thickness is measured after sintering has been completed.

Cross-linking - Reaction or formation of covalent bonds between chain-like polymer molecules or between polymer molecules and low-molecular compounds such as carbon black fillers. As a result of cross-linking polymers, such as thermosetting resins, may become hard and infusible. Cross-linking is induced by heat, UV or electron-beam radiation, oxidation, etc. Cross-linking can be achieved either between polymer molecules alone as in unsaturated polyesters or with the help of multifunctional cross-linking agents such as diamines that react with functional side groups of the polymers. Cross-linking can be catalyzed by the presence of transition metal complexes, thiols and other compounds.

Crystalline Form - See *Crystalline Phase*

Crystalline Melting Point - The temperature of melting of the crystalline phase of a crystalline polymer. It is higher than the temperature of melting of the surrounding amorphous phase.

Crystalline Phase - This is an organized structural arrangement for polymer molecules. In this arrangement, polymer chains are aligned into a closely-packed ordered state called *crystalline phase*.

Crystalline Plastic - See *Semicrystalline Plastic*.

Crystallization Temperature - Temperature (or range of temperature) at which polytetrafluoroethylene crystallizes. PTFE chains which were randomly distributed in the molten or gel state become aligned into a close-packed ordered arrangement during the crystallization process.

Crystallinity - Crystalline content of a polymer expressed in weight percent. Also see *Crystalline Phase*.

CTFE See *Chlorotrifluoroethylene*.

D

Dedusting Powder - Refers to dispersion polymerized polytetrafluoroethylene which is added to powdery material to prevent formation of dust during powder handling. The polymer can be added in the form of fine powder or dispersion.

Deflection Temperature under Load - See *Heat Deflection Temperature*.

Deformation Under Load - See *Creep*.

Deflagration - A violent reaction whereby tetrafluoroethylene is degraded into carbon and tetrafluoromethane.

Degradation - Loss or undesirable change in plastic properties as a result of aging, chemical reactions, wear, use, exposure, etc. The properties include color, size, strength, etc.

Density - The mass of any substance (gas, liquid or solid) per unit volume at specified temperature and pressure. The density is called *absolute* when measured under standard conditions, e.g., 760 mm Hg pressure and 0°C temperature. Note: For plastics, the weight in air per volume of the impermeable portion of the material measured at 23°C according to ASTM D792. Also called *mass density, absolute gravity, absolute density*.

Die Cone Angle - The angle that the wall of the convergent section of the die forms with the axis of the paste extruder barrel (parallel to extrusion direction).

Die Land - The part of the die (orifice) that is downstream from the convergent section where both cross-sectional area and shape are constant.

Dielectric Breakdown Strength or Voltage - The voltage (minimum) required to breakdown through the thickness of a dielectric (insulation material), i.e., create a puncture. ASTM D149 is used to measure dielectric breakdown strength of plastic insulation material.

Dielectric Constant - The dielectric constant of an insulating material is the ratio of the capacitance of a capacitor insulated with that material to the capacitance of the same capacitor insulated with a vacuum.

Dielectric Dissipation Factor - The ratio of the power dissipated in a dielectric to the product of the effective voltage and the current; or the cotangent of the dielectric phase angle; or the tangent of dielectric loss angle. Note: measured according to ASTM D150 for plastics. Also called *tan delta, permittivity loss factor, dissipation factor, dielectric loss tangent.*

Dielectric Loss Tangent - See *Dielectric Dissipation Factor.*

Differential Scanning Calorimetry - DSC is a technique in which the energy absorbed or produced is measured by monitoring the difference in energy input into the substance and a reference material as a function of temperature. Absorption of energy produces an endotherm; production of energy results in an exotherm. May be applied to processes involving an energy change, such as melting, crystallization, resin curing, and loss of solvents, or to processes involving a change in heat capacity, such as the glass transition.

Dip Coating - This method is the most popular way to coat cloth and fibers with polytetrafluoroethylene dispersion. Typically, the substrate is dipped in the dispersion and excess dispersion is removed by a device such as a doctor blade. The wet coated substrate is then further processed. Viscosity of the dispersion determines the initial thickness of the wet coating immediately after removal from the dip tank.

Direct Contact Hot Tool Welding - A form of heated tool welding in which the thermoplastic parts are pressed directly against the hot tool or plate. Part surfaces are heated until the melting or glass transition is reached. The hot tool is then removed, and the parts are pressed together until cooled. The hot tool can be coated with polytetrafluoroethylene to reduce melt sticking. See also *Heated Tool Welding, Hot Tool Welding.*

Dispersing Agent - See *Surfactant.*

Dispersion A dispersion is often defined as a uniform mixture of solids particles and a liquid. It may contain other agents such as a surfactant and a resin soluble in the liquid (solvent). An example of a dispersion is a house paint. A feature of most dispersions is stability which means little or no settling of the solid particles.

Dispersion Polymerization - This technique is a heterogenous regime where a significant amount of surfactant is added to the polymerization medium. Characteristics of the process include small uniform polymer particles which may be unstable and coagulate if they are not stabilized. Hydrocarbon oil is added to the dispersion polymerization reactor to stabilize the polytetrafluoroethylene emulsion. Temperature and agitation control are easier in this mode than suspension polymerization. Polytetrafluoroethylene fine powder and dispersion are produced by this technique.

Dissipation Factor - See *Dielectric Dissipation Factor.*

Dope - Any thick liquid or pasty preparation used for further processing such as coating or fiber spinning. A polytetrafluoroethylene (PTFE) dope prepared by the addition of viscose, to PTFE dispersion, is spun to manufacture yarn.

Drum Tumbler - A device capable of tumbling a drum to achieve mixing.

Dry Bag Isostatic Molding - A method of isostatic molding in which polytetrafluoroethylene powder is compressed between a hard part and a elastic bag. The compaction is achieved by introducing a high-pressure liquid behind the bag as opposed to submerging the closed system in a vessel containing a fluid and pressurizing the vessel. After completion of dwell time, the vessel is depressurized, the mold is removed and the preformed part is recovered. Also see *Wet Bag Isostatic Molding.*

DSC - See *Differential Scanning Calorimetry.*

Dual Laminate - FRP (fiberglass reinforced polyester) piping lined with fluoropolymer is manufactured by forming FRP over a fluoropolymer tube. The resulting structure is called a *dual laminate.*

DU Bearing Process - A process which was originally developed by Glacier Company (US Patent 2,689,380) in which fillers are added to polytetrafluoroethylene (PTFE) dispersion and the mixture is co-coagulated. An aluminum salt is added to the mixture to convert the surfactant in the dispersion to insoluble species and coagulate the mixture. The viscosity of the resulting paste (or mush) is adjusted by the addition of an organic solvent such as a toluene. This paste is calendared onto the surface of a steel-backed porous bronze strip. After sintering, the steel strip is rolled into bearing such that the filled PTFE would form its inside surface.

Durometer Hardness - Indentation hardness of a material as determined by either the depth of an indentation made with an indentor under specified load or the indentor load required to produce a specified indentation depth. The tool used to measure indentation hardness of polymeric materials is called a *durometer*, e.g., Shore-type durometer.

Dynodes - A name for glass rings positioned as insulation in between metal rings that make up the vacuum cylinder in electron beam irradiation equipment.

E

Elasticity - Property whereby a solid material changes its shape and size under action of opposing forces, but recovers its original configuration when the forces are removed.

Electron Beam - See *Electron Beam Radiation.*

Electron Beam Radiation - Ionizing radiation propagated by electrons that move forward in a narrow stream with approximately equal velocity. Also called *electron beam.*

Elongation - The increase in gauge length of a specimen in tension, measured at or after the fracture, depending on the viscoelastic properties of the material. Note: Elongation is usually expressed as a percentage of the original gage length. Also called *ultimate elongation, tensile negation, breaking elongation.*

Elongation at Break - The increase in distance between two gauge marks, resulting from stressing the specimen in tension, at the exact point of break. Measurement taken at the exact point of break according to ASTM D638.

Emulsion - See *Dispersion.*

Emulsion Polymerization - See *Dispersion Polymerization.*

Encapsulation - This term means to enclose as in a capsule. Polytetrafluoroethylene (PTFE) can be used to encapsulate metal articles to impart chemical resistance to them. Examples include encapsulated metal gaskets and butterfly valve gate. The metal provides mechanical strength and resistance to creep.

End Groups - The functional groups appear at the ends of polymer chains and in effect "end" the chain growth.

Environmental Stress Cracking - Cracking or crazing that occurs in a thermoplastic material subjected to stress or strain in the presence of particular chemicals or weather conditions or as a result of aging. Also called *ESC.*

Epoxides - Organic compounds containing three-membered cyclic group(s) in which two carbon atoms are linked with an oxygen atom as in an ether. This group is called an epoxy group and is quite reactive, allowing the use of epoxides as intermediates in preparation of certain fluorocarbons and cellulose derivatives and as monomers in preparation of epoxy resins.

Epoxy - See *Epoxy Resin.*

Epoxy Resin - Epoxy resins are a family of thermoset polyethers containing cross-linkable glycidyl groups. The largest group of epoxy resins is prepared by polymerization of bisphenol A and epichlorohydrin. These resins have a wide viscosity range, depending on their molecular weight, and are cured at room or elevated temperatures with catalyzed polyamines and/or anhydrides. Aliphatic and cycloaliphatic epoxy resins are produced by peroxidation of olefins with peracetic acid or epoxidation of polyols with epichlorohydrin. Novolak epoxy resins are prepared by reacting novolak phenolic resins with epichlorohydrin. Vinyl ester or acrylic epoxy resins are prepared by treating epoxy resins with unsaturated carboxylic acids such as acrylic acid. There are other specialty type epoxy resins such as halogenated epoxy and phenoxy resins. Bisphenol A epoxy resins exhibit excellent adhesion and very low shrinkage during curing. Additionally, cured novolak and cycloaliphatic resins have good UV stability and dielectric properties, while cured vinyl ester resins show high strength and chemical resistance and brominated epoxy resins show fire retardant properties. Some epoxy resins have poor oxidative stability. Processed by injection, compression, transfer, and structural foam molding, casting, coating, and lamination. Widely used as protective coatings, adhesives, potting compounds, and binders in laminates, flooring, civil engineering, electrical and electronic products. Also called e*poxy, EP, bisphenol A, epoxy resin.*

Etching - In adhesive and solvent bonding, a process used to prepare plastic surfaces for bonding. Exposure of the plastic parts to a reactive chemical, such as chromic acid, or to an electrical discharge results in oxidation of the surface and an increase in surface roughness by removal of surface material.

ETFE - See *Ethylene Tetrafluoroethylene Copolymer.*

Ethane - An alkane (saturated aliphatic hydrocarbon) with two carbon atoms, CH_3CH_3. A colorless, odorless, flammable gas. Relatively inactive chemically. Obtained from natural gas. Used in petrochemical synthesis and as fuel.

Ethylene - An alkene (unsaturated aliphatic hydrocarbon) with two carbon atoms, $CH_2=CH_2$. A colorless, highly flammable gas with sweet odor. Autoignition point 543°C. Derived by thermal cracking of hydrocarbon gases or from gas synthesis. Used as monomer in polymer synthesis, refrigerant, and anesthetic. Also called *ethene*.

Ethylene Polymers - Ethylene polymers include ethylene homopolymers and copolymers with other unsaturated monomers, most importantly olefins such as propylene and polar substances such as vinyl acetate. The properties and uses of ethylene polymers depend on the molecular structure and weight.

Ethylene Tetrafluoroethylene Copolymer - Thermoplastic comprised of an alternating copolymer of ethylene and tetrafluoroethylene. Has high impact resistance and good abrasion resistance, chemical resistance, weatherability, and electrical properties approaching those of fully fluorinated polymers. Retains mechanical properties from cryogenic temperatures to 356°F. Can be molded, extruded, and powder-coated. Used in tubing, cable and wire products, valves, pump parts, wraps, and tower packing in aerospace and chemical equipment applications. Also called *ETFE.*

Extrusion - Process for converting of an ingot or billet into lengths of uniform cross section by forcing material to flow plastically through a die orifice; a product form produced by this process. There are many variations of this process used widely in working metals and processing plastics.

F

Fatigue - Process of progressive localized permanent structural change occurring in a material subjected to conditions that produce fluctuating stresses and strains at some point or points and that may culminate in cracks or complete fracture after a sufficient number of fluctuations, according to ASTM E1150. See also *Flexural Fatigue, Tensile Fatigue.*

Fatigue Life - Number of loading-unloading cycles of a specified type that material specimen can endure before failing in a fatigue test.

Fatigue Strength - The limiting value of the median fatigue strength as the number of loading cycles sustained before failure becomes very large. Fatigue strength, here, is the maximum stress that can be sustained without failure at this number of loading cycles.

FEP - See *Fluorinated Ethylene Propylene Copolymer.*

Fibrillation - This phenomenon occurs when polytetrafluoroethylene fine powder particles are subjected to shear usually at above its transition point (19°C). For example, when fine powder particles rub against each other, groups of polymer chains are pulled out of crystallites. These fibrils can connect polymer particles together. They have a width of less than 50 nm.

Film - A product, e.g., plastic, that is extremely thin compared to its width and length. There are supported and unsupported films such as coatings and packagings, respectively.

Fine Cut PTFE - Particles of suspension polymerized polytetrafluoroethylene (PTFE) which has undergone size reduction (usually by cutting) to less than 50 μm average size.

Fine Powder PTFE - Polytetrafluoroethylene (PTFE) polymerized by dispersion polymerization method.

Finishes - Highly formulated dispersions of polytetrafluoroethylene containing a variety of fillers such as pigments, resins, extenders, and others. Finishes are used to coat different surfaces such as cookware, houseware and industrial equipment.

Fishtail Guide - A guide used in calendaring of polytetrafluoroethylene (fine powder) bead into tape. It allows safe feeding of the bead into the nip point of calendar rolls. It is called *fish tail* guide because of its similarity to the tail of a fish.

Flame Retardant - A substance that reduces the flammability of materials such as plastics or textiles in which it is incorporated. There are inorganic flame retardants such as antimony trioxide (Sb_2O_3) and organic flame retardants such as brominated polyols. The mechanisms of flame retardation vary depending on the nature of material and flame retardant. For example, some flame retardants yield a substantial volume of coke on burning, which prevents oxygen from reaching inside the material and blocks further combustion. Also called *fireproofing agent*.

Flame Treatment - In adhesive bonding, a surface preparation technique in which the plastic is briefly exposed to a flame. Flame treatment oxidizes the surface through a free radical mechanism, introducing hydroxyl, carbonyl, carboxyl, and amide functional groups to a depth of ~ 4–6 nm, and produces chain scissions and some crosslinking. Commonly used for polyolefins, polyacetals, and polyethylene terephthalate, flame treatment increases wettability and interfacial diffusivity.

Flash - In welding thermoplastics, molten plastic that seeps out of the joint area during the welding process. In molding, surplus plastic attached to the molding along the parting line. Flash must usually be removed before parts are considered to be finished.

Flax - Flax is a blue flowered plant that is grown for its fiber and its oily seeds. Its fiber is the source of linen.

Flex Life - Flex fatigue life is the total number of cycles that a specimen can be "flexed" in a prescribed manner before failure occurs. Failure is defined as physical breakdown of the specimen material. A number of methods such as MIT, Ross and De Mattia are used to measure flex life. See also *Fatigue Life.*

Flexural Fatigue - Progressive localized permanent structural change occurring in a material subjected to cyclic flexural stress that may culminate in cracks or complete fracture after a sufficient number of cycles.

Flexural Fatigue Strength - Maximum stress that can be sustained for a specified number of bending cycles without failure.

Flexural Modulus of Elasticity - The ratio, within the elastic limit, of the applied stress on a test specimen in flexure to the corresponding strain in the outermost fibers of the specimen.

Flexural Properties - Properties describing the reaction of physical systems to flexural stress and strain.

Flexural Strain - The tensile elongation on the surface of a cross section opposite to that experiencing a locally impinging force in bending at any time of the test. See also *Strain, Compressive Strain, Tensile Strain.*

Flexural Strength - The maximum stress in the extreme fiber of a specimen loaded to failure in bending. Note: Flexural strength is calculated as a function of load, support span and specimen geometry. Also called *modulus of rupture, bending strength.*

Flexural Stress - The maximum fiber stress in a specimen at a given strain in a bending test. The maximum fiber stress is a function of load, support span, and specimen width and depth. It depends on the method of load application relative to the supports and on the specimen geometry. It has to be calculated. Note: Flexural stress is calculated as a function of load at a given strain or at failure, support span and specimen geometry.

Fluid Energy Mill - A mill that utilizes high speed air to reduce the size of solid particles.

Fluorinated Ethylene Propylene Copolymer - A random copolymer of tetrafluoroethylene and hexafluoropropylene:

$$\left[-CF_2-CF_2-CF_2-\underset{\underset{CF_3}{|}}{CF}-CF_2-CF_2-\right]_n$$

Fluorine Sheath - An analogy comparing the molecule of (PTFE) polytetrafluoroethylene with a wire. Fluorine atoms form a sheath around the carbon backbone of PTFE, rendering it impervious to chemicals, resembling the function of an insulation around a conductor.

Fluoroadditives - These are the finely divided low molecular weight polytetrafluoroethylene powders added to other products to impart some of the fluoropolymer properties to the host systems.

Fluoroalkenes - Unsaturated linear perfluorinated hydrocarbons containing at least one double bond are called fluoroalkenes.

Fluorohydrocarbon Resin - See *Fluoropolymer.*

Fluoroplastic - See *Fluoropolymer.*

Fluoroplastic Homopolymer - A fluoropolymer entirely compiled of one monomer is called fluoroplastic homopolymer. Examples include polytetrafluoroethylene, polyvinylidene fluoride and polyvinyl fluoride.

Fluoropolymer - Thermoplastics (mostly polyolefins) in which all or some of the hydrogen atoms replaced by fluorine. Characterized by excellent chemical resistance, antifriction properties, thermal stability, antiadhesive properties, low flammability, and weatherability. The disadvantages include low creep resistance and strength and difficulty of processing. The properties of fluoropolymers depend on fluorine content. The processing is achieved by extrusion and molding. The uses are chemical apparatus, bearings, films, coatings, and containers. Also called *polyfluorohydrocarbon, polyfluorocarbon, fluoroplastic, fluorohydrocarbon resin, fluorocarbon resin.*

Free Flow PTFE - Suspension polymerized PTFE (polytetrafluoroethylene) particles first undergo size reduction followed by granulation and drying which result in a free flowing powder. Free flow PTFE has a similar consistency to sugar contrasted with fine cut PTFE resembling flour.

Free Radical - An atom or group of atoms with an odd or unpaired electron. Free radicals are highly reactive and participate in free radical chain reactions such as combustion and polymer oxidation reactions. Scission of a covalent bond by thermal degradation or radiation in air can produce a molecular fragment named a free radical. Most free radicals are highly reactive because of their unpaired electrons, and have short half lives.

$$R - R´ \rightarrow R· + R´$$

FTIR - Fourier transform infrared spectroscopy is a spectroscopic technique in which a sample is irradiated with electromagnetic energy from the infrared region of the electromagnetic spectrum (wavelength ~0.7 to 500 mm). The sample is irradiated with all infrared wavelengths simultaneously, and mathematical manipulation of the Fourier transform is used to produce the absorption spectrum or "fingerprint" of the material. Molecular absorptions in the infrared region are due to rotational and vibrational motion in molecular bonds, such as stretching and bending. FTIR is commonly used for the identification of plastics, additives, and coatings.

G

Gamma Radiation - Ionizing radiation propagated by high-energy protons, e.g., emitted by a nucleus in transition between two energy levels.

Gamma Ray Irradiation - A technique for reduction in the molecular weight of polytetrafluoroethylene by exposing this polymer to gamma rays from a source such as ^{60}Co.

Generator - An electronic device that converts standard 120/240 volt, 50/60 Hz line voltage into high-frequency electrical energy.

Glass Fiber - Glass fibers are a large family of short (staple, chopped, milled) or continuous fiber reinforcement, used widely in both thermosets and thermoplastics for increased strength, dimensional stability, thermal stability, corrosion resistance, and dielectric properties. The fibers are made by the melt drawing of various grades (electric, chemical, high tensile strength) of glass and are comprised of strands of filaments (roving) that can be further processed by size reduction, twisting, or weaving into fabrics or mats. Used in molding compounds, sprayup processes, die molding, layup, and other lamination processes. Glass fibers are often surface modified, e.g., with coupling agents, to improve bonding with polymer matrix or to impart special properties such as electrical conductivity (by coating with nickel). Also called *glass roving.*

Glass Filler - Glass fillers are a widely used family of fillers in the form of beads, hollow spheres, flakes, or milled particles. They increase dimensional stability, chemical resistance, moisture resistance, and thermal stability of plastics.

Glass Roving - See *Glass Fiber*.

Granular Polytetrafluoroethylene - The name used to refer to the products of suspension polymerization of tetrafluoroethylene.

Granular PTFE - See *Granular Polytetrafluoroethylene.*

Graphite Fiber - See *Carbon Fiber.*

Graphite Filler - A crystalline form of carbon in powder form. Graphite occurs naturally and also is produced by heating petroleum coke, carbon black, and organic materials. Used as a lubricating filler for nylons and fluoropolymers. Also called *powdered graphite, plumbago, graphite powder, carbon graphite, black lead.*

Graphite Powder - See *Graphite Filler.*

Gravure Coating - A coating technique using engraved rolls capable of picking up a precise amount of coating (dispersion) from a trough and transferring it to a web that is in contact with the roll. The engravings on the surface act as small reservoirs for the dispersion.

H

Halogenated Solvents - Organic liquids containing at least one atom of a halogen (Cl, F, I, Br) are called halogenated solvents.

Hammer Mill - A mill often used in producing polytetrafluoroethylene filled compounds. It consists of a rotor equipped with a set of small hammers, rotating inside a basket made from mesh screen. The resin and filler blend are placed or fed continuously into the basket and subjected to the hammer action. After sufficient grinding, the mixture passes through the screen and is discharged.

Harshaw Process for Dustless Powder - A process for compounding dispersion polymerized polytetrafluoroethylene in powders to suppress dust formation. It consists of working the polymer and powder mixture in a blender or a slow turning mixer like a cement mixer at an elevated temperature. The polymer is added to the powder in the form of dispersion to insure uniform blending. The mixer should be operated at about 100°C.

HDT - See *Heat Deflection Temperature.*

Heat Affected Zone - In welding, the region of the part that is affected by heat used to melt the joining surface. Microstructure of the heat affected zone is an important determinant of the mechanical strength of the weld. Also called *HAZ.*

Heat Deflection Temperature - The temperature at which a material specimen (standard bar) is deflected by a certain degree under specified load. Also called *tensile heat distortion temperature, heat distortion temperature, HDT, deflection temperature under load.*

Heat Distortion Temperature - See *Heat Deflection Temperature.*

Heat Seal Temperature - Temperature of a thermoplastic film or sheet required to join two or more films or sheets in contact by fusion.

Heat Sealing - A method of joining plastic films by the simultaneous application of heat and pressure to the areas in contact. Heat can be applied using hot plate welding, dielectric heating, or radiofrequency welding.

Heat Stability - See *Thermal Stability.*

Heated Tool Welding - A method for joining thermoplastic parts in which a hot plate or hot tool is used to provide heat to melt the joining surfaces. The tool is then removed, and the parts are pressed together. While in the molten state, molecular diffusion across the joint interface occurs, and a homogeneous, permanent bond is formed after the parts are allowed to cool. A hot plate is used for flat surfaces and a hot tool in the shape of the joint for irregularly shaped surfaces. Also called *fusion bonding, heat sealing, hot shoe welding, butt fusion., hot plate welding.*

Hexafluoropropylene, HFP - $CF_3 — CF = CF_2$

HFP - See *Hexafluoropropylene.*

Hold Time - In welding, the length of time allotted for the melted plastic to solidify. In process engineering, the residence time of an individual ingredient in reaction vessel or other processing apparatus. Also called *holding time, holdup time.*

Homopolymer - A polymer that contains only a single type of monomer (i.e., propylene).

Hot Melt Adhesive - An adhesive that is applied in a molten state which forms a bond after cooling to a solid state. Acquires adhesive strength through cooling, unlike adhesives that achieve strength through solvent evaporation or chemical cure.

Hot Stamping - This is a process for obtaining a specific geometry in a polytetrafluoroethylene stock shape. Hot stamping is conducted using a stamp which is heated well above (>360°C) the polymer melt point. Pressure and time are the other variables to manipulate for obtaining the desired pattern.

Hot Tool Welding - See *Heated Tool Welding*.

Hourglassing - Refers to the change in the shape of a cylindrical object made of polytetrafluoroethylene during sintering. A reduction in the diameter of the cylinder near the middle and an increase in the bottom diameter creates a resemblance in the shape of the sintered part to an hourglass.

Hydrocarbon - A chemical compound that contains only hydrogen and carbon atoms.

Hydrocarbon Lubricant - A hydrocarbon liquid added to dispersion polymerized polytetrafluoroethylene powder (fine powder) to form a paste for extrusion.

Hydrofluoric Acid - HF is a highly corrosive acid.

Hydrophilic Surface - Surface of a hydrophilic substance that has a strong ability to bind, adsorb or absorb water; a surface that is readily wettable with water. Hydrophilic substances include carbohydrates such as starch.

Hysteresis Loop - Hysteresis means a retardation of the effect when the forces acting upon a body are changed. For example, the viscosity of shear thinning liquids tends to decrease as shear increases, but it will not increase to the similar value at a given shear if, at the end of the shear rate increase period, the shear rate is decreased back to the initial value. The plot obtained by plotting the viscosity against shear rate is called a *hysteresis loop*.

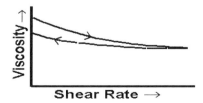

I

Impact Strength - The energy required to break a specimen, equal to the difference between the energy in the striking member of the impact apparatus at the instant of impact with the specimen and the energy remaining after complete fracture of the specimen. Also called *impact energy*.

Impregnation - A term applied to filling or saturation of porous material such as glass cloth and flax with polytetrafluoroethylene dispersion.

Infrared Oven - An oven equipped with infrared lamps where heat is generated by infrared rays.

Injection Molding - A molding procedure in which a heat-softened plastic material is forced from a cylinder into a cavity which gives the article the desired shape. It is used for all thermoplastics and some thermosets.

Interference Fit - A mechanical fastening method used to join two parts, such as a hub and a shaft, in which the external diameter of the shaft is larger than the internal diameter of the hub. This interference produces high stress in the material and must be determined carefully to avoid exceeding the allowable stress for the material. Stress relaxation can occur in interference fits, causing the joint to loosen over time. Also called *press fit*.

Intermolecular Forces - See *van der Waals Forces*.

Ionic Strength - This is a property of solutions containing ions. An increase in the concentration and the number of soluble salts in water increase the ionic strength of the solution. For a solution containing salts each with a molarity of m_i, ($i = 1, 2, ..., k$) and a valence of Z_i, ionic strength (I) is calculated from the following expression:

$$I = \sum_{i=1}^{k} \left(m_i Z_i^2\right)$$

Irradiation - See *Gamma Ray Irradiation*.

Isoparaffinic Hydrocarbon - A hydrocarbon which contains branches in its chemical structure.

Isotatic Molding - A molding method for producing parts (often with complex geometry) from polytetrafluoroethylene powder. The resin powder is compressed between a hard part and an elastic bag. Pressure for compaction is supplied by a high pressure fluid. The fluid can be introduced in pressure vessel by dry bag method or by submerging the mold in a vessel containing the fluid and pressurizing the vessel (wet bag). Also see *Dry* and *Wet Bag Isotatic Molding.*

Izod - See *Izod Impact Energy.*

Izod Impact Energy - The energy required to break a v-notched specimen equal to the difference between the energy in the striking member of the impact apparatus at the instant of impact with the specimen and the energy remaining after complete fracture of the specimen. For metals, it is measured according to ASTM E23. Also called *notched Izod strength, notched Izod impact strength, Izod v-notch impact strength, Izod strength, Izod impact strength, Izod, IVN.*

Izod Impact Strength - See *Izod Impact Energy.*

Izod Strength - See *Izod Impact Energy.*

Izod V-Notch Impact Strength - See *Izod Impact Energy.*

J

Joining - See *Adhesiveless Joining* and *Adhesive Bonding.*

L

Lab Joint - A joint in which one adherend is placed partly over the other adherend; overlapped areas are bonded together.

Lap Shear Test - See *ASTM D1002.*

Laser - A device used to produce an intense light beam with a narrow band width. Laser is an acronym for *light amplification by stimulated emission of radiation.*

Lewis Base - A substance that donates a pair of electrons in a chemical reaction to form a bond with another substance.

Linear Polyethylenes - Linear polyethylenes are polyolefins with linear carbon chains. They are prepared by copolymerization of ethylene with small amounts of higher alpha-olefins such as 1-butene. Linear polyethylenes are stiff, tough and have good resistance to environmental cracking and low temperatures. Processed by extrusion and molding. Used to manufacture film, bags, containers, liners, profiles and pipe.

Linings - Inserts, usually made from plastics to protect metallic or nonmetallic substrates. Linings or liners are either inserted or formed in-place and are usually thicker than coatings fabricated from a dispersion.

Lubricants - Also see *Hydrocarbon Lubricant.*

M

Macroscopic Flaws - Defects such as cracks or inclusions in fluoropolymer parts which can be detected visually or by the use of a simple magnifying glass.

Mandrels - Hard parts (normally metallic) which are inserted in mold to obtain desired part geometries. For example a solid metal mandrel is placed inside an isotatic cylindrical mold in order to obtain a polytetrafluoroethylene liner.

Mason and Pierce Equation - An equation for calculating viscosity of a suspension based on the volume fraction of solids.

$$\eta = -\frac{1}{\left[1-\left(\frac{\phi}{A}\right)^2\right]}$$

ϕ = Volume fraction of solids

A = A constant in the range of 0.64–0.68 for spherical solid particles. For other shapes, A is <0.64.

Matched Die Forming - A type of thermoforming process consisting of preheating the plastic sheet prior to forming and then subjecting it to simultaneous motion of the two mold halves.

Melt Creep Viscosity - A viscosity value measured for polytetrafluoroethylene using a dynamic mechanical analyzer (DMA).

Melting Point - The temperature at which the solid crystalline and liquid phases of a substance are in thermodynamic equilibrium. The melting point is usually referred to normal pressure of 1 atm.

Mica - Mica is a crystalline platy filler made by wet or dry grinding of muscovite or phlogopite, minerals consisting mainly of aluminum and potassium orthosilicates, or by chemical reaction between potassium fluorosilicate and alumina. Used as a filler in thermosetting resins to impart good dielectric properties and heat resistance, and in thermoplastics such as polyolefins to improve dimensional stability, heat resistance, and mechanical strength. Mica fillers also reduce vapor permeability and increase wear resistance. Mica fillers having increased flake size or platiness increase flexural modulus, strength, heat deflection temperature, and moisture resistance. Surface modified grades of mica are available for specialty applications.

Micron - A unit of length equal to 1×10^{-6} meter. Its symbol is Greek small letter mu (μ).

Microporosity - Defects such as small voids or inclusions in fluoropolymer parts which can be detected by a microscope or the use of a fluorescent dye.

Milled Glass Filler - Milled glass fillers are made by hammer milling continuous glass strands. Used as anticrazing reinforcing fillers for adhesives and in plastics. Also called *milled glass fiber*.

Mineral Filler - Mineral fillers are a large subclass of inorganic fillers comprised of ground rocks or natural or refined minerals. Some fillers, so-called commodity minerals, are relatively inexpensive and are used mostly as extenders. A good example of these is ground limestone. Other fillers, so-called specialty minerals, are usually reinforcing fillers. These are inherently small particle size fillers, such as talc, and surface chemically modified fillers. See also *Organic Filler*.

Modified PTFE - Also see *Modifier*.

Modifier - Generally a modifier refers to an additive which alters the properties of the host system. In case of PTFE (polytetrafluoroethylene), a modifier is a comonomer which modifies the properties of PTFE and is present at low concentration (<1%). An example is perfluorpropylvinylether.

Moisture Vapor Permeation - Refers to permeation of water vapor through films and membranes which can be measured by a number of standard methods (e.g., ASTM).

Mold Shrinkage - The difference between the dimension(s) of the mold cavity at 23°C and the dimension(s) of specimen molded. Measured for plastics after cooling the molding to room temperature according to ASTM D955.

Molecular Weight - The molecular weight (formula weight) is the sum of the atomic weights of all the atoms in a molecule (molecular formula). Also called *MW, formula weight, average molecular weight*.

Molecular Weight Distribution - The relative amounts of polymers of different molecular weights that comprise a given specimen of a polymer. It is often expressed in terms of the ratio between weight- and number-average molecular weights, Mw/Mn.

Monomer - The individual molecules from which a polymer is formed, i.e., ethylene, propylene.

Multifilament - In textiles, a fiber or yarn composed of several individual filaments, each of 75 denier or less, that are gathered into a single continuous bundle.

Multilayer Coating - A coating that is produced by multiple passes of the substrate through the coating process. After each pass the thickness of the coating increases. Multilayer coating is a means of overcoming critical cracking thickness when relative thick coatings are required.

Mush - The paste produced by co-coagulation in the DU Bearing Process is called "mush." Also see *DU Bearing Process.*

N

Nanometer - A unit of length equal to 1×10^9 meter. Often used to denote the wavelength of radiation, especially in UV and visible spectral region. Also called *nm*.

Nonpolar - In molecular structure, a molecule in which positive and negative electrical charges coincide. Most hydrocarbons, such as polyolefins, are nonpolar.

Notch Effect - The effect of the presence of specimen notch or its geometry on the outcome of a test such as an impact strength test of plastics. Notching results in local stresses and accelerates failure in both static and cycling testing (mechanical, ozone cracking, etc.).

Notched Izod Impact Strength - See *Izod Impact Energy.*

Notched Izod Strength - See *Izod Impact Energy.*

Nucleophile - Nucleophiles or Nucleophilic reagents are basic, electron-rich reagents. Negative ions and chemical groups can be nucleophiles, in addition to neutral compounds such as ammonia and water. Both ammonia and water molecules contain a pair of unshared electrons.

Nylon - Nylons are thermoplastic, mostly aliphatic polyamides prepared usually either by polymerization of dicarboxylic acid with diamine, or polymerization of amino acid, or ring opening polymerization of lactam. Nylons have good resistance to solvents, bases, and oils; to impact; abrasion; and creep. They have also high tensile strength and barrier properties, and a low coefficient of friction. The disadvantages include high moisture pickup, light degradation, and high mold shrinkage. Processing is achieved by injection, blow, and rotational molding; extrusion; and powder coating. Uses are automotive parts, electrical and electronic devices such as plugs, machine parts such as gears and pumps, housings for appliances and power tools, wire and cable jacketing, pipes, films, and fibers.

O

OI - See *Orientation Index*.

Optical Properties - The effects of a material or medium on light or other electromagnetic radiation passing through it, such as absorption, reflection, etc.

Organic Compound - A chemical compound that contains one or more carbon atoms in its molecular structure.

Organic Filler - Organic fillers are made from natural or synthetic organic materials. Natural material derived organic fillers include wood and shell flours. Synthetic material derived fillers include fluoropolymer spheres and milled polymer waste. Organic fillers are characterized by relatively low cost and low density. They might increase the flammability and decrease the moisture resistance of plastics. See also ***Mineral Filler***.

Organometallic Compound - A chemical compound in which carbon is bonded to a metal. Many organometallic compounds are used as catalysts in polymerization reactions.

Orientation - A process of drawing or stretching of as-spun synthetic fibers or hot thermoplastic films to orient polymer molecules in the direction of stretching. The fibers are drawn uniaxially and the films are stretched either uniaxially or biaxially (usually longitudinally or longitudinally and transversely, respectively). Oriented fibers and films have enhanced mechanical properties. The films will shrink in the direction of stretching, when reheated to the temperature of stretching.

Orientation Index - This is a factor that characterizes the orientation of polytetrafluoroethylene tube in machine and transverse directions. Orientation index (OI) provides a numerical value to monitor the disparity of orientation:

$$OI = 1 - \frac{\text{Yield strength in transverse direction}}{\text{Yield strength in machine direction}}$$

OI = 0 means balanced orientation.

OSHA - Occupational Safety and Health Administration

Oxygen Sensor - A sensor in automobiles senses the oxygen content of the exhaust gas.

Ozone - O_3

P

Paraffins - Linear saturated hydrocarbons with the general chemical formula of C_nH_{2n+2} are also called *paraffins*.

Pascal - An SI unit of measurement of pressure equal to the pressure resulting from a force of one newton acting uniformly over an area of one square meter. Used to denote the pressure of gases, vapors or liquids and the strength of solids. Also called *Pa*.

Paste Extrusion - Process by which dispersion polymerized polytetrafluoroethylene (PTFE) is extruded with the aid of a hydrocarbon lubricant. A paste of the PTFE and hydrocarbon is made and formed into a preform in a separate step. The preform is ram-extruded through a die, dried, and sintered.

PCTFE - See ***Polychlorotrifluoroethylene***.

PEEK - See ***Polyetheretherketone***.

PEK - See ***Polyetherketone***.

Peel Strength - The bond strength of a film adhered by an adhesive to a substrate is measured by different techniques and is called *peel strength*. An extensiometer can be used to measure peel strength.

Pendant Methyl Group - A methyl (CH_3) functional group attached to the main chain of a polymer molecule.

Perfluorinated Fluoropolymers - Polymer consisting of only carbon and fluorine (and an occasional oxygen atom) atoms are called *perfluorinated fluoropolymers*.

Perfluorinated Paraffins - Refers to a linear saturated hydrocarbon where all hydrogen atoms have been replaced with fluorine, with the general chemical formula of C_nF_{2n+2}.

Perfluoroalkoxy (PFA) Polymer - R_f = A perfluorinated alkyl group containing one or more carbon atoms, typically a maximum of four carbon atoms.

$$-CF_2-CF_2-\underset{\underset{R_f}{|}}{\underset{O}{\underset{|}{CF}}}-CF_2-$$

Perfluoroalkylvinylether (PAVE) -

$$R_f - O - CF = CF_2$$

where R_f is a perfluorinated alkyl group containing one or more carbon atoms, typically a maximum of four carbon atoms.

Perfluoroammoniumoctanoate - (C8)

$$CF_3-(CF_2)_6-\overset{\overset{O}{\|}}{C}-O-NH_4$$

PFA - See *Perfluoroalkoxy Polymer.*

PTFE - See *Polytetrafluoroethylene.*

Perfluoropolymer - See *Perfluorinated Fluoropolymer.*

Perfluoropropylvinylether (PPVE) -

$$CF_3-\underset{\underset{CF_3}{|}}{CF}-O-CF=CF_2$$

Permeability - The capacity of material to allow another substance to pass through it; or the quantity of a specified gas or other substance which passes through under specified conditions.

Permittivity Loss Factor - See *Dielectric Dissipation Factor.*

Persulfate Initiation - This is in reference to the action of persulfates such as ammonium persulfate in the free radical polymerization of tetrafluoroethylene.

Petroleum Solvents - A solvent that is derived from processing petroleum.

Plasma Arc Treatment - In adhesive and solvent bonding, a method for treating the surfaces of parts prior to solvent and adhesive bonding, in which an electrical current between two electrodes in a gas at low pressure excites the gas particles, producing free radicals. Contaminants are stripped from the surface of the part, and wettability is increased by reduction of the contact angle. Also called *plasma discharge, plasma treatment*. See also **Corona Discharge Treatment.**

Poisson's Ratio - This ratio is defined as the ratio of the change in the width per unit width of a material to the change in its length per unit length, as a result of strain in the length direction. At 23°C Poisson ratio of polytetrafluoroethylene has a value of 0.46.

Poker Chipping - Fracture of a ram extruded part such as a rod or tube at the interface of charges (weld) is called *poker chipping*.

Polar - In molecular structure, a molecule in which the positive and negative electrical charges are permanently separated. Polar molecules ionize in solution and impart electrical conductivity to the solution. Water, alcohol, and sulfuric acid are polar molecules; carboxyl and hydroxyl are polar functional groups.

Polyaramide - Commercial examples of these plastics include Kevlar® and Nomex®. Kevlar® has the following chemical structure:

$$\left[-\underset{\underset{O}{\|}}{C}-\underset{}{\bigcirc}-\underset{\underset{O}{\|}}{C}-\underset{\underset{H}{|}}{N}-\underset{}{\bigcirc}-\underset{\underset{H}{|}}{N}-\right]_n$$

Thermoplastic aromatic polyamide prepared by polymerization of aromatic diamine and aromatic diacid or its derivatives (e.g., anhydride). These plastics are characterized by high heat resistance, good retention of mechanical and dielectric properties at elevated temperatures, good dielectric properties, chemical resistance, high stiffness, and low flammability. High crystallinity results in good mechanical properties such as impact strength. Resistance to light is somewhat low. Processing is difficult because of the high melting point. Processed by solution casting, molding, and extrusion. Uses include film, fibers, and molded parts requiring good surface finish. Also called *PARA, aromatic polyamide, aramid.*

Polyarylene Plastics - Examples of these plastics include polyetherketone, polyetheretherketone, polyetherketoneketone, polyphenylenesulfide, and others.

Polyarylsulfone - Polyarylsulfone is a thermoplastic containing repeating sulfone and ether groups in its wholly aromatic backbone. It has excellent resistance to high and low temperatures, good impact strength, improved resistance to environmental stress cracking, good dielectric properties, rigidity, and resistance to acids and alkalies. Polyarylsulfone is nonflammable, but is attacked by some organic solvents. Processed by injection molding, compression molding, and extrusion. Used in high temperature electrical and electronic applications such as circuit boards and lamp housings, piping, and auto parts.

Polycarbonate - Polycarbonates are thermoplastics prepared by either phosgenation of dihydric phenols such as bisphenol A or by ester exchange between diaryl carbonate, usually diphenyl carbonate, and dihydric phenol. They are characterized by the presence of repeating carbonyldioxy groups in the backbone, but classified by some as polyesters. They have very good mechanical properties, especially impact strength, low moisture absorption, and good thermal and oxidative stability. They are self-extinguishing and some grades are transparent; but are attacked by strong acids and bases, soluble in organic solvents, and subject to stress cracking. Processed by injection and blow molding, extrusion, thermoforming, but require high processing temperatures. Used in telephone parts, dentures, business machine housings, safety equipment, nonstaining dinnerware, and food packaging. Also called *PC.*

Polychlorotrifluoroethylene (PCTFE) -

$[-CF_2-CFCl-]_n$

Thermoplastic prepared by radical polymerization of chlorotrifluoroethylene. It has good transparency and great barrier properties. Dielectric properties and resistance to solvents, especially chlorinated, of PCTFE are somewhat lower than those of perfluoropolymers, but tensile strength and creep resistance are higher. Processing is difficult, because of high melt viscosity, but possible by extrusion, injection molding, compression molding, and coating. Uses include chemical apparatus, cryogenic seals, films, and coatings. Also, PCTFE spheres are used as fillers and PCTFE oil is used as a lubricant in various plastics.

Polyetheretherketone - Polyetheretherketone is a partially crystalline thermoplastic containing repeating ether and keto groups in the backbone. Its systematic name is poly(oxy-1,4- phenyleneoxy-1,4-phenylenecarbonyl-1,4-phenylene). PEEK has superb chemical resistance, toughness, rigidity, load-bearing properties, thermal stability, radiation resistance, and very low flammability. Processed readily by injection molding, spinning, cold forming, and extrusion. Used in fibers, films, automotive engine parts, aerospace composites, and wire and cable insulation. Also called *PEEK.*

Polyetherketone - Thermoplastic with excellent high-temperature properties and thermal stability. Used in advanced composites, wire coating, chemical-resistant filters, integrated circuit wafer carriers, and bearings. Synthesized by polyarolyation or polyetherization. Also called *PEK.*

Polyethersulfone - Polyethersulfone is a thermoplastic containing repeating sulfone and ether groups in its wholly aromatic backbone. It has excellent thermal stability in air and water, good transparency, very low flammability and smoke emission, good dimensional stability, rigidity, and toughness. It has low mold shrinkage and is a good, heat resistant dielectric. Polyethersulfone has good resistance to acids and bases but is attacked by many organic solvents, has poor weatherability, and is subject to stress cracking. Processed by injection molding, extrusion, blow molding, and compression molding; but requires high processing temperatures. Used in high temperature electrical applications such as multipin connectors, bakery oven windows, medical devices, radomes, structural aircraft components, appliance housings, and auto parts. Also called *polyether polysulfone, PESV.*

Polyethylene - A family of polyolefins consisting of linear and branched polyethylenes. Polyethylenes are thermoplastics but can be cross-linked by irradiation or chemically and then show improved strength and dielectric properties. All linear polyethylenes, except the high density grade, are prepared by copolymerization of ethylene with higher olefins. Branched polyethylenes are prepared from ethylene alone or together with polar comonomers such as vinyl acetate. The density, melt index, crystallinity, degree of branching, molecular weight, polydispersity, and related properties of polyethylenes vary widely depending on the catalysts and methods of polymerization and on modifying comonomers and treatments. Polyethylenes have good impact resistance at low temperatures, good chemical resistance, and good moisture resistance, but high thermal expansion, poor weatherability, poor thermal stability and resistance to stress cracking. They are readily processible by all thermoplastic methods but are flammable and difficult to bond. Food grades are available. Processed by extrusion, blow and injection molding, thermoforming. Used very broadly as films, coatings, in containers and consumer goods, electrical insulation, and piping. Also called *PE, expandable polyethylene bead, EPE bead.*

Polymer - Polymers are high molecular weight substances with molecules resembling linear, branched, cross-linked, or otherwise shaped chains consisting of repeating molecular groups. Synthetic polymers are prepared by polymerization of one or more monomers. The monomers comprise low-molecular-weight reactive substances, often containing more than one reactive molecular bond or chemical bond. Natural polymers have molecular structures similar to synthetic polymers but are not man made, occur in nature, and have various degrees of purity. Also called *synthetic resin, synthetic polymer, resin, plastic.*

Polyolefin - Polyolefins are a large class of carbon-chain elastomeric and thermoplastic polymers usually prepared by addition (co)polymerization of olefins or alkenes such as ethylene. The most important representatives of this class are polyethylene and polypropylene. There are branched and linear polyolefins and some contain polar pendant groups or are halogenated. Unmodified polyolefin are characterized by relatively low thermal stability and a nonporous, nonpolar surface with poor adhesive properties. Processed by extrusion, injection molding, blow molding, and rotational molding. Other thermoplastic processes are used less frequently. This class of plastics is used more and has more applications than any other. Also called *olefinic resin, olefinic plastic.*

Polyphenylene Sulfide - PPS is a high-performance engineering thermoplastic. Its linear chains consist of benzene rings para-substituted with sulfur atom links. PPS has excellent thermal stability; good resistance to solvents and chemicals, except for chlorinated hydrocarbons; good radiation resistance; excellent dimensional stability; low water absorption; and good dielectric properties, but somewhat low impact strength. PPS is nonburning. Processing of PPS is hampered by its high melting point and is achieved at elevated temperatures by injection molding, compression molding, transfer molding, and extrusion. Used in hydraulic components, bearings, valves, electronic parts, small appliances such as hair dryers, and under-the-hood automotive parts. Also, PPS spheres are used as fillers in various plastics. Also called *polythiophenylene.*

Polyphenylene Sulfide Sulfone - Thermoplastic with good thermal stability, chemical resistance, and dielectric properties and low flammability and creep. Processed by injection molding. Used in electrical components. Also called *polyphenylenesulfidesulfone.*

Polypropylene - PP is a carbon chain thermoplastic comprised of propylene homopolymer prepared by stereospecific polymerization in the presence of Ziegler Natta catalysts. The majority of PP is isotactic. PP has low density and good flexibility and resistance to chemicals, abrasion, moisture, and stress cracking, but decreased dimensional stability, mechanical strength, and resistance to UV light and heat. PP is flammable. Processed by injection molding, spinning, extrusion, and film techniques. Used as films for pressure sensitive tapes, packaging, liners, and shrink films, and as fibers in textiles. Also called *PP homopolymer, PP, polypropylene homopolymer, expandable polypropylene bead, EPP bead.*

Polymer Fume Fever - A condition that occurs in humans as a result of exposure to degradation products of polytetrafluoroethylene and other fluoropolymers. The symptoms of exposure resemble those of flu and are temporary. After about twenty-four hours, the flu-like symptoms disappear.

Polysulfone - Polysulfones are a class of thermoplastics characterized by the presence of sulfone linkages in their repeating units. Most polysulfones are aromatic and also contain ether linkages to enhance oxidation resistance. The most important of those are polyarylsulfone and polyethersulfone. As a class, polysulfones show excellent thermal stability, low creep, good dielectric properties, transparency, self extinguishing properties, and resistance to chemicals, but are attacked by many solvents, have poor weatherability, are subject to stress cracking, and require high processing temperatures. Processed by injection molding, compression molding, blow molding, and extrusion. Used in appliances, electronics, automotive parts, and electric insulators. A high-performance engineering thermoplastic, called polysulfone in the trade, consists of diaryl sulfone groups linked by oxygen atoms and shares many properties typical for its class. It has good rigidity at elevated temperatures, good impact strength, and resists alkalies and acids but is attacked by polar solvents. Used for medical devices, in food processing equipment, electronics, chemical apparatus, and battery cells. Also called *PSU, PSO.*

Polytetrafluoroethylene - Thermoplastic prepared by radical polymerization of tetrafluoroethylene. It has low dielectric constant, superior chemical resistance, very high thermal stability, low friction coefficient, excellent antiadhesive properties, low flammability, and high weatherability. Impact resistance of PTFE is high, but permeability is also high whereas strength and creep resistance are relatively low. The very high melt viscosity of PTFE restricts its processing to sinter molding and powder coating. Uses include coatings for cooking utensils, chemical apparatus, electrical and nonstick items, bearings, containers. Also, PTFE spheres are used as fillers and PTFE oil is used as a lubricant in various plastics. Also called *TFE, PTFE, modified PTFE.*

Polytetrafluoroethylene Compounds - Material obtained by intimate mixing of fillers (metallic and nonmetallic) with polytetrafluoroethylene. One or more of polymer properties such as cold flow, wear and surface hardness are altered by the addition of fillers.

Polyvinyl Chloride - PVC is a thermoplastic prepared by free-radical polymerization of vinyl chloride in dispersion (emulsion), bulk, or suspension processes. A small amount of comonomer is sometimes added to enhance adhesion or other properties. Unmodified PVC is rigid and requires plasticizers to make it more flexible. The main end forms of PVC are rigid and flexible. The flexible form is often made from plastisols, suspensions of PVC in liquid plasticizers. PVC can be chlorinated to increase its heat deflection temperature and tensile strength and to reduce flammability and smoke generation. PVC is dimensionally stable, largely nonflammable, and resistant to weathering, but has limited thermal stability, high density, and is attacked by many solvents. Processed by injection molding, calendaring, extrusion, powder coating, blow molding, extrusion coating, and film techniques. Used very widely as films, fabric coatings, wire coatings, toys, bottles, and pipes. Also called *PVC.*

Polyvinyl Fluoride (PVF) -

$$[-CHF-CF_2-]_n$$

Thermoplastic prepared by free radical polymerization of vinyl fluoride. It is extruded into thin (<100 μm) films as a plastisol (dispersion in a polar solvent). PVF is known for its weather resistance, release, adherability and mechanical strength.

Polyvinylidene Fluoride (PVDF) -

$$[-CF_2-CH_2-]_n$$

Thermoplastic prepared from vinylidene fluoride. It has substantially higher strength, wear resistance, and creep resistance than other fluoropolymers but relatively high dielectric constant and loss factor. PVDF is nonflammable, resists most solvents, and has excellent weatherability. Its service temperature limit and chemical resistance are lower than those of perfluoropolymers. PVDF is processed readily by extrusion, injection molding, transfer molding, and powder or dispersion coating. Uses include electrical insulation, pipes, chemical apparatus, coatings, films, containers, and fibers. Also called *VF2, PVF2, PVDF.*

Porosity - Porosity is defined as the volume of voids per unit volume of a material or as the volume of voids per unit weight of material.

Preform - Also see *Preforming.* A shape produced by the compaction of a polymer powder (or its filled compounds) below the melting point of the polymer (e.g., PTFE). The preform is subsequently sintered in a separate step to produce a useful part.

Preforming - Also see *Preform.* The process of producing a preform which consists of filling a mold with the polymer powder, compaction of the resin by pressing it down in a press and finally removal of the preform from the mold. The mold is usually operated at the room temperature and is designed to allow easy escape of the air during the resin compaction.

Pressure Hoses - Reinforced polytetrafluoroethylene (PTFE) hoses comprised of a PTFE tube and reinforcement, usually in the form of single ply or multiplies of metal wire braiding, for super atmospheric pressure end uses.

Pressure Sensitive Adhesive - An adhesive that requires applied pressure on the parts for bonding to occur. Usually composed of a rubbery elastomer and modifying tackifier, pressure sensitive adhesives are applied to the parts as solvent-based adhesives or hot melts; curing does not usually occur. They adhere tenaciously under slight pressure and are highly thixotropic. Disadvantages include limited temperature capability and susceptibility to oxidative degradation. These adhesives do not undergo progressive viscosity increase like other adhesives but instead they are in a permanent tacky stage. They are usually coated on paper, plastic films, foam or cloth and applied with pressure to the adherend, as their name implies. Most pressure sensitive adhesives contain a blend of elastomers like SBR or natural rubber with low or medium tacky fibers. See also **Contact Adhesive.**

Primer - In adhesive bonding, a reactive chemical species dispersed in a solvent that is applied to the part surface by spraying or brushing. After the solvent is flashed off, the part surface may be bonded immediately, as in polyolefin primers for cyanoacrylates, or may require time to react with atmospheric moisture, as in silane and isocyanate-based primers used for silicone and polyurethane-based adhesives, respectively. Primers generally contain a multifunctional chemically reactive species capable of acting as a chemical bridge between the substrate and the adhesive. Primers are commonly used with acetals, fluoropolymers, polybutylene terephthalate, silicone, polyurethane, and polyolefins. In coatings, coatings applied on a substrate prior to subsequent coatings or topcoat in order to seal the pores, improve adhesion of the topcoat, improve corrosion protection, hide surface imperfections or color, etc. Usually based on polymers with functional additives. Applied by the same techniques as coatings. Also called *primer coating.*

PTFE Fiber - This is a polytetrafluoroethylene (PTFE) yarn produced by spinning of a blend of PTFE and viscose followed by chemical conversion, drying and sintering.

PVC - See *Polyvinyl Chloride.*

PVF - See *Polyvinyl Fluoride.*

PVDF - See *Polyvinylidene Fluoride.*

PV Limit - Wear of plastics is proportional to load (or pressure P) and velocity (V). Combinations of pressure and velocity are defined where the material can be used, thus a *PV* limit is defined.

R

Radiation Dosage - See *Radiation Dose.*

Radiation Dose - Amount of ionizing radiation energy received or absorbed by a material during exposure. Also called *radiation dosage, ionizing radiation dose.*

Radiation Resistant Materials - Materials that resist degradation on long- and medium-term or repeated exposure to ionizing radiation, e.g., steel grades designed for nuclear reactors. Radiation damage to materials includes swelling, radiolysis, blistering, changes in electrical and mechanical properties, etc. There are different mechanisms of radiation damage but most can be linked to free-radical reactions. The resistance of materials to radiation can be improved by stabilizing them with agents that can neutralize free radicals, such as dimethyl sulfoxide, carbohydrates and various reducing agents. Also called *radiation stabilized material.*

Radicals - See *Free Radicals.*

Radiography - An x-ray based technique used for detecting flaws in materials, for example voids and cracks in metals.

Ram Extrusion - Ram extrusion is the only continuous process for fabrication of parts from suspension polymerized (granular) polytetrafluoroethylene. All three required steps of processing are performed in one machine called *ram extruder.* The most common shapes are solid rods and tubes.

The basic steps of ram extrusion are compaction, sintering and cooling. Free flow powder is fed into a die and compacted against the previous charge by the action of a reciprocating ram which also advances the charge into the heated section of the die. After sintereing, the charge advances into the cooling zone of the die and finally exits.

Reduction Ratio - This factor is the ratio of cross sectional surface areas of the preform and the extrudate in paste extrusion. Different dispersion polymerized polytetrafluoroethylene powders accommodate different reduction ratio ranges. In general, reduction ratio of the resin decreases as molecular weight increases. Fracture and defects appear in the extrudate if it is processed at above its maximum reduction ratio.

Repro - This is short for "reprocessed" and is applicable to a great variety of scrap generated during processing of fluoropolymers. Scrap materials are treated in multiple step processes using acids and heat for purification. Afterwards they are ground, sized and consumed. A number of key properties of fluoropolymers decline as a result of purification process.

Repulsive Intermolecular Forces - Forces generated when atoms or molecules approach each other closely.

Rheology - A science that studies and characterizes flow of polymers, resins, gums and other materials.

Rietz Mill - A type of hammer mill supplied by Bepex Corp., Santa Rosa, CA. See *Hammer Mill*.

Rotolining - See *Rotational Lining*.

Rotational Lining - See also *Rotational Molding*. Rotational lining is a process by which a hollow object is lined with a plastic. The surface of the part, contrary to rotational molding process, is prepared to adhere the liner to the mold wall.

Rotomolding - See *Rotational Molding*.

Rotational Molding - Also known as rotocasting or rotomolding is a process for manufacturing hollow plastic parts. A typical procedure for rotational molding is as follows: Very fine plastic powder is placed in a mold and the closed mold is heated above the melting point of the powder while the mold is rotated in two planes at right angle to each other. The heating continues until the polymer powder fuses and melts to form a homogeneous layer of uniform thickness. The mold is rotated while it is cooled down to the removal temperature. At the end mold rotation is stopped and the part is removed.

S

Semicrystalline Plastic - A plastic material characterized by localized regions of crystallinity. See also *Amorphous Plastic*.

Shear - Displacement of a plane of a solid body parallel to itself, relative to other parallel planes within the body; deformation resulting from this displacement.

Shelf Life - Time during which a physical system such as material retains its storage stability under specified conditions. Also called *storage life*.

Shore A - See *Shore Hardness*.

Shore D - See *Shore Hardness*.

Shore Hardness - Indentation hardness of a material as determined by the depth of an indentation made with an indentor of the Shore-type durometer. The scale reading on this durometer is from 0, corresponding to 0.100" depth, to 100 for zero depth. The Shore A indenter has a sharp point, is spring-loaded to 822 gf, and is used for softer plastics. The Shore D indenter has a blunt point, is spring-loaded to 10 lbf, and is used for harder plastics. Also called *Shore D, Shore A*.

Silicone - Silicones are polymers, the backbone of which consists of alternating silicon and oxygen atoms. Pendant organic groups are attached to silicon atoms. They are usually made by hydrolyzing chlorosilanes, followed by poly condensation and cross-linking. Depending on the degree of cross-linking and the nature of pendant groups, silicones can be liquid, elastomeric, or rigid. Liquid silicones or silicone fluids such as dimethylsiloxane have very good anti adhesive properties, lubricity, resistance to heat and chemicals and are used as release agents, surfactants, and lubricants in plastics. As lubricants they improve wear resistance of plastics. Silicone elastomers, or rubbers, have high adhesion, resistance to compression set, flexibility, good dielectric properties, weatherability, low flammability, good moisture barrier properties, and thermal stability, but somewhat low strength. Optically clear grades are available. Processed by coating and injection molding. Used as optical fiber coatings, electronic connector encapsulants, printed circuit board coatings, seals, diaphragms, fabric coatings, medical products, adhesives, sealants, and glazing compounds. Rigid silicone resins offer good flexibility, weather ability, dirt release properties, dimensional stability, and are stronger and harder than silicone rubbers. The resins are attacked by halogenated solvents. Processed by coating, casting, injection molding, compression molding, and transfer molding. Used as coatings, adhesives, sealants, bonding agents, and molded parts. Also called siloxane, silicone rubber, silicone plastic, silicone fluid, SI, polysiloxane.

Sintering - Consolidation and densification of polytetrafluoroethylene particles above its melting temperature is called *sintering*. Also see **Coalescence.**

Skiving - This is a popular method for producing films and tapes of polytetrafluoroethylene (PTFE). Skiving resembles peeling of an apple where a sharp blade is used at a low angle to the surface of a billet (cylinder) of PTFE. A similar method is used in the production of wood veneer from trees.

Sliding Velocity - The relative speed of movement of one body against the surface of another body (counterbody) without the loss of contact as in a sliding motion during wear and friction testing of materials. In the sliding motion, the velocity vectors of the body and the counterbody remain parallel and should be unequal if they have the same direction.

Slitting - This is a process to cut film and webs into narrower width than the starting material. A roll of the wide film is unwound and passed over sharp knives positioned to obtain the required cut widths. The narrower films are wound up on separate cores. Special machines are available for slitting films.

Sodium Etching - Sodium etching is a process by which the surface of fluoropolymers is rendered adherable. During etching the fluoropolymer surface is brought into contact with a sodium solution (1:1) in naphthalene dissolved in tetrahydrofuran or an anhydrous solution of sodium in liquid ammonia.

Sodium Naphthalenide - See *Sodium Etching.*

Softening Point - Temperature at which the material changes from rigid to soft or exhibits a sudden and substantial decrease in hardness.

Solubility - The solubility of a substance is the maximum concentration of a compound in a binary mixture at a given temperature forming a homogeneous solution. Also called *dissolving capacity.*

Solubility Parameter - Solubility parameter characterizes the capacity of a substance to be dissolved in another substance, e.g., of a polymer in a solvent. It represents the cohesive energy of molecules in a substance and determines the magnitude and the sign of the heat of mixing two substances in given concentrations. The magnitude and the sign of the heat of mixing determine the sign of the free energy of mixing. The solution occurs when the sign of the free energy of mixing is negative.

Spark Tester - A machine in which a wire or a thin film is passed through a narrow gap and subjected to an electric voltage to find the defective areas in the film. Defects and voids spark out at an appropriately high voltage.

Specific Wear Rate - Also known as wear factor, specific wear rate is defined as the volume of material worn away per unit of sliding distance and per unit of load.

Spherulite - In plastics, a rounded aggregate of radiating lamellar crystals with amorphous material between the crystals. Has the appearance of a pompom. Spherulites exist in most crystalline plastics and usually impinge on one another to form polyhedrons. Range in size from a few tenths of a micron in diameter to several millimeters.

Standard Specific Gravity - SSG or standard specific gravity is determined for polytetrafluoroethylene using a specially molded and sintered sample. SSG is used as a means of assessing molecular weight of this polymer. Molding and sintering are procedures specified by ASTM F4894.

Steric Hindrance - A spatial arrangement of the atoms of a molecule that blocks reaction of the molecule with another molecule.

Stick Slip - This is a jerking action that occurs in a moving part such as a bearing in overcoming a higher static coefficient of friction than a dynamic coefficient of friction before movement begins.

Strain - The per unit change, due to force, in the size or shape of a body referred to its original size or shape. Note: Strain is nondimensional but is often expressed in unit of length per unit of length or percent. Also called *mechanical strain*. See also ***Flexural Strain, Compressive Strain, Tensile Strain.***

Stress Cracking - Appearance of external and/or internal cracks in the material as a result of stress that is lower than its short-term strength. Also see ***Environmental Stress Cracking.***

Stress Relaxation - Time-dependent decrease in stress in a solid material as a result of changes in internal or external conditions. Also called *stress decrease*.

Surface Energy - See ***Surface Tension.***

Surfactant - Derived from *surface active agent*. Defined as substances which aggregate or absorb at the surfaces and interfaces of materials and change their properties. These agents are used to compatibilize two or more immiscible phases such as water and oil. In general, one end of a surfactant is water soluble and the other end is soluble in an organic liquid.

Surface Roughening - In adhesive bonding, a commonly used surface preparation technique in which the substrate surface is mechanically abraded. The roughened surface increases bondability by dramatically increasing the number of sites available for mechanical interlocking.

Surface Roughness - The closely spaced unevenness of a solid surface (pits and projections); can be quantified by various methods, e.g., by using a profilometer in coatings.

Surface Tension - The surface tension is the cohesive force at a liquid surface measured as a force per unit length along the surface or the work which must be done to extend the area of a surface by a unit area, e.g., by a square centimeter. Also called *free surface energy*.

Suspension Polymerization - Refers to a heterogeneous polymerization regime in which the product of the reaction is a solid forming a suspension in the liquid medium of reaction. Little or no surfactant is added to the reaction medium. Characteristics of the process include high agitation rate and poor particle size control. An advantage of this reaction is high purity of the polymer product as compared to that of the dispersion method.

T

Taber Abrasion Resistance - The weight loss of a plastic or other material specimen after it was subjected to abrasion in a Taber abraser for a prescribed number of specimen disk rotations, usually 1000. A Taber abraser consists of an idling abrasive wheel, designated depending on the type and grit of abrasive used as CS-10F, H 22, etc., and a rotary disk with the specimen mounted on it. The load is applied to the wheel. The produced motion simulates that of rolling with slip.

Talc - Talc is a filler made by dry or wet grinding of mineral magnesium silicate. Talc improves stiffness, dimensional stability, flexural modulus, creep resistance, flow, surface smoothness, moisture resistance, tensile strength, and wear resistance of plastics. It also increases heat deflection temperature and decreases vapor permeability. Can be used as a film antiblock agent. Used mainly in polypropylene but also in thermoplastic and unsaturated polyesters and epoxy resins at low levels. Surface-modified grades are available.

Tensile Elongation - See ***Elongation.***

Tensile Fatigue - Progressive localized permanent structural change occurring in a material subjected to cyclic tensile stress that may culminate in cracks or complete fracture after a sufficient number of cycles. See also ***Fatigue, Flexural Fatigue.***

Tensile Heat Distortion Temperature - See ***Heat Deflection Temperature.***

Tensile Properties - Properties describing the reaction of physical systems to tensile stress and strain.

Tensile Strain - The relative length deformation exhibited by a specimen subjected to tensile force. See also ***Flexural Strain, Strain.***

Tensile Strength - The maximum tensile stress that a specimen can sustain in a test carried to failure. Note: The maximum stress can be measured at or after the failure or reached before the fracture, depending on the viscoelastic behavior of the material. Also called *ultimate tensile strength, tensile ultimate strength, tensile strength at break*.

Tensile Strength at Break - The maximum load per original minimum cross-sectional area of the plastic specimen in tension within the gage length when the maximum load corresponds to the break point. Measured according to ASTM D638. See also ***Tensile Strength.***

Tensile Strength at Yield - The maximum load per original minimum cross-sectional area of the plastic specimen in tension within the gage length, when the maximum load corresponds to the yield point. Measured according to ASTM D638.

Tensile Stress - The force related to the smallest original cross-section of the specimen at any time of the test.

Thermal Conductivity - The time rate of heat transfer by conduction across a unit area of substance at unit thickness and unit temperature gradient.

Thermal Expansion Coefficient - The change in volume per unit volume resulting from a change in temperature of the material. The mean coefficient of thermal expansion is commonly referenced to room temperature.

Thermal Properties - Properties related to the effects of heat on physical systems such as materials and heat transport. The effects of heat include the effects on structure, geometry, performance, aging, stress-strain behavior, etc.

Thermal Recycling - A plastics recycling method in which mixed plastic waste undergoes controlled combustion, producing heat that can be used as a substitute for oil, gas, and coal or for the generation of energy at power plants.

Thermal Stability - The resistance of a physical system such as material to decomposition, deterioration of properties, or any type of degradation in storage under specified conditions. Also called *oven stability, heat stability*.

Thermoforming - The process of heating a thermoplastic sheet to a point at which it softens and flows, then applying differential pressure to make the sheet conform to the shape of a mold or die.

Thermoplastic - Thermoplastics are resin or plastic compounds which, after final processing, are capable of being repeatedly softened by heating and hardened by cooling by means of physical changes. There are a large number of thermoplastic polymers belonging to various classes such as polyolefins and polyamides. Also called *thermoplastic resin*.

Thermoplastic Resin - See *Thermoplastic.*

Thermoset - Thermosets are resin and plastic compounds which, after final processing, are substantially infusible and insoluble. Thermosets are often liquids at some stage in their manufacture or processing and are cured by heat, oxidation, radiation, or other means often in the presence of curing agents and catalysts. Curing proceeds via polymerization and/or cross-linking. Cured thermosets cannot be resoftened by heat. There are a large number of thermosetting polymers belonging to various classes such as alkyd and phenolic resins. Also called *thermosetting resin, thermoset resin*.

Thermoset Resin - See *Thermoset.*

Thermosetting Elastomer - A large class of polymers that can be stretched at room temperature to at least twice their original length and, after having been stretched and the stress removed, return with force to approximately their original length in a short time. To attain this elastic property the rubbers must be cross-linked or vulcanized, usually by heating in the presence of various cross-linking agents and catalysts. There are natural and synthetic rubbers. The most important synthetic rubber families are olefinic rubbers, dienic rubbers (nitrile, butadiene, neoprene), silicone rubbers, and urethane rubbers. Used often as impact modifiers/fillers in plastics. Also called *synthetic rubber, rubber*.

Thixotropic Liquids - These liquids exhibit lower viscosity as shear rate increases. A practical example is house paint which appears thinner when stirred. See also **Hysteresis Loop.**

Tornado Mill - See also **Hammer Mill.** A type of hammer mill supplied by Tornado Engineering Division of Solus Industries, Niagara Falls, New York.

Torsional Damping - One of the methods to measure the crystallinity of polytetrafluoroethylene is torsional damping. Other methods include infrared spectroscopy, x-ray diffraction and ultrasonics.

Transition Point - See *Transition Temperature.*

Transition Temperature - This is a temperature (19°C for 100% homopolymer of tetrafluoroethylene) at which the unit crystalline cell of polytetrafluoroethylene changes from triclinic to hexagonal.

Tribological Characteristics - These characteristics deal with friction or contact related phenomenon in materials. Coefficient of friction and wear rate are the most important tribological characteristics of a material.

Tubular Ovens - Ovens that have round cylindrical shapes.

Two Roll Coating - Any coating technique in which two rolls are involved can be generally called two roll coating. One roll is in contact with the liquid coating and transfers the liquid to a web that is wrapped around the second roll. The first roll is called the applicator roll and the second roll is called the back up roll.

U

Ultraviolet Light - See *Ultraviolet Radiation.*

Ultraviolet Radiation - Electromagnetic radiation in the 40–400 nm wavelength region. Sun is the main natural source of UV radiation on the earth. Artificial sources are many, including fluorescent UV lamps. Ultraviolet radiation causes polymer photodegradation and other chemical reactions. Note: UV light comprises a significant portion of the natural sun light. Also called *UV radiation, UV light, ultraviolet light.* See also ***Ultraviolet Radiation Exposure.***

Ultraviolet Radiation Exposure - In adhesive bonding, a surface preparation technique in which the substrate is irradiated with high intensity UV light. Exposure to UV radiation results in chain scissions, cross-linking, and oxidation of the polymer surface. The effectiveness of this technique is dependent on the wavelength of radiation used. It is commonly used for polyolefins. Also called *UV exposure.*

V

Vacuum Forming - Vacuum forming is a type of thermoforming process consisting of preheating the plastic sheet prior to forming. The sheet is formed into the female mold by application of vacuum through holes in the mold.

van der Waals Attraction - See *van der Waals Forces.*

van der Waals Forces - Weak attractive forces between molecules, weaker than hydrogen bonds and much weaker than covalent bonds.

V-Cone Blender - See *V-Shaped Blender.*

Vinyl Chloride (VC) - Monomer for polyvinylchloride, $CH_2 = CHCl$

Vinylidene Fluoride (VF$_2$) - Monomer for polyvinylidene fluoride, $CH_2 = CF_2$.

Vinyl Fluoride (VF) - Monomer for polyvinylfluoride, $CH_2 = CF_2.$

Vinylidene Fluoride Hexafluoropropylene Copolymer - Thermoplastic copolymer of vinylidene fluoride and hexafluoropropylene. Has better thermal stability; antistick, dielectric, and antifriction properties; and chemical resistance, but lower mechanical strength at room temperature and creep resistance, compared to incompletely fluorinated fluoropolymers. Processing by conventional thermoplastic techniques is difficult due to its high melt viscosity. Uses include chemical apparatus, containers, films, and coatings. Also called *FKM.*

Vinylidene Fluoride Hexafluoropropylene Tetrafluoroethylene Terpolymer - Thermosetting elastomeric polymer of vinylidene fluoride, hexafluoropropylene, and tetrafluoroethylene having good chemical and heat resistance and flexibility. Used in auto parts. Also called FKM/TFE viscoelasticity A property of a material that exhibits both elastic and viscous behavior. Viscoelastic materials have both solid-like characteristics—elasticity, strength, and stability of form—and liquid-like characteristics, such as flow that depends on time, temperature, and stress. All plastics exhibit some degree of viscoelasticity.

Viscosity - The internal resistance to flow exhibited by a fluid, the ratio of shearing stress to rate of shear. A viscosity of one poise is equal to a force of one dyne/square centimeter that causes two parallel liquid surfaces one square centimeter in area and one centimeter apart to move past one another at a velocity of one cm/second.

Voids - See *Porosity.*

V-Shaped Blender - (Also see *Attrition Bar*) A V-Shaped blender consists, in general, of two cylindrical halves that are attached at one end, thus the name V-shaped. The blender can rotate around an axis that is horizontal and allows tumbling of the material inside the blender. Some of these blenders are equipped with an attrition bar to intensify the mixing action.

W

Wafer Carrier - A basket, often made from fluoropolymers, for transportation of silicon wafers during the processing and shipping.

Warpage-See *Warping*.

Warping - Dimensional distortion or deviation from the intended shape of a plastic or rubber article as a result of nonuniform internal stress, e.g., caused by uneven heat shrinkage. Also called *warpage*.

Wear - Deterioration of a surface due to material removal caused by any of various physical processes, mainly friction against another body.

Wear Factor - See *Specific Wear Rate*.

Wear Rate - See *Tribological Characteristics*.

Weep Test - This is a test to determine the integrity of a high pressure hose. In this procedure (MIL-S-3136, type III and MIL-H-25579) unbraided samples of the tube are pressurized by filling with a specified fuel (70% i-octane + 30% toluene) to a calculated initial pressure. Pressure is incrementally (0.035 MPa) increased until seepage occurs. This pressure is recorded as weep pressure.

Weld Factor - The ratio of weld strength to the strength outside the welded zone, typically determined by tensile stress tests.

Welding - Also see *Heated Tool Welding*. A method used for joining thermoplastic parts without an adhesive. Heat, pressure and time are variables used to weld together plastic parts. A simple method consists of clamping the parts together and heating the joint with a heater band.

Welding by Distance - A form of heated tool welding in which mechanical stops on the hot tool and holding fixtures are used to control the dimensions of the welded parts. See also *Welding by Pressure, Heated Tool Welding*.

Welding by Pressure - A form of heated tool welding in which flow of the molten plastic after heating is regulated by application of specific pressures to the hot tool and parts. Accurate pressure control is necessary, and final part dimensions may vary due to variations in melt thickness and melt viscosity. See also *Welding by Distance, Heated Tool Welding*.

Welding Pressure - The pressure applied to the parts during welding to hold the parts in place and provide intimate contact for bonding. Welding pressure varies depending on the welding process used and the composition of the parts being joined.

Welding Time - A processing parameter in welding. Definition varies depending on the welding process—can refer to the time vibrations are applied to the parts being joined (ultrasonic welding), the amount of time parts are held together under pressure for cooling and solidification (heated tool welding), or both the spin time and cooling time (spin welding). Welding time is dependent on other processing parameters, such as weld amplitude, weld frequency, weld pressure, power level, temperature, etc.

Weld Strength - Strength of a welded plastic part at the seam that it has been welded is called weld strength. It is measured by methods similar to those for measuring the strength of adhesive bonds. Also see *Adhesive Bond Strength.*

Wet Bag Isostatic Molding - Also see *Dry Bag*. A method of isostatic molding in which polytetrafluoroethlene (PTFE) powder is compressed between a hard part and an elastic bag. The compaction is achieved by submerging the filled mold in fluid inside a high pressure vessel and pressurizing the vessel.

Wettability - The rate at which a substance (particle, fiber) can be made wet under specified conditions. See also *Wetting*.

Wetting - The spreading out (and sometimes absorption) of a fluid onto (or into) a surface. In adhesive bonding, wetting occurs when the surface tension of the liquid adhesive is lower than the critical surface tension of the substrates being bonded. Good surface wetting is essential for high strength adhesive bonds; poor wetting is evident when the liquid beads up on the part surface. Wetting can be increased by preparation of the part surface prior to adhesive bonding.

Y

Yellowing - Developing of yellow color in near-white or near-transparent materials such as plastics or coatings as a result of degradation on exposure to light, heat aging, weathering, etc. Usually is measured in terms of yellow index.

Yield Deformation - The strain at which the elastic behavior begins, while the plastic is being strained. Deformation beyond the yield deformation is not reversible.

Young's Modulus of Elasticity - In the elastic region, the relationship between stress and strain of a polymer, undergoing tensile or compressional strain, is linear (Hooke's Law). In this relationship stress is proportional to strain. The coefficient of proportionality in this stress-strain relationship is called *Young's Modulus of Elasticity*.

Z

Zero Strength Time - Zero strength time is defined as the time required to hold a sample of polychlorotrifluoroethylene at 250°C, under conditions specified by ASTM D1430, until the sample loses its strength.

ZST - See *Zero Strength Time*.

Index

A

Abrasion resistance 210
Abused particles 137
Acceleration tube 195
Aclar® 47, 60, 311
Aclon® 189
Acrysol® 174
Acyl fluoride 192
Adhesive bond strength 278
Adhesiveless joining 281
Adhesives 277
Agglomerates 191, 214
Agglomeration of PTFE 37
Agglomeration processes 37
Aircraft wiring 293
Algoflon® 60
Aluminum surface coatings 179
Ammonia 278
Amorphous locking 162
   temperature range 163
Amorphous phase 154
Amorphously locked PTFE 164
Annealing 95
   PTFE 276
Applications
   of dispersions 168
   fine powder PTFE 138
   fluoroadditives 199
   fluoropolymers 302
   PTFE 6, 135
Asbestos 178
ASTM Method D4895 156
Atomic oxygen 236
Attractive intermolecular forces 14
Attrition bar 208
Autopolymerization inhibitors
   TFE 25

B

Backpressure 121, 124, 125
Bag
   design 112
   flexible 115
   sealing 114
   thickness 111
Ball mill 196
Bead extrusion 158
Beaker production 105
Beam current 195
Beam power 195
Bearing pads 294

Bearings 293
   filled 182
Biaxially oriented sheets 213
Billets 78
   cracking 88
   molding 79
Bisulfite initiators 33
Bleaching
   thermal 180
Blenders 141
Blending room 140
Bond strength 278, 281
Bonding
   PCTFE 189
   problem 277
   techniques 281
Bonds
   C-F and C-H 9, 10
   failure 279
   polarity 10
Bottle blending 140
Braiding 153
Brake 119, 124
Break elongation 243, 244
Bronze 207
Buckling
   of bags 116
Building code requirements 293
Bulk properties 200
Burst strength 156

C

C8 169
Calendar guides 160
Calendared tape
   properties 161
Calendaring 157, 212
   beads 158
   starting up 160
Capture velocity 297
Carbon
   electronegativity 10
Carbon disulfide 180
Carbon filled compounds 207
Carbonyl fluoride 296
Carboxylic acid 192
Cast films 181
Casting 180
Charge-to-charge welds 123
Charging
   automatic 97
Chemical contamination 289
Chemical priming 180

Chemical process industry 285
Chemical resistance 223
Chlorine 12
Chlorotrifluoroethylene 22, 23
   properties 30
   synthesis 30
Cleanrooms
   molding 79, 83
Clothing
   protective 301
Co-coagulation 182, 214
Coagulated dispersion 33
   powder 135
Coagulation 214
   of PTFE particles 169
Coalescence 76
   of polytetrafluoroethylene 147
Coatings
   on aluminum surfaces 179
   filled 179
   on glass cloth 175
   industrial 201
   processes 170
   unfilled 179
Codes
   building 293
Coefficient of friction 255
Cold flow 97, 206, 210
Cold plasma 279
Coloring PTFE 141
Commercial fluoroplastic homopolymers 59
Compaction 121, 142
   pressure 118
   process 109
Compounding techniques 208
Compression
   PTFE particles 84
Compression molding 76, 97, 185, 187, 214
   PTFE 78
Compressive loading 254
Conductivity of dispersions 174
Connectors
   high frequency 293
Contact adhesives 277
Contact angle 173
Contamination 289
   of process streams 285
Continuous use temperature
   ETFE 5
Coolants 276
Cooling 95, 127
   quenching 155
Copolymerization
   PTFE 34
Copolymers
   of CTFE 23

   definition 19
   of TFE 23
   of VDF 23
Core pin 151, 152
Corrosion products 128
Corrosion resistance 285
Cracking 88, 155, 193
   critical thickness 174, 181
Creep 97, 206, 210, 249
   variables 249
   viscosity 92
Crosslinking 239
Crystalline phases 239
Crystalline PTFE 135
Crystallinity 95, 127, 154, 186, 243, 254, 267
   measurement 245
   PTFE 11, 147, 154
Crystallization temperature 95
CTFE
   copolymers 23

D

Dancer roll 160
De-dusting powders 182
De-molding 97
Deagglomeration 191
Decomposition
   of fluoropolymers 296
   products 300, 303
Decomposition rate 259
Defluorination depth 278
Deformability
   of tape 157
Deformation
   under load 213
Degassing 88
Degradation 259
   of polytetrafluoroethylene 191
   products 296, 297
Degradation temperatures 11, 191
   CTFE 185
Degreasing 180
Delrin® 152, 160
Density
   expanded part 163
Detection methods 244
Die cone angle 158
Die land 146
Dielectric breakdown resistance
   of PTFE 147
Dielectric breakdown strength 85, 267
Dielectric constant value 258
Dielectric strength 243
Dies
   extrusion 131
Dip-coating 170

Dip tank 175
Dispersing agent 36
Dispersion polymerization 33, 35
  PTFE 209
Dispersions
  applications 168
  coagulated 214
  copolymers of TFE
  PCTFE 60, 189
  process 22
  properties of 173
  PTFE 23, 59, 168
  storage 169
Disposal
  fluoropolymer 303
Drilling
  PTFE 277
Dripping 182
Dry bag 107
  mold 113
Dry lubricants 203
Drying 180
  ovens 147
  process 147
DU Bearing Process 182
Dual laminate 286
Duplicate measurements
  averages 245
Durability of insulation 258
Dust particulates 300
Dwell time 88, 118
Dynodes 195

E

Ekonol® 209
Einstein's equation 172
Elastomers 200
Electrical cables
  tape 157
Electrical properties
  PCTFE 267
Electron beam irradiation 193
Electronegativity 10
Electrophilic attack 12
Electrostatic charge dissipation 210
Elongation 243
Emulsion polymerization 33
Encapsulation 283
End groups 192
Energy utilization coefficient 195
Engineering polymers 210
Environmental stress cracking 241
Etching
  aluminum surfaces 179
ETFE
  molecular structure 5
Ethylene tetrafluoroethylene
  molecular structure 4
Excessive microporosity 244
Exhaust air 297
Expansion
  and strength 165
Expansion process 162
Exposure hazards 300
Exposure limits 297
Extruders 119, 128, 144
  drive system 130
  large 154
Extrusion 189
  aids 135, 138
  of bead 158
  conditions 151, 158
  dies 131
  parameters 125
  pressure 138, 144, 151, 158, 166
  ram 130
  rate 123

F

FDA compliance 203
FEP
  degradation curves 11
FEP and PTFE
  properties of 11
Fibrillation 135, 144, 209
  of fine powder particles 137
  pressure 166
Fibrils 136
Fick's First Law
  mass transfer 237
Filled bearings 182
Filled coatings 179
Filled compounds 206
Fillers 210, 288
  material 206
  separation 208
  variables 226
Film casting 180
Fine cut PTFE 36, 59
Fine cut resins 76
Fine powder
  applications 138
  processing 137
  PTFE 22, 35, 59, 135, 209
  resins 33
Finishes
  PTFE 168
Fire hazards 301
Fire retardants 293
Fishtail guide 160

Flammability 147
  solvents 138
  tetrafluoroethylene 25
Flax 178
Flex life 154, 254
Flexibility characteristics 254
Flexible bag 115
Fluid energy mill 196, 197
Fluid handling components 290
Fluon® 60
Fluorinated ethylene propylene polymer (FEP) 4
  molecular structure 4
Fluorine 236
  abstraction from PTFE 12
  electronegativity 10
  in fluoropolymers 296
  sheath 5
Fluoroadditives 191, 197, 199, 201, 302
  irradiated 203
Fluoroalkenes 296
Fluoroplast® 60
Fluoroplastic homopolymers
  commercial 59
Fluoropolymers 4, 236, 239, 242
  applications 233
  classification 19
  discovery of 3
  dispersions 301
  disposal of 303
  environmental stress cracking 241
  extreme properties 9
  fundamental property 233
  handling and processing 296
  history 5
  manufacturers 60
  processing 1, 242
  thermoplastic 286
Fluorotech® 279
Food contact 302
Free flow compounds 208
Free flow powder 206
Free flow PTFE 59
Free flow resins 76, 121
Free radical reaction 33
Friction 14, 124, 207, 223, 226
Fume fever 297

G

Gamma ray irradiation
  PCTFE 268
Gas chromatography 239
Gaskets 210, 288
Gel state 155
Glass cloth 175

Glass cloth coating 175
Glass fabric backing 286
Glass fibers 206, 209
Glow discharge 279
Gore-Tex GR® 289
Gore-Tex® 162
Granular powders 206
Granular PTFE 35, 36, 59, 76, 191
  properties 216
Granular resins 33
Graphite impregnation 178
Gravure coating 170
Greases 201
Grinding 196
  compounds 276
Guides
  calendar 160
Gylon® 289

H

Halogenated solvents 233
Hammer mills 196, 208
Harshaw Process for Dustless Powders 182
Health hazards 297
Heat of fusion 155, 243
Heat of polymerization 36
Heated console ovens 147
Heating rate 92, 93
Helical conformation
  PTFE 15
Hexafluoropropylene
  as comonomer 27
  properties of 27
HFP
  yields 27
Hiflon® 60
High frequency
  connectors 293
Hold temperature 94
Hoods
  ventilation 297
Horizontal extrusion 154, 155
Hostaflon® 60, 208
Hot stamping 283
Hot-tool welding 282
Hourglassing 88, 125
Hydrocarbon lubricant 135
Hydrocarbon olefins 35
Hydrofluoric acid 301
Hydrogen
  electronegativity 10
Hydrostatic molding 97
Hysteresis loop 171

I

Impact strength 254
Impregnation
    asbestos 178
    of graphite 178
    porous metals 178
Impurities 243
Infrared ovens 155
Infrared spectroscopy 245
Initiator concentration 36
Injection molding 187
Ink printing 283
Inks 200
Inorganic compounds 302
Insulated wire 138
Insulation 211, 301
    wire and cable 290
Interfacial forces 173
Intermolecular forces 14
Ionic strength
    of dispersions 174
Irradiation 192, 203
    distribution 195
    electron beam 193
    PCTFE 268
Isopar® 69, 138, 147, 157, 158
Isoparaffin lubricant 22
Isoparaffinic hydrocarbons 138
Isostatic mold design 116
Isostatic molding 97, 103, 104, 109, 288
Isostatic process 105, 116

J

Jar blending 140
Jet mills 196, 197
Joining
    adhesiveless 281

K

Kapton® 180
Kel F® 47
Kevlar® 153
Kneader 193

L

Lapping and grinding compounds 276
Lewis bases 233
Linings 285, 286
    complex parts 289
Liquid oxygen contact
    surface condition 237
Liquid pigments 141
Load removal 254
Low flow PTFE 59, 208

Low flow resin 206
Lubricants 135, 138, 211, 300
    extraction 160
    insulation 147
    isoparaffin 22
Lubricating oils 201
Lubrication 154

M

Machining 276
    high speed 300
    PCTFE 189
    PTFE 276
Macroscopic flaws 244
Mandrels 105, 125
Marine paints 201
Mason and Pierce equation 172
Mass spectroscopy 239
Matched-die forming 283
Material Safety Data Sheets 301
Media mills 196
Medical applications 302
Melt creep viscosity 92
    PTFE 34
Melt processes 300
    definition 19
Melt-processible
    PTFE 5
Melt viscosity
    FEP vs PTFE 11
Melting point 302
    PTFE 147
Melting temperature
    PTFE 121
    PTFE vs PE 10
Metals
    porous 178
Mica particles 207
Microporosity 243, 244
    measuring 245
Micropowders 191, 197, 200, 201
Microvoids 244
Milling 208
Mills 196, 197
Modified PTFE 1, 253
Modifier molecules 281
Moisture vapor permeation 239
Molded parts 276
Molding 97, 187, 214
    areas 79
    complex shapes 109
    isostatic 103, 288
    pressure 82, 118
    techniques 76
    temperature 82
    wet- and dry-bag 107

Molds
　configuration 111
　design 79, 111
　filling 109
　sealing 114
Molecular structure 239
Molecular weight 189, 243
　micropowders 191
　PCTFE 185
　PTFE 193
Monomers 22
　polytetrafluoroethylene 25
Motor oils 201
Multilayer coating techniques 170
Mush 182

N

Natrosol® 174
Necking 165
Neoflon® 47, 71, 72
Non-stick surfaces 179
Nucleophile 12
Nucleophilic attack 12

O

Oils 201
Organic contaminants 302
Orientation 243
　polymer molecules 211
　after sintering 156
Orientation Index 155, 156, 157
Oriented tape 157
OSHA 297
Ovens 147
　lubricant removal 161
Oxidation products 296
Oxygen sensors 295
Oxygen uptake 236
Ozone 303
　reactive substance 236
　resistance of polymers 236

P

Paints 201
Paraffins 36, 97
Particle sizes
　PTFE 36
Particulate contamination 289
Particulates 297
Paste extrusion 135, 144, 300
PCTFE 12, 237, 267. *See also*
　Polychlorotrifluoroethylene
　amorphous 267
　bonding 189
　consumption 60
　crystal structure 16
　degradation of 187
　dispersions 189
　films 268
　glass transition temperature 16
　machining 189
　stress 186
　thermal properties 267
Peel strength 280
Pendent group 281
Perfluorinated fluoropolymers 259
Perfluorinated paraffin 97
Perfluoroalkoxy
　molecular structure 4
Perfluoroalkylvinylethers
　properties of 29
Perfluoroammoniumoctanoate 169
Perfluoropolymers 4
Perfluoropropylvinylether 29
Permeated species 237
Permeation 237
　inhibiter 238
　process 233
　simplest form 237
Permeation rate
　affecting factors 238
　measurement 239
Personal protective equipment 300
Persulfate initiation 33
Petroleum solvents 138
PFF 297
Pigment dispersions 141
Pigment filled coatings 179
Pigmenting 154
Pipes
　sealing tape 157
Piping 285, 286
Plane slippage 84
Plasma treatment 279
Plastics
　properties 19
Plenum cables 293
Plunkett, Roy 3
Poisson's ratio 165
Poker chipping 118, 121, 125
Polarity
　C-F bonds 10
Pollution abatement 161
Polyaramide 178
Polyarylene plastics 210
Polychlorotrifluoroethylene 1, 185, 237, 267
　crystal structure 16
　density 16
　molecular structure 4
Polyethylene 241
Polyflon® 60, 208

Polymer
 chain movements 238
 chemical characteristics 238
 cracking 241
 fume fever 297
 physical characteristics 238
 weakness 241
Polymerization
 particles 191
 polytetrafluoroethylene 35
 pressure conditions 35
 regimes 33
 tetrafluoroethylene 15
Polymerization reaction
 inhibiters 36
 tetrafluoroethylene 36
Polymerization technologies 35
Polytetrafluoroethylene 3, 233, 236
 coefficient of expansion 263
 coefficient of friction 256
 degradation 191
 design of parts 249
 dispersions 169
 electrical stability 258
 fillers 255
 good wear properties 256
 heat deflection temperature 263
 key properties 243
 manufacturers 60
 modified 281
 monomers 25, 33
 phase diagram 16
 polymerization 35
 transition temperature 259
 unsintered 157
 vacuum 259
 viscosity 171
Polytetrafluoroethylene compounds
 fillers 263
Polytetrafluoroethylene parts
 standard test methods for inspection 265
Polytetrafluoroethylene resins
 processing conditions 243
Polyvinyl fluoride 1
 molecular structure 4
Polyvinylidene fluoride
 molecular structure 4
Pores
 micro 162
Porosity 245
Porous metals
 impregnation 178
Powders 206
 coating 286
 de-dusting 182
 low density 36

 metal processing 97
 modified 59
 selection criteria 116
 shrinkage 113
Pre-sintered resins 121
Preform
 pressure 88
 PTFE 92
 release 88
 shape 107, 109
Preforming 82, 142, 154
 and sintering 80
Prepressing 144
Presses
 molding 80
Pressure
 compaction 118, 121, 142
 decay 103
 for fibrillation 166
 for welding 282
 forming 283
 hoses 153, 155
 molding 97
Pressure sensitive adhesives 277
Primers 180
Printing inks 200
Protective clothing 301
PTFE 3, 10. *See also* Polytetrafluoroethylene
 additives 201
 amorphously locked 164
 applications 6
 auto-ignition 236
 chain stability 236
 color of 13
 coloring 141
 crystalline form 135
 decomposition rates 259
 degradation 128, 259
 density 163
 dielectric strength 258
 enthalpy 263
 fabrication 135, 243
 fiber strength 180
 fine powder 22
 finishes 168
 granular 35, 59, 76
 hardness 255
 helical conformation 15
 limiting oxygen index 236
 machining 276
 melting point 19, 147, 170, 276
 melting temperature 121
 molecular weight 33, 34, 193
 organic and inorganic compounds 233
 parts 244
 Poisson's ration 249

preform 92
process conditions 243
properties 6, 9
radiation stability 264
reagent effects 233
resins 254
scrap 302
seals 288
self ignition temperature 301
sintering temperatures 206
solvents 13
specific heat 263
stress-strain behavior 246
structure of 13
surface tension 138
suspension polymerized 206
thermal decomposition 191
triclinic 16
van der Waals forces 33
viscosity 19, 172
yarn 180
PTFE and FEP
properties of 11
PTFE dispersions 23, 206, 214
pH 173
Pulmonary edema 297
Pulverization 196
Pumps 288
PV convention 256
PV limits 226, 256
Pyrolysis 193

Q

Quenching 95, 127

R

Radiation
exposure 192
of PCTFE 268
Radicals 192
Radiography 244
Ram 130
pressure 144
speeds 84
Ram extruder 118, 144
design 128
Ram extrusion 121
powders 59
vertical 125
Reactive elements 12
Reactor deflagaration 36
Recycling 302
Reduction ratio 59, 135, 144, 148, 149, 158, 166
Reinforced fine powder PTFE 210
Reinforced material 211

Repeat distance
PTFE 10
triclinic 15
Repro 302
Repulsive intermolecular forces 14
Residence time 123
Resins
compression 142
extrudates 128
fine powder 33
flow 76
for isostatic molding 116
for ram extrusion 121
Reverse roll gravure coating 170
Rheology 171
Rietz mill 208
Rod-like cylinders 15
Rods
PCTFE 187
Roller mills 196
Rolls
diameter 160
Rotational lining 286
Rotolining 286, 289

S

Safety requirements 293
Sandblasting 180
Sawing
PTFE 277
Scotch-Weld® 280
Scrap material 302
Scrubbers 286
Sealing
mold and bag 114
Seals 210, 288
Semiconductor industry 290
Semicrystalline polymer 267
Shear damage 135
Shear stress 127, 137
Shearing
PTFE 277
Sheeting 211
Shrinkage 80, 113, 127
preform 88
Silicones
stress cracking 237
Single layer coating methods 170
Sintering 121, 147, 179, 216, 300
cycle profiles 92
cycles 97
ovens 81
sheet 213
time 123
Sintering temperatures 121
PTFE 206

Sizing agents 175
Skin formation 127, 128
Skiving 277
Slide bearings 294
Slide coating techniques 170
Slitting 162
Sodium etching 278
Sodium naphthalenide 278
Solubility
   PTFE 11
   PTFE and FEP 13
Solvents
   flammability data 138
   lubricants 138
   sorption of 13
Spaghetti tubing 151
Spark tester 144, 147
Specific gravity 186, 243
Specific wear rate 226
Spillage 301
Spinning
   yarn 180
Springback 121, 130
Stack gas treatment 194
Static charge 153, 210
Steel powder 207
Steel vessels 286
Stick-slip 293
Storage
   of dispersions 174
   granular PTFE 78
Strength
   by expansion 165
Stresses 95, 186
   cracking 155
   reduction 186
   relaxation 254
Stretch rates 161, 163
Stretch Void Index 155, 156
Styrene
   heat of polymerization 26
Surface energy 14
Surface finish 110
Surface modification
   for adesion 277
Surface properties 200
Surface science 173
Surface tension 14, 173
   PTFE 138
Surfactants 169, 174
   perfluorinated carboxylic ammonium 35
Suspension polymerization 33, 35, 36
Suspension polymerized PTFE 206
Suspension process 22
Swelling 13

T

Tanks 286
Tape
   oriented 157
   properties 161
   slitting 162
   unsintered 157
Tapping
   PTFE 277
Tear strength 200
Tedlar® 22
Teflon® 3, 60, 208
Teflon® AF 5
Tefzel® 236
Tensile properties 244
   degree of orientation 244
Tensile strength 165, 243, 244, 246
Tetra-etch® 279
Tetrafluoroethylene 1, 25, 192
   copolymers 253
   homopolymers 243
   molecular structure 4
   monomer 33
   polymerization 15, 33, 36
   properties 26
   yields 25
Tetrafluoromethane 296
TFE
   autopolymerization inhibitors 25
   copolymers 23
   heat of polymerization 26
   preparation of 25
Thermal conductivity 263, 276
Thermal cracking 193
Thermal degradation 11, 264, 300
   PCTFE 267
Thermal exposure 259
Thermal stability 296
Thermoforming 283
Thermogravimeric analysis techniques 259
Thermoplastics 286, 302
Thickeners 172
Thixotropic liquids 171
Threading
   PTFE 277
Threads
   sealing tape 157
Tie layer 181
Tool speeds 276
Torlon® 210
Tornado® mill 208
Torsional damping 245
Toxic gases 301
Toxicity
   of fluoropolymers 296

Transition point
  PTFE 135
Transition temperature
  PTFE 15, 137
Tribological characteristics
  of polymers 200
  of PTFE compounds 226
Triton® 45, 174, 178, 179, 180, 214
Tubes
  compression molding 187
Tubing
  categories 150
  extrusion 151
  jacketed 295
  pressure hoses 152
Tubular oven 147
Tumbler 208
Turbula® 141
Two-roll coating 170
Tyvek® 301

U

U.S. Food and Drug Administration 302
Ultrapure fluids 290
Ultrasonics 245

V

V-cone blenders 140
V-shape blenders 208
Vacuum forming 283
Valve diaphragms 254
Valves 288
van der Waals attraction 136
van der Waals forces 14
Ventilation 297
Vertical extrusion 154, 155
Vespel® 226
Vessels 286
Vinyl chloride
  heat of polymerization 26
Vinyl fluoride 22, 23
Vinylidene fluoride 22
  copolymers 23
  in PCTFE 60
Virgin 302
Viscosity
  equation 172
  PCTFE 185
Viton® 202
Voids 124, 156, 254
  closure 34
  in PTFE 82

W

Wafer carriers 289
Waste fluoropolymer 302
Water
  purified 36
  removal 180
Wear 207, 226
  rate 223
Weep Test 155, 156
Weld factor 281
Weld strength
  between charges 121
Welding 281, 282, 301
Wet bag 107
Wet processing equipment 289
Wetting behavior 173
Wire
  coating 144
  insulation 138, 211
  quality 149
Wood pulp process 180

X

X-ray diffraction 245

Y

Yarns
  bleaching 180
Yield deformation 246
Young's Modulus of Elasticity 165

Z

Zero strength time 60, 185, 188
Zonyl® 202